Springer Series in Reliability Engineering

Series editor

Hoang Pham, Piscataway, USA

More information about this series at http://www.springer.com/series/6917

Hoang Pham

Editor

Quality and Reliability Management and Its Applications

 Springer

Editor
Hoang Pham
Department of Industrial and Systems
 Engineering
Rutgers University
Piscataway
USA

ISSN 1614-7839   ISSN 2196-999X   (electronic)
Springer Series in Reliability Engineering
ISBN 978-1-4471-6776-1   ISBN 978-1-4471-6778-5   (eBook)
DOI 10.1007/978-1-4471-6778-5

Library of Congress Control Number: 2015952761

Springer London Heidelberg New York Dordrecht

Printed on acid-free paper

Springer-Verlag London Ltd. is part of Springer Science+Business Media (www.springer.com)

To Michelle, Hoang Jr., David

Preface

Most of the products that affect our daily lives are becoming more complex. Quality and Reliability Management—which integrates processes, policies, and reliability predictions from the beginning of the product development life cycle to ensure high levels of product performance and safety—helps companies to address the challenges of increasingly complex systems and globally widespread processes in today's competitive marketplace.

This book consists of 15 chapters, organized into four parts: Quality Management, Reliability Management, Maintenance Management, and Design, Applications and Practices. It aims to present both theoretical research and practical aspects of product quality and reliability management with particular emphasis on system design, processes, modeling, and its applications. The topics covered include:

Design for reliability
Fatigue reliability assessment
Data fusion
Machine condition monitoring
Multivariate process capability
Random replacement and policies
Random environments and uncertainty
Safety inspection
Maintenance inspection
System failure behavior
Systemability
Ultrasonic inspection
Supplier quality evaluation
Multi-state system reliability
Product development life cycle
Applications in automotive industry

Change-point models
Failure mechanisms and failure modes
Health monitoring
Multi-criteria analysis
Network reliability
Reliability management
Statistical quality management
System reliability and availability
Surveillance system
Soft computing
Sensor fusion
Warranty policy
Maintenance and risk management
Performance monitoring healthcare
Product durability design
Case studies

Each chapter has been written by active researchers and/or experienced practitioners with international reputations in the field and with a hope of bridging the gap between the theory and practice in the area of quality and reliability management. The book consists of four parts.

Part I—*Quality Management*—contains three chapters, focuses on the aspects of supplier quality multicriteria management approach, risk-adjusted monitoring in healthcare quality management, and the area of multivariate statistical process control. Choosing the right supplier quality always plays an important role in decisions—making organizations profitable. The first chapter by Al Salem, Awasthi, and Wang discusses a new multi-criteria multistep approach to evaluate the quality of large data sets of suppliers based on both qualitative and quantitative criteria using the analytic hierarchy process (AHP) among several other techniques. The proposed approach enables customers to deal with suppliers' large data sets and identify the quality of the suppliers. "Risk-Adjusted Performance Monitoring in Healthcare Quality Control" by Zeng provides in detail an overview of existing studies on risk-adjusted monitoring in healthcare aspects. He also discusses several case studies that illustrate the use of some techniques such as Bayesian for related problems in risk-adjusted monitoring. Some potential research problems in risk-adjusted monitoring using data mining techniques to select significant patient risk factors have also been discussed. "Univariate and Multivariate Process Capability Analysis for Different Types of Specification Limits" by Chakraborty and Chatterjee gives a brief overview of the process capability index (PCI) for both the bilateral and unilateral specification limits. It is worth noting that to compute the PCI of a process, practitioners need to make sure, or assume, that the quality characteristic under consideration follows normal distribution and the process is under statistical control. The chapter also presents some recent studies in the area of multivariate process capability indices.

Part II—*Reliability Management*—containing three chapters, focuses on the aspects of reliability modeling of complex systems and its applications in practice with the uncertainty of operating environments. In today's highly competitive world, reliability and maintainability are the two most important measures that determine the quality of the product. "Modeling and Analyzing System Failure Behavior for Reliability Analysis Using Soft Computing-Based Techniques" by Garg discusses an optimization reliability model considering the reliability, availability, and maintainability aspects of complex systems to obtain the optimal values of mean time between failures and mean time to repair for each of its constituent components in the system. Production managers often look at ways to identify the most sensitive stations in order to increase the reliability of the entire manufacturing networks. "System Reliability Evaluation of a Multistate Manufacturing Network" by Lin, Chang, and Huang discusses graphical transformation and decomposition probability models in order to determine the reliability of multi-state manufacturing networks (MMN) with multiple production lines in parallel and multiple reworking actions. The production managers can use the transformed MMN and decomposed paths approach to develop a decision-making strategy to assign the amount of output that each production line should produce to fulfill the demand. Reliability is defined as the ability of a system or component to perform its required functions under stated operating conditions for a specified period of time. In reality, the operating environment is often unknown and different from the laboratory or testing environments. "Systemability: A New Reliability Function for Different Environments" by Persona, Sgarbossa, and Pham provides a literature review of systemability—a concept of reliability with consideration of operating environments. It also discusses recent studies on systemability of age replacement maintenance policy. The chapter also discusses several real-world applications in the automatic packaging machines for beer production, gear component, and motorcycle drive-system to illustrate the systemability in practice.

Part III—*Maintenance Management*—containing four chapters, focuses on the aspects of maintenance modeling and inspection design policy programs of complex systems and its maintenance applications in practice. "Innovative Maintenance Management Methods in Oil Refineries" by Bevilacqua et al. aims to discuss an innovative maintenance program applied to the turnaround management and two methodologies based on the risk analysis and the application of criticality index to evaluate the criticalities of equipment and plants. These approaches can be used to optimize the use of economical, human and instrumental resources needed for the refinery maintenance activities. If the age of an operating unit is always known and its failure rate increases with age, it may be wise to replace it before failure on its age. A commonly considered age replacement policy for such a unit is made if the unit is replaced at a total operating time T after its installation or at failure, whichever occurs first. "Age Replacement Models with Random Works" by Zhao and Nakagawa discusses four age replacement models for an operating unit where it works successively for jobs with random working cycles. Optimal policies for each model that minimizes the expected cost rate are analytically discussed.

"Availability of Systems with or Without Inspections" by Hwang and Mi provides an overview of availability of systems subject to different inspection policies where failures are subject to either self-announcing or not self-announcing. This chapter also discusses explicitly expressions of the steady-state availability, limiting average availability, and the instantaneous availability of systems with and without inspections.

The modeling of the surveillance systems has recently received wide attention in various applications especially in the security areas. "Reliability and Maintenance of the Surveillance Systems Considering Two Dependent Processes" by Zhang and Pham discusses the existing works related to surveillance system modeling including sensor deployment, intelligent surveillance system design involving data mining and computer automation techniques, and the attack-defense model that quantifies the interaction behavior between the defender and adversary. The chapter further discusses several recent works in the field of surveillance system reliability modeling with considerations of two stochastic processes.

Finally Part IV of the book contains five chapters, on *Design, Applications and Practices*.

Reliability management is responsible for the oversight of reliability activities. In general there are two basic approaches to managing product reliability: reactive and proactive. "Reliability Management" by Schenkelberg discusses the difference between reactive and proactive reliability programs and provides an introduction to the reliability maturity matrix and how to take specific steps to move the organization to proactively managing reliability. Reactive organizations respond to each field failure, to each product testing failure, and to each vendor component failure. Proactive organizations design and build products with an acceptable reliability and anticipate the type and number of field failures. Reliability, time to market, and cost are the three most important factors that determine whether a product is successful in the marketplace or not. "Design for Reliability and Its Application in Automotive Industry" by Yang discusses effective designs for reliability (DFR) process and techniques, and the integration of DFR into the product life cycle especially in the automobile industry. It describes phases of the product life cycle including product planning phase, design and development phase, design verification and process validation phase, production phase, field deployment phase, and the disposal phase as the terminal phase of a product in the life cycle. A practical application in the automobile industry is discussed to illustrate how DFR improves reliability and robustness.

"Product Durability/Reliability Design and Validation Based on Test Data Analysis" by Wei et al. discusses several practices in product durability and reliability designs. It also discusses the concepts and approaches on five major aspects which are essentially the procedures of newly developed durability and reliability analysis and design methods. The five aspects are: failure mechanisms and modes, linear data analysis, design curve construction, Bayesian statistics for sample size reduction, and accelerated testing. These approaches can serve as a practical guide for product design engineers and testing managers in their test planning and validation analysis. "Turbine Fatigue Reliability and Life Assessment Using Ultrasonic Inspection: Data Acquisition, Interpretation, and Probabilistic Modeling" by Guan et al. presents a systematic method and procedure for assessing fatigue reliability of steam turbines using ultrasonic nondestructive inspections. The uncertainties from ultrasonic inspections, flaw characterization, and fatigue model parameters are also discussed. Based on the inspection information, a probabilistic of detection model using a classical log-linear model coupling the actual flaw size and the NDE reported flaw size is developed in order to quantify the uncertainties from flaw sizing and model parameters. An application of steam turbine rotor integrity assessment with actual ultrasonic inspection data is used to demonstrate the overall method. "Fusing Wavelet Features for Ocean Turbine Fault Detection" by Duhaney, Khoshgoftaar, and Wald focuses on employing feature-level sensor fusion to enable machine learners to detect changes in the operational state of the dynamometer. The authors discuss a machine condition monitoring system that allows for automated detection of changes in the state of a machine being monitored. This chapter also discusses several case studies to show the performance of feature level fusion.

All the chapters are written by more than 35 leading experts in the field with a hope to provide readers the gap between theory and applications and to trigger new research challenges in quality and reliability management in practice.

I am deeply indebted and wish to thank all of them for their contributions and cooperation. Thanks are also due to the Springer staff for their editorial work. I hope that the readers including engineers, teachers, scientists, postgraduates, researchers, managers, and practitioners will find this book a state-of-the-references survey and a valuable resource for understanding the latest developments in quality and reliability management and its applications in process, design, and development of products.

Piscataway, New Jersey   Hoang Pham
December 2014

Contents

Part III   Maintenance Management

Part IV   Design, Applications and Practices

Editor and Contributors

About the Editor

Hoang Pham is Distinguished Professor and former Chairman (2007–2013) of the Department of Industrial and Systems Engineering at Rutgers University, New Jersey, USA. He is the author or coauthor of five books and has published over 140 journal articles, and edited 10 books including *Springer Handbook in Engineering Statistics* and *Handbook in Reliability Engineering*. Dr. Pham has served as editor-in-chief, editor, associate editor, guest editor, and board member of many journals. He is the editor of *Springer Book Series in Reliability Engineering*, editor of *World Scientific Book Series on Industrial and Systems Engineering*, and has served as Conference Chair and Program Chair of over 30 international conferences and workshops. His numerous awards include the 2009 IEEE Reliability Society *Engineer of the Year Award*. He is a Fellow of the IEEE and IIE.

Contributors

Waheed A. Abbasi Siemens Energy, Inc, Pittsburgh, PA, USA

Aqeel Al Salem CIISE-EV 7.640, Concordia University, Montreal, QC, Canada

Anjali Awasthi CIISE-EV 7.640, Concordia University, Montreal, QC, Canada

M. Bevilacqua Dipartimento di Energetica, Università Politecnica della Marche, Ancona, Italy

Ashis Kumar Chakraborty SQC & OR Unit, Indian Statistical Institute, Kolkata, India

Ping-Chen Chang Department of Industrial Management, National Taiwan University of Science & Technology, Taipei, Taiwan

Moutushi Chatterjee SQC & OR Unit, Indian Statistical Institute, Kolkata, India

F.E. Ciarapica Dipartimento di Energetica, Università Politecnica della Marche, Ancona, Italy

Janell Duhaney Computer and Electrical Engineering and Computer Science, Florida Atlantic University, Boca Raton, Florida

Harish Garg School of Mathematics, Thapar University Patiala, Patiala, Punjab, India

G. Giacchetta Dipartimento di Energetica, Università Politecnica della Marche, Ancona, Italy

Xuefei Guan Siemens Corporation, Corporate Technology, Princeton, NJ, USA

Cheng-Fu Huang Department of Industrial Management, National Taiwan University of Science & Technology, Taipei, Taiwan

Kai Huang Department of Mathematics and Statistics, Florida International University, Miami, FL, USA

Taghi M. Khoshgoftaar Computer and Electrical Engineering and Computer Science, Florida Atlantic University, Boca Raton, Florida

Dmitri Konson Tenneco Inc., Grass Lake, MI, USA

Burt Lin Tenneco Inc., Grass Lake, MI, USA

Yi-Kuei Lin Department of Industrial Management, National Taiwan University of Science & Technology, Taipei, Taiwan

Limin Luo Tenneco Inc., Grass Lake, MI, USA

B. Marchetti Università degli Studi ECampus, Novedrate, Como, Italy

Jie Mi Department of Mathematics and Statistics, Florida International University, Miami, FL, USA

Toshio Nakagawa Aichi Institute of Technology, Toyota, Japan

C. Paciarotti Dipartimento di Energetica, Università Politecnica della Marche, Ancona, Italy

Alessandro Persona Department of Management and Engineering, University of Padova, Vicenza, Italy

Hoang Pham Department of Industrial and Systems Engineering, Rutgers, The State University of New Jersey, Piscataway, NJ, USA

El Mahjoub Rasselkorde Siemens Energy, Inc, Pittsburgh, PA, USA

Fred Schenkelberg Reliability Engineering and Management Consultant, FMS Reliability, California, USA

Fabio Sgarbossa Department of Management and Engineering, University of Padova, Vicenza, Italy

Randall Wald Computer and Electrical Engineering and Computer Science, Florida Atlantic University, Boca Raton, Florida

Chun Wang CIISE-EV 7.640, Concordia University, Montreal, QC, Canada

Zhigang Wei Tenneco Inc., Grass Lake, MI, USA

Fulun Yang Tenneco Inc., Grass Lake, MI, USA

Guangbin Yang Chrysler Group, Auburn Hills, Michigan, USA

Li Zeng Department of Industrial and Manufacturing Systems Engineering, The University of Texas at Arlington, Arlington, TX, USA

Yao Zhang Rutgers University, New Brunswick, USA

Xufeng Zhao Qatar University, Doha, Qatar; Aichi Institute of Technology, Toyota, Japan

S. Kevin Zhou Siemens Corporation, Corporate Technology, Princeton, NJ, USA

Part I
Quality Management

A Multicriteria Multistep Approach for Evaluating Supplier Quality in Large Data Sets

Aqeel Al Salem, Anjali Awasthi and Chun Wang

List of Symbols

AD	Affinity Diagram
AHP	Analytical Hierarchy Process
CA	Cluster Analysis
VIKOR	Vlse Kriterijumska Optimizacija Kompromisno Resenje

1 Introduction

Most companies today depend on outsourcing to build their products. Outsourcing strategy has shown its effectiveness in increasing organizational profits through the development of better products when outsourced from the right supplier. The decision to outsource is made by a company's procurement or purchasing department. The decision involves many factors, and it gets more complex as the number of factors increases.

In this chapter, a modeling framework for analyzing the quality of a large number of suppliers from different environments is proposed. The review of the literature pertaining to supplier quality evaluation has not revealed any previous study for large sets. Most researchers have applied their model on a small set of suppliers. Some have evaluated suppliers based on very few criteria and in some cases; criteria may not be carefully evaluated. Unfortunately, most of their models

A. Al Salem · A. Awasthi (✉) · C. Wang
CIISE-EV 7.640, Concordia University, 1515 Ste Catherine Street West,
Montreal, QC H3G2W1, Canada
e-mail: awasthi@ciise.concordia.ca

A. Al Salem
e-mail: knhnh@hotmail.com

C. Wang
e-mail: cwang@ciise.concordia.ca

© Springer-Verlag London 2016
H. Pham (ed.), *Quality and Reliability Management and Its Applications*,
Springer Series in Reliability Engineering, DOI 10.1007/978-1-4471-6778-5_1

do not provide a mechanism for efficient analysis of a large number of suppliers. It is commonly known that as the number of suppliers and criteria increase, the problem of evaluation becomes more difficult and needs more time to be resolved. Therefore, this chapter proposes to develop a comprehensive and efficient model to analyze this type of problem for tracking or monitoring the quality performance of suppliers.

The proposed modeling framework integrates three methods that have heretofore been used separately for the purpose of evaluating supplier quality. Each of the methods was adopted for its strengths and advantages with respect to the problem under study. The first method is based on affinity diagram (AD) and the analytic hierarchy process (AHP) and concentrates on determining criteria and their weights. AHP has the ability to handle qualitative and quantitative criteria, simplifies the problem through building hierarchy, and is widely used and approved by many researchers and consultants for the purpose of prioritizing criteria.

The second method based on cluster analysis (CA) is used to manage large supplier data sets in such a way that suppliers with similar attributes are grouped together in clusters. Cluster analysis has the ability to group similar objects—in this case, suppliers—into clusters. Suppliers in a given cluster are more alike in many aspects than those in other clusters. CA technique was chosen for its ability to handle a large number of data efficiently and to guarantee that the best suppliers are not eliminated at least at the initial levels (Holt 1996).

The third method based on VIKOR (Vlse Kriterijumska Optimizacija Kompromisno Resenje) technique is used to rank suppliers and select the best supplier(s) based on the overall criteria. The VIKOR method was selected for its ability to find the compromise solution that is closest to the ideal solution. The compromise solution is most likely to be accepted by decision makers since it was developed on the basis of "the majority of criteria" rule (Opricovic and Tzeng 2004).

Integrating these methods confers their respective advantages upon the model and enables it to handle the supplier quality evaluation problem in different ways: managing large data sets, evaluating or analyzing them, and ranking them quickly and efficiently. Moreover, this model can be used to monitor selected suppliers' performance after a period of cooperation through comparison of results at different stages and under different situations. The strength of the proposed model is that it works with both small and large sets of supplier data, however, its chief purpose is to analyze large data sets. In short, this model is capable of handling the multicriteria problem on any scale of information.

The rest of the chapter is organized as follows. In Chap. 2, we present the problem definition. A literature review on supplier quality evaluation criteria and methods is presented in Sect. 3. In Sect. 4, the proposed model is set out. Section 5 presents a numerical application of the proposed approach. Finally, we provide the conclusions and future works in Sect. 6.

2 Problem Definition

The problem this chapter addresses consists of evaluating a large number of alternatives (suppliers) under a given set of criteria (quantitative or qualitative). According to Zanakis et al. (1998), most existing methods of supplier evaluation and selection are not suitable for application to a large number of alternatives, since these methods tend to generate inconsistencies. For this reason, the large data sets of suppliers must be treated in a way that overcomes this problem. To this end, the model will solve the following challenges:

1. How do buyers deal with large numbers of suppliers in heterogeneous business environments, that is, under different geographical location, product type, and product volume conditions?
2. Which criteria should buyers use for supplier quality evaluation?
3. How should buyers rank criteria or decide criteria weights?
4. How should buyers deal with qualitative and quantitative criteria?
5. How should buyers generate supplier quality rankings?

3 Literature Review

In the literature, many methods have been applied to solve multicriteria supplier selection and evaluation. Some papers handle the problem using the single method of Chan and Chan (2004), who used AHP to select supplier that matches with the company's strategies. Choy et al. (2005) applied case-based reasoning (CBR) method for outsource manufacturing. Barla (2003) conducted simple multi-attribute rating technique (SMART) model for a manufacturing work under lean philosophy to reduce supplier base. In addition, more authors like Talluri and Narasimhan (2003), Sarkis and Talluri (2002), and Karpak et al. (2001) applied different single methods for supplier quality evaluation. On the other hand, some papers find that integrating methods lead to better results. For example, Chen and Yang (2011) integrated fuzzy AHP and fuzzy TOPSIS to solve multicriteria problems for supplier selection. Jain et al. (2004) integrated fuzzy with genetic algorithm (GA). Rezaei and Ortt (2012) present a multivariable approach to supplier segmentation.

AHP has been widely used for the purpose of supplier quality evaluation (Liu and Hai 2005). The AHP method involves breaking down a complex problem into different levels. Once these levels have been identified, pairwise comparison is performed to find the interrelationships among them (Lam et al. 2010). The AHP method has been combined with other methods such as fuzzy theory, linear programming, goal programming, and data envelopment analysis (Vaidya and Kumar 2006) for the purpose of supplier selection. Vaidya and Kumar (2006) conducted an overview of applications that had used AHP. From 150 papers, they found that most

Table 1 Dickson's supplier quality evaluation criteria

Rank	Factor	Mean rating	Evaluation	Rank	Factor	Mean rating	Evaluation
1	Quality	3.508	Extreme importance	12	Desire for business	2.256	Average importance
2	Delivery	3.417	Considerable importance	13	Management and organization	2.216	
3	Performance history	2.998		14	Operating controls	2.211	
4	Warranties and claim policies	2.849		15	Repair service	2.187	
5	Production facilities and capacity	2.775		16	Attitude	2.12	
				17	Impression	2.054	
6	Price	2.758		18	Packaging ability	2.009	
7	Technical capability	2.545		19	Labor relations record	2.003	
8	Financial position	2.514		20	Geographical location	1.872	
9	Procedural compliance	2.488	Average importance	21	Amount of past business	1.597	
10	Communication system	2.426		22	Training aids	1.537	
11	Reputation and position in industry	2.412		23	Reciprocal arrangements	0.61	Slight importance

of the researchers used this method for selection and evaluation purposes. The applications of these papers were in engineering, personal, and social categories. Moreover, many researchers such as Narasimhan (1983), Nydick and Hill (1992), and Partovi et al. (1989), suggested using AHP for supplier evaluation and selection because of its ability to deal with qualitative and numerical attributes. However, AHP is more efficient when the pairwise comparisons at each level are reasonably small (Partovi 1994). Saaty (1980) suggests that each level should be limited to nine pairwise comparisons.

Supplier quality evaluation criteria were first proposed by Dickson (1966), where he listed 23 criteria for supplier quality evaluation based on a survey of purchasing agents and managers. Dickson's criteria for supplier quality evaluation are presented in Table 1.

It can be seen in Table 1 that quality is the most important criterion for supplier quality evaluation followed by delivery. However, this survey was conducted 45 years ago, in 1966. Nowadays, many salient features of supply and production have changed with globalization and technological progress. However, most of these criteria are still valid for evaluation purposes. Weber et al. (1991) studied all the literature pertaining to supplier quality evaluation criteria that had been published from Dickson's paper until 1991. They found that each of the 74 articles has

Table 2 Mentioned criteria in JIT's articles as mentioned by Weber et al. (1991)

Criteria	Number of mentioned out of 13	Ranked in Dickson's table
Quality	13	1
Delivery	13	2
Price	8	5
Geographical location	7	20
Production facilities and capacity	6	6

at least one of the criteria that Dickson mentioned. Moreover, 64 % of these articles mentioned at least two of Dickson's criteria. Weber et al. (1991) also studied 13 articles related to JIT (just in time) philosophy in order to see which of Dickson's criteria were mentioned in them. Their results are listed in Table 2.

Notice that even after 45 years these criteria still have relevance. Dickson's table ranked geographical location as 20th out of the 23 criteria (average importance). When it comes to supplier quality evaluation using JIT philosophy criteria, it might be considered in a more advanced position than Dickson's ranking.

Huang and Keskar (2007) proposed comprehensive metrics for supplier quality evaluation of original equipment manufacturers (OEMs). They came up with a list of metrics for seven categories under three divisions: "reliability, responsiveness and flexibility" in the product-related division; "cost and financial" and "assets and infrastructure" in the supplier-related division; and safety and environment in the society division. Additionally, they considered three types of products in their construction of the metrics: make to stock (MTS), make to order (MTO), and engineer to order (ETO). They came up with a total of 101 metrics for supplier quality evaluation for OEMs. For the list of metrics, the reader may refer to the original paper by Huang and Keskar (2007).

4 The Proposed Methodology

The proposed model consists of three stages. The first stage is devoted to determining the evaluation criteria and their weights. The second stage focuses on prequalifying suppliers and grouping them based on similar characteristics. The final stage deals with evaluating supplier quality and finding the best solution. These stages are summarized in detail as follows.

4.1 Determining the Criteria and Their Weights

To generate the evaluation criteria, affinity diagram technique is used. Affinity diagram technique was proposed by Kawakita Jiro (Foster 2010) and is used to generate groupings of data under uncertainty through brainstorming or analyzing verbal data gathered through meetings, surveys, or interviews. The supplier quality evaluation criteria generated from affinity diagram are presented in Fig. 1.

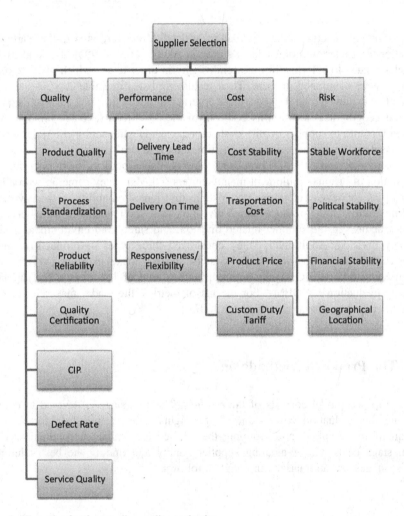

Fig. 1 Hierarchy for the supplier quality evaluation

The description of criteria and sub-criteria is presented as follows:

I. Quality

This is the most important criterion for any organization that is looking to build a strong reputation through satisfying its customers' needs. It can be measured using the following sub-criteria:

1. Product quality (C01)

The quality of the product fits in with customer regulation, as the organization seeks to gain their customers' satisfaction about the product. In short, product quality is the essence of what the customers need.

2. Process standardization (C05)

This is a binary factor; either the supplier has process standardization or it does not. Process standardization pertains to the use of standard methods, techniques, and components.

3. Product reliability (C06)

This sub-criterion represents the robustness of the product, in terms of its number of failures and the likelihood of its durability in retaining the same performance and efficiency.

4. Quality certification (C07)

It involves obtaining quality certificate in any kind of quality that has been gained through satisfying certificate requirements, such as ISO 9000 or any other quality certification.

5. Continuous improvement program (CIP) (C09)

This includes presence of a program or initiative whereby the organization continuously tries to improve the quality of product or production process or adapts to new technology. Friedl and Wagner (2012) study the importance of supplier development for continuous quality improvement in organizations.

6. Defect rate (C14)

This is the rate at which products are rejected by customers because of defects.

7. Service quality (C04)

The service quality level is measured in terms of empathy, ease of communication, and user friendliness.

II. Cost

Cost or price is also a significant factor in supplier selection. Customers are always looking for the minimum product cost so they can maximize their profit or the value of their purchase. The sub-criteria related to this criterion are:

1. Cost stability (C08)

Cost stability refers to how often the supplier changes its product cost. Put another way, this is a measure of whether the customer has a long-term agreement with the supplier.

2.Transportation cost (C15)

The assumption of transportation cost variations depends on the supplier's location. It is different for local, international, and global suppliers.

3. Product price (C17)

It is the purchase price of product expressed in dollars.

4. Custom cost/Tariff (C18)

This sub-criterion applies to global suppliers for their customs charges.

III. Performance

The performance of the supplier is its ability to react to and meet the customer's needs within the agreement period or as quickly as possible. This criterion can be measured through the following attributes:

1. Responsiveness/Flexibility (C03)

This is the ability of the supplier to respond to any change from the customer in terms of any increase in the product quantity or an urgent order.

2. Delivery on time (C02)

This is the ability of the supplier to deliver a shipment at the right time.

3. Delivery lead time (C16)

It is the time from ordering the item until it arrives at the point of sales. For example, this is assumed to be between 2 and 4 days for local suppliers, 3 and 7 days for national suppliers, and between 12 and 20 days for global suppliers.

IV. Risk

Risk is an important factor that buyers should consider and study carefully, especially when dealing with global suppliers. It can affect the ability to meet the customer's expectations, such as receiving late shipment or low-quality products. The following sub-criteria are related to risk:

1. Workforce stability (C10)

This represents the satisfaction of the employees with their job and the environment that they work in.

2. Political stability (C11)

This is an important factor, especially when the supplier is international. Political change in a given country can change business policies and practices, and therefore affect the long-term partnership between supplier and buyer.

3. Financial stability (C12)

The financial status of the supplier is important for a long-term partnership with buyers. It is the backbone that gives supplier the ability to improve, adapt to new technologies, and survive among its competitors.

4. Geographical location (C13)

The geographical locations of suppliers are classified as local, national, and global, respectively. Bayo-Moriones et al. (2011) study the effect of supplier localization on quality assurance practices in the global supply chain.

Table 3 summarizes the different values assumed for the various criteria and sub-criteria presented above for study purposes. The qualitative criteria are of two types: nominal and binary. The nominal value is par value, where a specific value assigns to specific expression of word. However, binary value is either 0 (not present) or 1 (present).

To assign weights to the generated criteria and sub-criteria, AHP is used. The AHP method consists of four steps: first, define the problem; next, build the hierarchy; then, perform pairwise comparisons; and finally, evaluate the weights.

Table 3 Sub-criteria assumption

Code	Sub-criteria	Scale	Objective	Data type
C01	Product quality	[1–7]	Maximize	Nominal
C02	Delivery on time	[1–7]	Maximize	Nominal
C03	Responsiveness/flexibility	[1–7]	Maximize	Nominal
C04	Service quality	[1–7]	Maximize	Nominal
C05	Process standardization	[0–1]	Maximize	Binary
C06	Product reliability	[1–7]	Maximize	Nominal
C07	Quality certification	[0–1]	Maximize	Binary
C08	Cost stability	[0–1]	Maximize	Binary
C09	CIP	[0–1]	Maximize	Binary
C10	Stable workforce	[0–1]	Maximize	Binary
C11	Political stability	[1–7]	Maximize	Nominal
C12	Financial stability	[1–7]	Maximize	Nominal
C13	Geographical location	[1–2–3]	Minimize	Nominal
C14	Defect rate	[2–15]/100,000 items	Minimize	Continuous
C15	Transportation cost	L/N [1000–1750], G [1500–2500]	Minimize	Continuous
C16	Delivery lead time	L [2–4], N [3–7] and G [12–20]	Minimize	Continuous
C17	Product price	L [$250–$350], N [$200–$300] and G [$100–$200]	Minimize	Continuous
C18	Custom cost/tariff	10 % of C17	Minimize	Continuous

L, N, and G stand for Local, National, and Global, respectively

In our case, the criteria/sub-criteria information is provided by the affinity diagram and the hierarchy is generated using AHP. The first level or level one contains the problem objective as stated in step one. Level two contains the main criteria. Level three contains the sub-criteria associated with the main criteria. At the last level are the alternatives for evaluation.

To generate criteria weights, a pairwise comparison is conducted for each element at the same level and with respect to the one above it using the principle of AHP (Saaty 1980). For example, a pairwise comparison should be done between the main criteria at first. Then, another comparison should be done to the set of sub-criteria below each of the main criteria. Saaty (2008) suggests that the pairwise comparison be done through the use of a scale. This scale is shown in Table 4. The next step is using the pairwise matrix to rank the priorities of criteria using the eigenvector approach. In this approach, the matrix is multiplied by itself and then each row is summed. After that, the summed rows will be normalized. This will be done again to the last matrix by repeating the same procedure. Then the results will be compared to the previous one. If the results nearly match, the process stops; otherwise, the process will be repeated until no differences between two consecutive calculations appear.

Table 4 AHP pairwise comparison scale adopted from (Saaty 2008)

Intensity of importance	Definition	Explanation
1	Equal importance	Two activities contribute equally to the objective
2	Weak or slight	
3	Moderate importance	Experience and judgment slightly favor one activity over another
4	Moderate plus	
5	Strong importance	Experience and judgment strongly favor one activity over another
6	Strong plus	
7	Very strong or demonstrated importance	An activity is favored very strongly over another; its dominance demonstrated in practice
8	Very, very strong	
9	Extreme importance	The evidence favoring one activity over another is of the highest possible order of affirmation
Reciprocals of above	If activity *i* has one of the above nonzero numbers assigned to it when compared with activity *j*, then *j* has the reciprocal value when compared with *i*	A reasonable assumption

Table 5 Random index for each matrix size (adopted from Saaty 1982)

Matrix size (n)	1	2	3	4	5	6	7	8	9	10
Random index (RI)	0	0	0.58	0.9	1.12	1.24	1.32	1.41	1.45	1.49

Since the decision maker's judgment could be subjective or random, an evaluation of the outputs of the pairwise comparisons performed in the previous step should be done to check the inconsistency of the results. Saaty (1982) recommended that the value of consistency ratio should be equal to or less than 10 % in order to accept the inconsistency. Otherwise, a revision should be done. To check for inconsistencies, a consistency ratio (CR) should be applied as follows:

1. First, find the eigenvalue (λ_{Max}) by multiplying the pairwise matrix with the weight matrix.
2. Then, divide the result over its corresponding weight. The eigenvalue is the average of the results.
3. After finding the eigenvalue, calculate the consistency index (CI) as $CI = (\lambda_{Max} - n)/(n - 1)$ where n is the matrix size.
4. The final step is to calculate the consistency ratio (CR) using the formula $CR = CI/RI$, where RI is the random index. Saaty (1982) suggested some values for the random index and they are listed in Table 5.

4.2 Clustering the Suppliers into Groups

Before subjecting suppliers to quality evaluation, we perform clustering to group suppliers with similar characteristics together. This step acts as a prequalifying step in the sense that all poor quality suppliers will be grouped together and therefore we can eliminate them from subjecting to our next level of analysis. Following steps are performed in this stage:

4.2.1 Normalizing Data

Since the variables have different units, a normalization process has to be done to make the data dimensionless and bring in the range between 0 and 1. This process allows the variables to contribute equally to the dissimilarity or similarity measure when applied in cluster analysis (Romesburg 1984). A number of formulas have been proposed in literature for normalization. Some of them, as listed by Milligan and Cooper (1988) are presented as follows:

$$z_1 = (x - \bar{x})/s$$
$$z_2 = x/s$$
$$z_3 = x/\max_j x_{ij}$$
$$z_4 = x/(\max_j x_{ij} - \min_j x_{ij})$$
$$z_5 = (x - \min_j x_{ij})/(\max_j x_{ij} - \min_j x_{ij})$$

where x is the data value, $\bar{x}$ is the average, and s is the standard deviation. Each of these formulas has its advantages and disadvantages. For example, z_1 "may not perform properly if there are substantial differences among the within-cluster standard deviations" (Milligan and Cooper 1988). Milligan and Cooper (1988) conclude that z_4 and z_5 are better than the others when they are used for cluster analysis because they form the original cluster structure better than the others. Moreover, they indicate that using z_4 or z_5 in Euclidean distances would provide the same results if applied on same data.

4.2.2 Multiplying Sub-criteria Weights by Normalized Data

The purpose of this step is to make use of the criteria/sub-criteria weights given to criteria in the clustering process so that the suppliers with the most similar pro-prieties fall in the same cluster. If this step is neglected, there can be no guarantee that the best suppliers will occur within one or two clusters. Integrating the sub-criteria weights to the data leads to better investigation of all the suppliers. For example, if the weights are not considered and an element (supplier in our case) is good in 10 variables, the results might change when weights are considered. The element might show as good in only 3 or 4 criteria instead of 10 during the clustering process, and vice versa. This explains why it is important to consider the weights at this stage.

4.2.3   Determine the Number of Clusters

To prequalify the suppliers, the number of clusters needs to be known. Since determining the number of clusters is critical, researchers have proposed a number of methods to find the value of k. But none of the proposed methods so far has been commonly agreed upon by the researchers, and therefore the problem still exists. One of the ways to determine the value of k or the number of clusters is through the dendrogram, which is obtained from hierarchical clustering. The best cut is used to find the distinct clusters inherent within it (Holt 1996). The best cut has been defined by Romesburg (1984) to be the largest width of range between two joint distances. It is clarified in Fig. 2. This figure shows that the best cut is between distance 0.1583 and 0.1449, since the width of range between the two joint distances is the higher among the others (0.0024, 0.0048, and 0.0124). In this case, the suitable number of clusters is four ($k = 4$), as seen in Fig. 3.

4.2.4   Applying k-Means Clustering to Qualify the Suppliers

k-means clustering is a very well-known technique due to its ability to deal efficiently with large data sets as long as the initial number of (k) clusters is known. This is why it is has been considered for use in this stage of the process. k-means clustering is a type of partitional clustering. It has the ability to deal with large sets of data more efficiently than hierarchical clustering. It uses an iteration procedure to

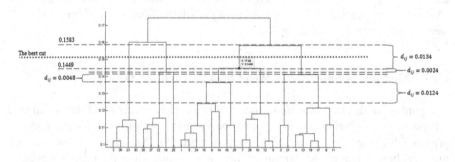

Fig. 2 Dendogram showing best cut

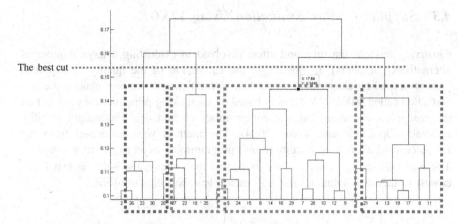

Fig. 3 Dendrogram showing best cut and cluster memberships

form the clusters. *k*-means cluster analysis consists of five steps (Khan and Ahmad 2004):

Step 1   Determining the number of clusters (*k*)

Step 2   Choosing initial seeds or centroids for each cluster

Step 3   Determining the distance from centroid to each object

Step 4   Grouping objects based on minimum distance and

Step 5   If the clusters are stable (the position of objects are not changed from the previous iteration), end. Otherwise, start over from Step 2 and update the cluster centroids

The most common distance measure used with *k*-means clustering is Euclidean distance (Mu-Chun and Chien-Hsing 2001).

To execute this technique, SPSS software was used.

4.2.5   Analyzing the Clustering Results

To find which cluster has the best group of suppliers, an analysis needs to be performed. The analysis will be done based on the center of each cluster. This center is the weighted average of the different criteria centers present in each cluster. The cluster that has the highest mean will have the best suppliers (Holt 1996). Since the sub-criteria weights are already integrated with the data, the cluster with the highest mean will have the best suppliers satisfying those weights.

4.3 Supplier Quality Evaluation Using VIKOR

Clustering reduces the time and effort involved in evaluating a large number of alternatives (suppliers). In this stage, the evaluation of the quality of suppliers present in the best cluster is performed using multicriteria decision-making method (MCDM) called VIKOR. VIKOR is based on outranking principle and used to find the compromise ranking list, the compromise solution, and the weight stability intervals (Opricovic and Tzeng 2004). The method was developed from the L_p—metric which is used in compromise programming as an aggregation function. The method uses L_p—metric concepts to find the compromise solution that is the closest to the ideal solution. The L_p—metric has the following form:

$$ L_{p,j} = \left\{ \sum_{i=1}^{n} [w_i (f_i^* - f_{ij})/(f_i^* - f_i^-)]^p \right\}^{1/p} ; \quad 1 \le p \le \infty, \, j = 1, 2, \ldots, J $$

Figure 4 shows the relationship between a compromise solution (F^c) and the ideal solution (F^*) in the L_p—metric.

VIKOR method has the following steps:

1. Find the best of f_i^* and the worst of f_i^- of all criterion $i = (1, 2, \ldots, n)$ as follows:

   a. If i represents a benefit, then $f_i^* = \max_j f_{ij}$ and $f_i^- = \min_j f_{ij}$
   b. If i represents a cost, then $f_i^* = \min_j f_{ij}$ and $f_i^- = \max_j f_{ij}$

2. Compute linear normalization for all alternatives $d_{ij} = (f_i^* - f_{ij})/(f_i^* - f_i^-)$

3. Find the values $S_j = \sum_{i=1}^{n} w_i d_{ij}$ and $R_j = \max_i w_i d_{ij}$ of all alternatives $j = (1, 2, \ldots, J)$. Where w_i is the weight of criterion i.

Fig. 4 Ideal and compromise solutions adopted from (Opricovic and Tzeng 2004)

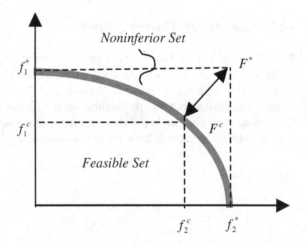

4. Compute the values $Q_j = v(S_j - S^*)/(S^- - S^*) + (1 - v)(R_j - R^*)/(R^- - R^*)$ of all alternatives, where $S^* = \min_j S_j$, $S^- = \max_j S_j$, $R^* = \min_j R_j$, $R^- = \max_j R_j$, and v is the weight of the strategy of "the majority of criteria."
5. Rank the alternatives, sorting by the values $S, R,$ and Q, in decreasing order.
6. The compromise solution is the first-ranked alternative (a') by Q if a'

   a. has the acceptable advantage that $Q(a'') - Q(a') \geq 1/(J - 1)$, (Note: if $J \leq 4$ then $1/(J - 1) = 0.25$ (Chen and Wang 2009) where a'' is the alternative with second position in the ranking list by Q; and
   b. has the acceptable stability in decision making, so that it will be the best ranked by S or/and R.
   If a' is not satisfied by point (b) then, the compromise solutions are a' and a''. But if a' is not satisfied by point (a), then the compromise solutions consist of $a', a'', \ldots a^M$, where a^M is determined by the relation $Q(a^M) - Q(a') < 1/(J - 1)$ for maximum M.

5 Numerical Example

Consider an organization ABC who is dealing with 625 suppliers in a heterogeneous business environment. These suppliers consist of 130 local suppliers, 195 national suppliers, and 300 global suppliers. First, clustering the suppliers will be based on three criteria: product type, supplier location, and product volume. The results of the first level of clustering are 250 suppliers, which consist of 48 local suppliers, 88 national suppliers, and 114 international suppliers. This step was performed to reduce computational complexity. Now, detailed analysis of quality of these 250 suppliers will be performed using the proposed model. Eighteen sub-criteria will be used. Thirteen of them are qualitative and the other five are numerical data. The criteria data were randomly generated using Excel. The criteria are a mixture of qualitative and quantitative variables. To treat the problem in a numerical way, the qualitative data have been quantified using the following scale:

For nominal variables, the criteria are quantified into seven scales: very low (VL) quantify to 1, low (L) to 2, and so on as seen in Table 6.

However, for binary variables 1 represents YES or present, and 0 represents NO or absent. For the geographical location criterion, 1 has been assigned to local suppliers, 2 to national suppliers, and 3 to global suppliers. The data have been

Table 6 Linguistic scale

Very low (VL)	Low (L)	Medium low (ML)	Medium (M)	Medium high (MH)	High (H)	Very high (VH)
1	2	3	4	5	6	7

normalized so that all criteria will be unitless and in the range of 0 and 1. For the normalization process, the following formulas have been used:

$$x_{ij}^* = \frac{x_{ij} - \min_j x_{ij}}{\max_j x_{ij} - \min_j x_{ij}} \tag{1}$$

if the variable needs to be maximized and

$$x_{ij}^* = \frac{(1/x_{ij}) - \min_j(1/x_{ij})}{\max_j(1/x_{ij}) - \min_j(1/x_{ij})} \tag{2}$$

if the variable needs to be minimized.

5.1 Criteria/Sub-criteria Weight Generation Using Analytic Hierarchy Process

The purpose of using this technique is to find the appropriate weight for each criterion along with the weights to be given to their sub-criteria. This method was run in Excel. The pairwise comparison results for the four main criteria and their sub-criteria are shown in Tables 7, 8, 9, 10, and 11.

Table 7 The evaluation of main criteria

	Quality	Service	Cost	Risk	Weight
Quality	1	4	3	8	0.55
Performance	1/4	1	1/2	5	0.154
Cost	1/3	2	1	7	0.253
Risk	1/8	1/5	1/7	1	0.043

Table 8 The evaluation of sub-criteria with respect to quality criterion

	C01	C05	C06	C07	C09	C14	C04	Weight
C01	1	3	3	6	5	2	5	0.327
C05	1/3	1	1/2	5	2	1/3	6	0.127
C06	1/3	2	1	5	4	1/2	4	0.169
C07	1/6	1/5	1/5	1	1/2	1/6	3	0.044
C09	1/5	1/2	1/4	2	1	1/3	3	0.067
C14	1/2	3	2	6	3	1	5	0.233
C04	1/5	1/6	1/4	1/3	1/3	1/5	1	0.033

Table 9 The evaluation of sub-criteria with respect to performance criterion

	C03	C02	C16	Weight
C03	1	1/2	2	0.311
C02	2	1	2	0.493
C16	1/2	1/2	1	0.196

Table 10 The evaluation of sub-criteria with respect to cost criterion

	C08	C15	C17	C18	Weight
C08	1	2	1	4	0.344
C15	1/2	1	1/2	6	0.233
C17	1	2	1	5	0.36
C18	1/4	1/6	1/5	1	0.063

Table 11 The evaluation of sub-criteria with respect to risk criterion

	C10	C11	C12	C13	Weight
C10	1	1/2	1/2	3	0.197
C11 ·	2	1	1/2	3	0.28
C12	2	2	1	5	0.443
C13	1/3	1/3	1/5	1	0.081

Table 12 Consistency test for the criteria

	λ_{MAX}	CI	RI	CR
Main criteria	4.131	0.043655	0.9	0.0485
Quality sub-criteria	7.481	0.080201	1.32	0.0608
Performance sub-criteria	3.054	0.026811	0.58	0.0462
Cost sub-criteria	4.129	0.043013	0.9	0.0478
Risk sub-criteria	4.065	0.021602	0.9	0.024

To check that the results are consistent, verification was performed. For the results to be considered valid, the consistency ratio should be less than 0.1 or 10 % of all pairwise comparisons. The results of this test are shown in Table 12. It can be seen that all the CRs < 0.1 for each of the pairwise comparison results, therefore the results are consistent.

After all the weights are verified to be consistent, the next step is to multiply the main criteria weight with their corresponding sub-criteria weight to make the sum of the weights of all sub-criteria equal to 1. Table 13 shows the sub-criteria weights generated using AHP.

5.2 Generating Supplier Groups Using Cluster Analysis

In this stage, dendrogram obtained from hierarchical cluster analysis will be used to determine the number of clusters. Once the number of clusters (k) is determined, then k-means cluster analysis will be performed to find the groups. Before performing the analysis, the clusters need to be guaranteed to have similar attributes for suppliers. Therefore, the weights determined from AHP will be multiplied with their corresponding normalized value to make sure that the best suppliers fall in the same group respecting the criteria weights. The dendrogram obtained from SPSS

Table 13 Weights of all sub-criteria

		Sub-criteria W	Criteria W × sub-criteria W
C01	Product quality	0.327	0.18
C02	Delivery on time	0.493	0.076
C03	Responsiveness/flexibility	0.311	0.048
C04	Service quality	0.033	0.018
C05	Process standardization	0.127	0.07
C06	Product reliability	0.169	0.093
C07	Quality certification	0.044	0.024
C08	Cost stability	0.344	0.087
C09	CIP	0.067	0.037
C10	Stable workforce	0.197	0.009
C11	Political stability	0.28	0.012
C12	Financial stability	0.443	0.019
C13	Geographical location	0.081	0.003
C14	Defect rate	0.233	0.128
C15	Transportation cost	0.233	0.059
C16	Delivery lead time	0.196	0.03
C17	Product price	0.36	0.091
C18	Customs cost/tariff	0.063	0.016

using Euclidean distance measures and the within-group linkage method, is shown in Fig. 5.

The dendrogram shows that the number of clusters is 4, where the best cut is occurred. Now that the analysis has produced a clear idea of the possible number of $k = 4$, k-means clustering can be applied. The results of the k-means cluster analysis are shown in Table 14.

The next step is to analyze these clusters and find the best cluster among them. This will be done by considering centers for the 18 sub-criteria in each cluster. Then, an average cluster center will be calculated using the weighted average of the 18 sub-criteria cluster centers. This step will be performed for each cluster. The cluster that has the highest average cluster mean is considered to be the best. Table 15 lists the center of 18 sub-criteria and average cluster center for the four clusters. It can be deduced that cluster number one is the best cluster among the group. Please note that the data used in clustering was normalized using Eqs. (1) and (2).

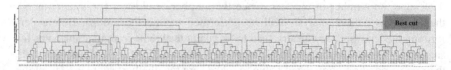

Fig. 5 Dendrogram for hierarchical analysis

Table 14 Clusters' memberships

	Cluster 1			Cluster 2	Cluster 3					Cluster 4			
1	S004	S077	S181	S002	S003	S057	S124	S176	S250	S001	S086	S165	S227
2	S007	S084	S186	S020	S008	S059	S125	S177		S005	S087	S166	S229
3	S012	S085	S199	S033	S009	S060	S127	S180		S006	S089	S169	S231
4	S013	S091	S200	S037	S010	S063	S128	S182		S016	S090	S170	S232
5	S014	S094	S202	S064	S011	S067	S132	S184		S018	S093	S179	S238
6	S017	S095	S207	S065	S015	S070	S134	S190		S022	S098	S185	S239
7	S023	S102	S210	S079	S019	S074	S136	S191		S025	S101	S188	S242
8	S024	S106	S212	S082	S021	S092	S137	S193		S039	S103	S189	S246
9	S026	S116	S220	S088	S027	S096	S138	S194		S048	S105	S192	
10	S028	S117	S224	S107	S029	S097	S140	S195		S049	S108	S197	
11	S032	S129	S226	S114	S030	S099	S141	S196		S050	S110	S201	
12	S036	S131	S230	S130	S031	S100	S145	S198		S051	S112	S203	
13	S038	S133	S233	S139	S034	S104	S146	S205		S052	S121	S204	
14	S040	S143	S234	S154	S035	S109	S147	S211		S061	S123	S206	
15	S043	S148	S235	S158	S041	S111	S151	S214		S066	S126	S208	
16	S044	S150	S236	S167	S042	S113	S152	S216		S069	S135	S209	
17	S056	S155	S240	S171	S045	S114	S157	S222		S071	S142	S213	
18	S058	S156	S241	S178	S046	S115	S160	S228		S073	S144	S215	
19	S062	S161	S243	S183	S047	S118	S162	S237		S076	S149	S217	
20	S068	S172	S245	S187	S053	S119	S164	S244		S078	S153	S218	
21	S072	S173	S249	S219	S054	S120	S168	S247		S081	S159	S221	
22	S075	S174		S225	S055	S122	S175	S248		S083	S163	S223	

5.3 Evaluating Suppliers in the Best Cluster Using VIKOR

Now that the best cluster has been found, it can be seen that only a few data will be dealt with instead of many for supplier quality evaluation. Cluster 1 groups the best 65 suppliers from 250. Evaluating 65 suppliers is much easier than evaluating 250 suppliers.

To use the VIKOR technique, the data acquired from 65 suppliers was normalized using the normalization method suggested by its founders (Sect. 4). Table 16 presents the results for the first 15 suppliers ranked by the VIKOR method. These results were computed using Excel Spreadsheet.

In this calculation v is assumed to be 0.5, and since there are 65 suppliers, $1/(J-1) = 1/(65-1) = 0.0156 \leq Q(a'') - Q(a') = 0.090 - 0.027 = 0.063$.

According to this, condition one is satisfied. However, condition two is not satisfied because supplier 230 is not ranked first in S and/or R. Consequently, a compromise solution should be considered. Such a compromise would be to select both supplier 230 and supplier 85. Note that the rectangle under column R means that all the suppliers have the same value.

Table 15 Criteria and cluster centers

	C01	C02	C03	C04	C05	C06	C07	C08	C09	C10	C11	C12	C13	C14	C15	C16	C17	C18	Average	Cluster center
Cluster 1 65	0.132	0.038	0.026	0.009	0.032	0.049	0.013	0.087	0.018	0.005	0.008	0.010	0.001	0.027	0.022	0.009	0.029	0.005	0.029	
Cluster 2 22	0.034	0.037	0.025	0.008	0.032	0.032	0.013	0.020	0.025	0.005	0.008	0.009	0.001	0.101	0.022	0.010	0.037	0.006	0.024	
Cluster 3 89	0.027	0.038	0.023	0.009	0.038	0.049	0.013	0.045	0.019	0.005	0.008	0.010	0.001	0.016	0.025	0.009	0.030	0.005	0.021	
Cluster 3 74	0.134	0.035	0.023	0.008	0.036	0.042	0.011	0.000	0.024	0.005	0.007	0.010	0.001	0.020	0.025	0.007	0.032	0.005	0.024	

Table 16 Top 15 suppliers from VIKOR method

	Main criteria weights
	Quality = 0.55, Performance = 0.154, Cost = 0.235, Risk = 0.043

	Q		S		R	
1	0.03	S230	0.23	S014	0.06	S085
2	0.09	S085	0.24	S230	0.06	S094
3	0.11	S058	0.26	S058	0.06	S230
4	0.13	S094	0.3	S085	0.06	S007
5	0.15	S014	0.3	S026	0.07	S072
6	0.18	S007	0.31	S094	0.07	S181
7	0.21	S072	0.31	S173	0.07	S058
8	0.22	S181	0.32	S044	0.07	S133
9	0.27	S026	0.33	S072	0.07	S200
10	0.27	S077	0.34	S040	0.07	S012
11	0.28	S012	0.34	S007	0.08	S077
12	0.28	S129	0.34	S181	0.08	S106
13	0.29	S133	0.35	S077	0.08	S129
14	0.3	S068	0.35	S161	0.08	S241
15	0.31	S062	0.35	S068	0.08	S249

6 Conclusions

Choosing the right supplier plays an important role in making organizations profitable and keeping them focused on their potential strengths. Therefore, evaluating the quality of suppliers carefully is vital for any company. The purpose of supplier quality evaluation, however, can differ from one company to another. Some companies might have a large set of suppliers and consequently might want to reduce the number of suppliers so that they can manage them more efficiently and focus on building long-term relationships with only preferred suppliers. On the other hand, some companies might be looking for new suppliers to deal with, therefore, they may use different rationale for supplier quality evaluation.

In this chapter, we propose a three-stage model for supplier quality evaluation. The first stage focuses on selecting evaluation criteria and assigning weight to each criterion using affinity diagram and analytical hierarchy process techniques. This is an essential step in the evaluation of supplier quality.

The second stage addresses the challenge of handling large supplier data sets and reducing complexity by conducting clustering. In this stage, the weights assigned in the first stage are integrated with suppliers' ratings. This step is important for the formation of the right clusters, and the determination of which cluster should have the best suppliers. Considering the weights in the clustering process ensures that similar suppliers are grouped together, since clustering suppliers without considering criteria weights will not consider the trade-offs between criteria.

In the third and the last stages, outranking technique VIKOR is applied to select the best supplier(s) in the supplier cluster obtained from stage 2. The purpose of selecting suppliers is not always to find the best supplier; sometimes its purpose is to reduce the number of suppliers or to choose a specific number of suppliers. By ranking the results, the user can have a clearer appreciation of each supplier and its relative position. Thus, this method enables customers to evaluate suppliers much more easily than before, and will save time and effort in evaluation of large data sets of suppliers.

The proposed model enables one to deal with suppliers' large data sets and simplifies the MCDM problem to the point of dealing with a reduced number of suppliers with a variety of variables (qualitative, quantitative). To assure customers about the quality of the supplier, a careful, efficient, and reliable evaluation must be performed. The proposed model offers such an evaluation, since it identifies the best group of suppliers as dominant over the others, and its last stage gives buyers the chance to choose the number of suppliers they need and keep a list of others for future references or as a backup.

This research has some limitations, as some of the data was generated by Excel which does not reflect the real data. Therefore, there is still scope for verifying and validating model results by considering real data.

References

Barla, S. B. (2003). A case study of supplier selection for lean supply by using a mathematical model. *Logistics Information Management, 16*(6), 451–459.

Bayo-Moriones, A., Bello-Pintado, A., & Merino-Díaz-de-Cerio, J. (2011). Quality assurance practices in the global supply chain: the effect of supplier localisation. *International Journal of Production Research, 49*(1), 255–268.

Chan, F. T. S., & Chan, H. K. (2004). Development of the supplier selection model—a case study in the advanced technology industry. *Proceedings of the Institution of Mechanical Engineers Part B - Journal of Engineering Manufacture, 218*(12), 1807–1824.

Chen, L. Y., & Wang, T. (2009). Optimizing partners' choice in IS/IT outsourcing projects: the strategic decision of fuzzy VIKOR. *International Journal of Production Economics, 120*(1), 233–242.

Chen, Z., & Yang, W. (2011). An MAGDM based on constrained FAHP and FTOPSIS and its application to supplier selection. *Mathematical and Computer Modeling, 54*(11–12), 2802–2815.

Choy, K. L., Lee, W. B., Lau, H. C. W., & Choy, L. C. (2005). A knowledge-based supplier intelligence retrieval system for outsource manufacturing. *Knowledge-Based Systems, 18*(1), 1–17.

Dickson, G. W. (1966). An analysis of vendor selection systems and decisions. *Journal of Purchasing, 2*(1), 5–17.

Foster S. T. (2010). Managing quality: integrating the supply chain, 4th edn. Prentice Hall.

Friedl, G., & Wagner, S. M. (2012). Supplier development or supplier switching? *International Journal of Production Research, 50*(11), 3066–3079.

Holt, G. D. (1996). Applying cluster analysis to construction contractor classification. *Building and Environment, 31*(6), 557–568.

Huang, S., & Keskar, H. (2007). Comprehensive and configurable metrics for supplier selection. *International Journal of Production Economics, 105*(2), 510–523.

Jain, V., Tiwari, M. K., & Chan, F. T. S. (2004). Evaluation of the supplier performance using an evolutionary fuzzy-based approach. *Journal of Manufacturing Technology Management, 15*(8), 735–744.

Karpak, B., Kumcu, E., & Kasuganti, R. R. (2001). Purchasing materials in the supply chain: Managing a multi-objective task. *European Journal of Purchasing and Supply Management, 7*(3), 209–216.

Khan, S. S., Ahmad, A., (2004). Cluster center initialization algorithm for K-means clustering. *Pattern Recognition Letters, 25*(11).

Lam, K. C., Tao, R., & Lam, M. C. K. (2010). A material supplier selection model for property developers using Fuzzy Principal Component Analysis. *Automation in Construction, 19*(5), 608–618.

Liu, F.-H. F., & Hai, H. L. (2005). The voting analytic hierarchy process method for selecting supplier. *International Journal of Production Economics, 97*(3), 308–317.

Milligan, G. W., & Cooper, M. C. (1988). A study of standardization of variables in cluster analysis. *Journal of Classification, 5*(2), 181–204.

Mu-Chun, S., & Chien-Hsing, C. (2001). A modified version of the K-means algorithm with a distance based on cluster symmetry. *IEEE Transactions on Pattern Analysis and Machine Intelligence, 23*(6), 674–680.

Narasimhan, R. (1983). An analytical approach to supplier selection. *Journal of Purchasing and Materials Management, 19*(4), 27–32.

Nydick, R. L., & Hill, R. P. (1992). Using the analytic hierarchy process to structure the supplier selection procedure. *Journal of Purchasing and Materials Management, 25*(2), 31–36.

Opricovic, S., & Tzeng, G. H. (2004). Compromise solution by MCDM methods: a comparative analysis of VIKOR and TOPSIS. *European Journal of Operational Research, 156*(2), 445–455.

Partovi, F. Y. (1994). Determining what to benchmark: an analytic hierarchy process approach. *International Journal of Operations & Production Management, 14*(6), 25–39.

Partovi, F. Y., Burton, J., & Banerjee, A. (1989). Application of analytic hierarchy process in operations management. *International Journal of Operations and Production Management, 10*(3), 5–19.

Rezaei, J., & Ortt, R. (2012). A multi-variable approach to supplier segmentation. *International Journal of Production Research, 50*(16), 4593–4611.

Romesburg, H. C. (1984). *Cluster analysis for researchers*. Belmont: Lifetime Learning.

Saaty, T. L. (1980). *The analytic hierarchy process*. New York: McGraw-Hill.

Saaty, T. L. (1982). *Decision making for leaders*. Belmont: Lifetime Learning.

Saaty, T. L. (2008). Decision making with the analytic hierarchy process. *International Journal of Services Sciences, 1*(1), 83–98.

Sarkis, J., & Talluri, S. (2002). A model for strategic supplier selection. *Journal of Supply Chain Management, 38*(1), 18–28.

Talluri, S., & Narasimhan, R. (2003). Vendor evaluation with performance variability: a max–min approach. *European Journal of Operational Research, 146*(3), 543–552.

Vaidya, O. S., & Kumar, S. (2006). Analytic hierarchy process: An overview of applications. *European Journal of Operational Research, 169*(1), 1–29.

Weber, C. A., Current, J. R., & Benton, W. C. (1991). Vendor selection criteria and methods. *European Journal of Operational Research, 50*, 2–18.

Zanakis, S. H., Solomon, A., Wishart, N., & Dublish, S. (1998). Multi-attribute decision making: A simulation comparison of select methods. *European Journal of Operational Research, 107*(3), 507–529.

Risk-Adjusted Performance Monitoring in Healthcare Quality Control

Li Zeng

1 Introduction

With the growing emphasis and concerns on quality of health care, performance monitoring of care providers has received much attention recently. Performance measures used as monitors are typically clinical outcomes, utilization of health services, and cost. By monitoring these measures continuously, changes in the performance of care providers can be detected promptly to avoid serious consequences as well as provide valuable information on the care delivery system for quality improvement.

One critical challenge in performance monitoring in medical contexts is the need to adjust for patient case mix, called *risk adjustment* (Iezzoni 1997). Unlike products in manufacturing processes which are relatively homogeneous in nature, patients vary a lot in their characteristics or risk factors, which may affect the performance monitors. For example, sicker patients tend to experience worse outcomes, even with excellent care, than their healthier counterparts. The affecting patient risk factors must be taken into account in the monitoring to fairly assess the performance of care providers. Performance monitoring with such considerations is referred to as risk-adjusted (RA) monitoring in the literature.

Two basic problems are involved in RA monitoring, as illustrated in Fig. 1: *establishing risk adjustment models,* which includes identifying the appropriate performance measures to monitor and associated patient risk factors, constructing statistical models that characterize the dependency of the performance measures on the risk factors, and *change detection based on the established models,* which includes estimating baseline parameters of the risk adjustment models and detecting deviations from them. The former has been a focus of physicians and medical

L. Zeng (✉)
Department of Industrial and Systems Engineering,
Texas A&M University, College Station, TX 77843, USA
e-mail: lizeng@tamu.edu

© Springer-Verlag London 2016
H. Pham (ed.), *Quality and Reliability Management and Its Applications,*
Springer Series in Reliability Engineering, DOI 10.1007/978-1-4471-6778-5_2

quality researchers, while the later has attracted the attention of statisticians, including those industrial statisticians who are extending their statistical process control (SPC) research in industrial contexts to medical contexts. The objective of this chapter is to review the main developments on the two problems. A case study will also be provided to demonstrate the use of a powerful method, Bayesian approaches, for RA monitoring. It is worth mentioning that, unlike previous reviews of this topic which merely focus on the change detection problem, establishment of risk adjustment models is also considered in this chapter to provide a broader view of the RA monitoring problem which can help readers to understand the techniques for change detection better as well as enable the identification of potential collaboration opportunities between researchers in different areas.

2 Risk Adjustment Models

As the basis and first step in risk-adjusted monitoring, a statistical risk adjustment model must be constructed based on domain knowledge of the application of interest and historical data available. As shown in Fig. 1, a risk adjustment model consists of three components: performance measures to monitor, patient risk factors that may affect the performance measures, and statistical models that characterize the dependency of performance measures on the risk factors. Such models vary from one application to another. Typical examples and considerations in constructing these models will be introduced in this section.

2.1 Performance Measures and Patient Risk Factors

Risk adjustment models have been developed in many critical areas in health care in the past two decades, such as thoracic and cardiac surgeries (e.g., Brunelli et al. 2007; Daley et al. 2001; Krumholz et al. 1999; Pinna-Pintor et al. 2002; Shroyer et al. 2003; Sousa et al. 2008; Tu et al. 1995), public mental health (e.g., Hendryx et al. 1999; Hermann et al. 2007), home health care (e.g., Murtaugh et al. 2007),

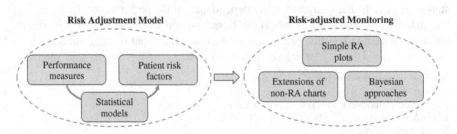

Fig. 1 The framework of risk-adjusted monitoring

Table 1 Examples of performance measures and risk factors in some applications

Application	Key performance measures	Important patient risk factors
Surgeries	30-day mortality survival time	Age, sex, race, previous myocardial infarction, cerebrovascular disease, renal failures requiring
	Presence of major complications	Dialysis
Public mental health	Inpatient length of stay inpatient cost patient satisfaction change in functioning	Age, illness severity, income, prior utilization of mental health services, amount of social support received
Home health care	Improvement in bathing stabilization in speech acute care hospitalization	Age, socioeconomic factors, prior service use, sensory status, diagnosis severity
Nursing home care	Decline in functional status worsening pressure ulcers mortality	Age, sex, medical conditions, physical restraints, decubiti at admission
General hospital care	Presence of readmission inpatient mortality occurrence of patient safety events	Age, sex, principal diagnosis, number of major chronic conditions and significant comorbidities

nursing home care (e.g., Mukamel and Spector 2000; Zimmerman 2003), and general hospital care (e.g., Benbassat and Taragin 2000; Forthman et al. 2010). Cardiac surgeries is the most popular area for such studies, a fact that is due largely to the motivation of well-publicized reports of surgical outcomes and cases where high rates of surgical complications remained undetected for an undue length of time (Steiner et al. 2000). Examples of key performance measures and important affecting risk factors in the above-mentioned applications are listed in Table 1.

The performance measures are typically clinical outcomes, especially adverse events, such as mortality and morbidities, service utilization, length of stay in intensive care unit, patient satisfaction, and cost that are commonly recognized indicators of quality of care in terms of its six main dimensions including effectiveness, efficiency, timeliness, patient-centeredness, equality, and safety (Reid 2005). The most widely used performance measure is mortality. The use of this measure dates back to the releasing of patient mortality records by the Health Care Financing Administration (HCFA) in the 1980s, which aroused much criticism then, but has stimulated discussions on how to measure quality of care (DesHarnais et al. 1991).

There are many affecting patient risk factors associated with a chosen performance measure, ranging from patients' demographic characteristics, diagnosis, severity of illness, medical conditions to socioeconomic status and social support received. To provide convenience in implementing risk adjustment in practice, the effects of different risk factors are often combined into a risk score such as the Parsonnet score and APACH score that have been widely used in cardiac surgeries and intensive care (Iezzoni 1997).

2.2 Statistical Models

With data of performance measures and patient risk factors, statistical models are
built to characterize their relationship. The data are typically obtained from
administrative claims database or patient medical records. As popular performance
measures in health care fall into two categories, binary measures (e.g.,
death/survival, presence/absence of certain complications) and continuous measures
(e.g., survival time following a surgery, length of hospital stay, cost), statistical
models for the two types of measures have been developed in existing studies.

Specifically, the logistic regression model is commonly used for binary measures

$$y \sim \text{Bernoulli}(p)$$
$$\log \frac{p}{1-p} = \beta_0 + \beta_1 x_1 + \beta_2 x_2 + \cdots + \beta_p x_p \tag{1}$$

where y is the performance measurement which takes value 1 or 0 corresponding to
the occurrence or not of the concerned adverse event such as death, p is the probability of the occurrence of the event, which is the parameter of the Bernoulli
distribution that generates y, $x_1, \ldots, x_p$ are the affecting patient risk factors, and $\beta_1, \ldots,$
β_p are the corresponding coefficients in the model which represent the effects of the
risk factors on the performance measure.

There are two types of continuous performance measures, regular ones (e.g.,
cost), and time to event (e.g., survival time following a surgery). For the former,
standard linear regression models are used

$$y = \beta_0 + \beta_1 x_1 + \beta_2 x_2 + \cdots + \beta_p x_p + \varepsilon \tag{2}$$

where ε is the error term assumed to follow a normal distribution. For the latter,
different survival models are used such as the accelerated failure time model

$$\log y = \beta_0 + \beta_1 x_1 + \beta_2 x_2 + \cdots + \beta_p x_p + e \tag{3}$$

where the distribution of the error term e can take different forms defined on $[0, \infty)$,
such as normal distribution, logistic distribution, and extreme value distribution
leading to log-normal, log-logistic, and Weibull distributions for y. More complex
models may also be used such as the Cox proportional hazard models.

In constructing risk adjustment models, two issues need to be considered: First,
variable selection. Selection of significant variables, a basic task in statistical
regression analysis, is especially critical in constructing risk adjustment models
because a large pool of patient characteristics is typically available in healthcare
databases which may contain a lot of irrelevant or redundant information.
Moreover, models involving too many variables may also pose difficulties to data
collection in practice in terms of time and cost. Simple methods, such as stepwise
selection procedures, have been used in the existing studies for this purpose.

Second, dealing with multiple performance measures. Multiple performance measures normally exist to characterize the quality of care in different aspects. The measures may bear complex correlations which need to be taken into account in risk adjustment (DesHarnais et al. 1991). Despite a common recognition of this issue, however, no formal analysis has been done about it in the current literature than simply combining multiple performance measures into one single measure.

3 Risk-Adjusted Performance Monitoring

With the established risk adjustment model, the performance of care providers will be characterized by the parameters of the model, and then monitoring will be conducted to detect changes in the parameters. An example of data used in RA monitoring is shown in Fig. 2, where the upper panel displays a stream of performance measurements in cardiac surgeries, i.e., patient mortality status within 30 days after the surgery (D/S denoting death/survival), and the lower panel shows the associated Parsonnet scores of patients, which indicate their preoperative risk of death.

Before introducing the techniques for RA monitoring, definitions of important concepts on performance monitoring are provided below to facilitate understanding of those techniques:

Phase I versus Phase II monitoring These are the two basic types of monitoring considered in SPC research. Phase I monitoring, also referred to as retrospective monitoring, aims to detect changes in performance during a fixed time period, while Phase II monitoring, also referred to as prospective monitoring, aims to detect deviation of performance from a baseline whenever a data point becomes available. In Phase I monitoring, the data sequence is fixed, and the goal is to

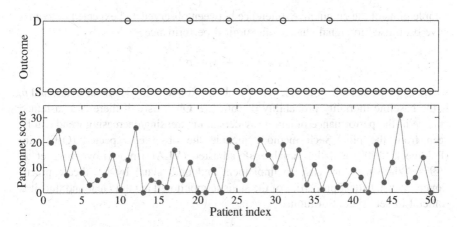

Fig. 2 A typical data set used in risk-adjusted performance monitoring

identify change points in the data as accurately as possible. The baseline is normally unknown, and needs to be estimated in Phase I monitoring. In Phase II monitoring, the data are obtained sequentially, and the goal is to capture a change as soon as possible. A baseline estimated from historical data is typically available, and a change is defined to be a deviation from the baseline. There is also a hybrid type of monitoring, called self-starting monitoring, which takes the online detection scheme in Phase II monitoring, but without a known baseline as in Phase I monitoring.

Grouped monitoring versus continuous monitoring In grouped monitoring, patients are divided into groups of similar sizes, and some aggregate summaries of each group will be monitored. This is consistent with the concept of "subgroup" in industrial SPC research. An alternative way is to monitor each patient individually, called continuous monitoring in the literature. Essentially, the continuous monitoring scheme represents 100 % inspection, which is supposed to be a better scheme for performance monitoring in medical contexts where prompt detection of changes is particularly desired (Woodall 2006).

Grigg and Farewell (2004a, b), Woodall (2006), and Cook et al. (2008) give excellent reviews on methods and techniques for RA performance monitoring. In this chapter, a brief review of the basics and advantages/disadvantages of popular RA monitoring techniques will be given in a unified way, with an emphasis on techniques developed after those reviews and, particularly, Bayesian approaches. These methods can be divided into three categories: simple RA plots, extensions of non-RA SPC control charts, and Bayesian approaches. All the techniques in the first two categories focus on Phase II monitoring except the last one in the second category, while Bayesian approaches can be used for both Phase I and Phase II monitoring.

3.1 Simple Risk-Adjusted Plots

Simple plots of the cumulative difference between observed and expected outcomes have been used to signal changes in surgical performance

$$C_t = C_{t-1} + (y_t - p_t)$$

where C_t is the statistic at time t, y_t is the observed outcome (death/survival), and p_t is the baseline mortality probability of patient t. Obviously, if there is a sustained change in the performance of care providers, an increasing/decreasing trend can be seen from the plot. Such methods include the observed-expected (O-E) plot (Poloniecki 1998) and the variable life-adjusted (VLAD) plot (Lovegrove et al. 1997). While they are easy to implement and understand by healthcare practitioners, the statistical properties of the statistic monitored are not clear and thus it is difficult to set up control limits.

3.2 Extensions of Non-risk-Adjusted SPC Control Charts

As SPC is a well-studied area in other contexts especially industrial applications, various non-risk-adjusted control charts are available in the literature. These techniques have been extended to medical contexts by incorporating risk adjustment. Popular extensions of these techniques will be briefly introduced as follows.

3.2.1 Risk-Adjusted p-Chart

p-chart is a basic SPC technique to monitor binary data such as defectiveness or non-defectiveness of products in industrial process control where defective rate is an important concern. To apply this technique in Phase II monitoring, subgroups of data need to be collected, and a normal distribution is assumed when the sample size of subgroups is adequately large

$$\hat{p}_i = \frac{\sum_{t=1}^{n_i} y_{it}}{n_i} \sim N\left(p_0, \frac{p_0(1-p_0)}{n_i}\right)$$

where n_i is the sample size of subgroup i, $\hat{p}_i$ is the corresponding average defective rate, y_{it} is the measurement of product t in subgroup i, and p_0 is the baseline defective rate estimated from historical data. 3-sigma control limits can be obtained based on this distribution.

To extend this method to medical contexts, patients in consecutive time periods of same length, e.g., 6 months, are grouped, and the distribution used in the monitoring becomes

$$\hat{p}_i = \frac{\sum_{t=1}^{n_i} y_{it}}{n_i} \sim N\left(\frac{1}{n_i}\sum_{t=1}^{n_i} p_{0ti}, \frac{1}{n_i}\sum_{t=1}^{n_i} p_{0ti}(1-p_{0ti})\right)$$

where p_{0ti} is the baseline mortality probability of patient t in group i. The simple idea here is to represent the mortality probability of each subgroup by averaging the mortality probabilities of patients in that group, which is reasonable when the number of patients in the group is large enough. Cockings et al. (2006) and Cook et al. (2003) use such charts to monitor mortality in intensive care.

The RA p-chart bears the merits of the simple plots, that is, easiness in implementation and interpretation, and also provides a convenient way to set up control limits. However, the need for grouping patients during considerably long periods may lead to delay in capturing changes in performance, which makes it not so appealing as techniques designed for continuous monitoring.

3.2.2 Risk-Adjusted Set Method

The set method monitors the time between events of interest (e.g., death) by counting
the number of events (e.g., survival) between. Specifically, letting C_t be the current
set number, i.e., count of events following the occurrence of an interested event,
$C_t = C_{t-1} + 1$, that is, this statistic will increase by 1 if the tth observation is not the
interested event. This continues until an interested event occurs, and then the set
number will be reset to 0. An alarm is signaled when $C_t \leq T$ happens n times, where
(T, n) is a pair of thresholds determined through simulation.

An extension of this method to incorporate risk adjustment has been proposed by
Grigg and Farewell (2004a, b). The basic idea is to weigh each event by the
baseline mortality probability of the patient. Specifically, the set number will be
calculated by

$$C_t = C_{t-1} + \frac{p_{0t}}{\bar{p}_0}$$

where p_{0t} is the baseline mortality probability of patient t, and $\bar{p}_0$ is the average
baseline mortality probability of all patients, which can also be termed as the
baseline mortality probability of an "average" patient. Here the average patient is
used as a benchmark to assess the normality of each observation, and patients with a
higher baseline mortality probability than the average patient can be assigned a
higher weight.

The set method also provides a graphical representation, called grass plot, to
assist decision making. The drawbacks of this method lie in the complexity in
determining the paired thresholds and inference based on the time between events
rather than individual observations, which may cause delay in change detection.

3.2.3 Risk-Adjusted CUSUM Chart

Cumulative sum (CUSUM) control charts is a popular SPC technique due to their
optimal properties. In the general setting, such charts aim to test the following
hypotheses:

$$H_0 : \theta = \theta_0 \quad H_1 : \theta = \theta_1$$

where θ denotes the parameter of the risk adjustment model, θ_0 is the baseline
which is typically known, and θ_1 is a hypothesized value of interest. The following
statistic is monitored

$$C_t = \max(0, C_{t-1} + W_t)$$

where W_t is the CUSUM score assigned to the tth observation. A control limit H will be found through simulation to achieve a specified in-control average run length (ARL$_0$), and an alarm is signaled when $C_t > H$.

The CUSUM score is given by the log-likelihood ratio of the two hypotheses

$$W_t = \log\left(\frac{L(\theta_1|y_t)}{L(\theta_0|y_t)}\right) \tag{4}$$

where y_t is the tth observation, and $L(\theta|y_t)$ is the likelihood function of the risk adjustment model. For example, for binary data following a Bernoulli distribution with parameter $\theta = p$, the likelihood function is

$$L(\theta|y_t) = \theta^{y_t}(1 - \theta)^{1-y_t}$$

CUSUM charts based on the above likelihood have been used widely to monitor defective rate of products in industrial processes.

For binary performance measures in healthcare which follow the logistic regression model in (1), the likelihood function is

$$L(\theta|y_t) = \left(\frac{Rp_{0t}}{1 - p_{0t} + Rp_{0t}}\right)^{y_t}\left(\frac{1 - p_{0t}}{1 - p_{0t} + Rp_{0t}}\right)^{1-y_t}$$

where the parameter of interest $\theta = R$, R is the odds ratio, and p_{0t} is the baseline mortality probability of patient t. The resulting CUSUM charts are the risk-adjusted version of CUSUM since p_{0t}, which depends on patient risk factors, is taken into account in the statistic.

When the performance measure is time to event such as survival times, and the accelerated failure time model in (3) is used, the likelihood function depends on the assumed distribution for the survival times. For example, if the Weibull distribution is used, the likelihood function is

$$L(\theta|y_t) = \frac{\alpha}{\theta}\left(\frac{y_t}{\theta}\right)^{\alpha-1}\exp\left[-\left(\frac{y_t}{\theta}\right)\alpha\right]$$

where $\theta = \lambda$, λ is the scale parameter, and α is the shape parameter which is often assumed to be fixed and can be estimated from historical data. A special issue in dealing with survival data is censoring where the observation y_t is either the survival time of patient t or the censoring time. In this case, a likelihood function taking into account censoring will be used.

Risk-adjusted CUSUM charts for binary performance measures were first proposed by Steiner et al. (2000) in monitoring 30-day mortality in cardiac surgeries, and then applied in other applications such as liver transplant to monitor 1-year mortality (Leandro et al. 2005) and coronary artery bypass surgeries to monitor adverse outcomes (Novick et al. 2006). RA CUSUM charts for time to event were developed by Sego et al. (2009), Gandy et al. (2010), and Biswas and Kalbfeisch

(2008), who use different models for the survival time. These charts, like their non-risk-adjusted counterparts in industrial contexts, are powerful in detecting small changes in performance, but their use is limited by the perceived difficulty of interpretation by healthcare practitioners (Cook et al. 2011; Pilcher et al. 2010).

3.2.4 Risk-Adjusted EWMA Chart

Like CUSUM charts, the exponentially weighted moving average (EWMA) charts are a popular and widely used SPC technique. The statistic monitored in these charts takes the following form:

$$C_t = \gamma S_t + (1 - \gamma) C_{t-1}$$

where S_t is the EWMA score assigned to the tth observation and $0 < \gamma \le 1$ is a smoothing constant. Essentially, the statistic is a linear combination of all the observations with higher weights assigned to recent observations. With the linearity in the statistic, its distribution can be obtained analytically and, consequently, control limits can be specified based on that.

There are different definitions for the EWMA score depending on the types of data monitored. For binary performance measures, S_t can be the baseline mortality probability or the difference in the observed and baseline mortality probability (Cook et al. 2008; Cook et al. 2011). For time to event performance measures, S_t can be the likelihood ratio scores as used in RA CUSUM charts (Steiner and Jones 2010).

The RA EWMA charts have similar performance to CUSUM charts in detecting small changes. Its main advantage over CUSUM charts lies in its intuitive interpretation as the EWMA statistic can be viewed as an estimate of the current level of the process. Moreover, the influence of previous observations is removed in the statistic gradually by adjusting the weights rather than resetting the statistic as CUSUM does, which is a more natural way to conduct monitoring and easier to accept by healthcare practitioners (Cook et al. 2011).

3.2.5 Likelihood Ratio Test for Phase I Monitoring

The above control charts are all designed for Phase II monitoring. Kamran et al. (2012) propose a control chart based on likelihood ratio test for Phase I monitoring. This method is built on the change-point setting which tests the following hypotheses:

$$H_0 : y_i \sim \mathrm{LG}(y_i | \theta_0) \qquad\qquad i = 1, \ldots, m$$

$$H_1 : y_i \sim \begin{cases} \mathrm{LG}(y_i | \theta_0) & i = 1, \ldots, K \\ \mathrm{LG}(y_i | \theta_1) & i = K+1, \ldots, m \end{cases} \qquad (4)$$

where $\mathrm{LG}(\cdot | \theta)$ is the logistic regression model with parameter θ, m is the total number of observations available, and K, $1 \le K \le m - 1$, is the change point at which the model parameter changes from θ_0 to θ_1, $\theta_1 \neq \theta_0$. A likelihood ratio statistic can then be constructed as

$$\Lambda(\tau) = \log \left(\frac{L(\hat{\theta}_0^1, \hat{\theta}_1^1 | y_1, \ldots, y_m)}{L(\hat{\theta}_0^0 | y_1, \ldots, y_m)} \right)$$

where $\hat{\theta}_0^1, \hat{\theta}_1^1$ and $\hat{\theta}_0^0$ are maximum likelihood estimates of parameters under the two hypotheses. Control limits will be determined through simulation.

3.3 Bayesian Approaches for RA Monitoring

Bayesian approaches have been used for process monitoring and change detection in various applications. Recently, such approaches are developed for different RA monitoring problems, including Phase I monitoring (Assareh et al. 2011a, b; Assareh and Mengersen 2012), estimating the location where change in performance occurs (Assareh et al. 2011c), and self-starting performance monitoring (Zeng and Zhou 2011). This section will first present the Bayesian framework for change detection in RA monitoring, and then its applications in different specific problems. Its advantages and disadvantages over the abovementioned non-Bayesian techniques will be summarized in the end.

3.3.1 Bayesian Framework for Change Detection

Assume the data monitored follow a change-point model as in (4)

$$y_i \sim \mathrm{CP}(y_i | \theta_0, K, \theta_1) = \begin{cases} \mathrm{LG}(y_i | \theta_0) & i = 1, \ldots, K \\ \mathrm{LG}(y_i | \theta_1) & i = K+1, \ldots, m \end{cases}$$

In the Bayesian framework, the unknown change point K is treated as a parameter of the change-point model $\mathrm{CP}(\cdot)$. Here this model is characterized by three sets of parameters, the pre-change parameter θ_0, the change point K, and the post-change parameter θ_1. Correspondingly, any inference regarding this model relates to finding the posterior distribution of these parameters:

$$p(\theta_0, K, \theta_1 | y_1, \ldots, y_m) = \pi(\theta_0, K, \theta_1) \cdot f(y_1, \ldots, y_m | \theta_0, K, \theta_1) \qquad (5)$$

where $p(\cdot | y_1, \ldots, y_m)$ is the posterior, $\pi(\cdot)$ is the prior, and $f(\cdot)$ is the sampling density as follows:

$$f(y_1, \ldots, y_m | \theta_0, K, \theta_1) = \prod_{i=1}^{K} \mathrm{LG}(y_i | \theta_0) \cdot \prod_{i=K+1}^{m} \mathrm{LG}(y_i | \theta_1)$$

Samples from the posterior distribution can be obtained through Markov chain Monte Carlo (MCMC) algorithms, and summaries of these samples will be used for decision making in performance monitoring.

One critical step in obtaining the posterior samples is to specify the priors. Assuming that the three sets of parameters are independent, their priors can be specified separately, that is,

$$\pi(\theta_0, K, \theta_1) = \pi(\theta_0) \cdot \pi(K) \cdot \pi(\theta_1)$$

For θ_0 and θ_1, specifying their priors is equivalent to specifying the prior for the logistic regression model in (1), where the parameter $\theta = [\beta_0, \ldots, \beta_p]'$. This problem has been considered in many studies, and appropriate choices depend on the availability of prior information such as historical data and expert knowledge. When there is prior information, the prior of θ can be either estimated from historical data or elicited from expert knowledge; otherwise regular priors such as flat priors, normal priors, and conjugate priors can be used. Zeng and Zhou (2011) propose ideas on specifying priors for θ in both cases in RA monitoring. For the change point K, a uniform prior on $\{1, 2 \ldots, m - 1\}$ is commonly used.

3.3.2 Bayesian Estimation of Change Points

As suggested by Assareh et al. (2011a, c), Bayesian approaches can be used in conjunction with the non-Bayesian control charts such as RA CUSUM charts to estimate the location of the change point when a change is detected using those charts. Summaries, such as mean, median, and mode of the posterior samples can be used as estimates of the change point.

3.3.3 Bayesian Phase I Monitoring

The central task of Phase I monitoring is to determine if there is any change point in the performance during the studied time period. This can be formulated as testing the following hypotheses regarding the parameter K:

$$H_0 : K \leq \tau_L \text{ or } K \geq \tau_U \quad H_1 : \tau_L < K < \tau_U \tag{6}$$

where τ_L and τ_U, $\tau_L > 1$, $\tau_U < m - 1$ are the specified lower and upper bounds. Such bounds are needed in decision making because very little evidence could exist to support K being at the very beginning or end of the data sequence.

This problem can be solved by using the Bayes factor (BF), which is a popular Bayesian tool for model comparison. Essentially, the Bayes factor compares the marginal likelihoods under the two hypotheses to determine the plausibility of one against another

$$\mathrm{BF}(H_1 : H_0) = \frac{P(y_1, \ldots, y_m | H_1)}{P(y_1, \ldots, y_m | H_0)}$$

A value of BF larger than a chosen threshold, η, $\eta > 1$, means that H_1 is more strongly supported by the data. BFs can be obtained from the posterior samples (Zeng and Zhou 2011).

In practice, there are possibilities that multiple change points may exist in the data. A simple binary segmentation strategy can be applied, that is, we first try to capture one change point in the dataset; if a change point is detected, the data will be broken into two segments by the identified change point, and then the procedure will be applied to each segment to capture one change point in that segment. This repeats until all the change points are identified. A more advanced way is to explicitly represent the number of change points and the locations of change points as random variables in the change-point model, and then find posterior distribution of the expanded parameter space.

3.3.4 Bayesian Phase II Monitoring

In Phase II monitoring, the number of observations (i.e., m) increases over time. For each value of m, the Bayes factor will be calculated and decision will be made on whether some change has occurred. If not, the monitoring will continue; otherwise, we will stop and estimate the change point. Since a baseline is normally available for Phase II monitoring, the pre-change parameter θ_0 in the change-point model will be known and, consequently, the posterior distribution in (5) contains only two parameters, θ_1 and K. Obviously, this scheme can also be used for self-starting monitoring where the baseline is unknown.

3.3.5 Advantages/Disadvantages of Bayesian Approaches

Compared to non-Bayesian RA monitoring techniques, Bayesian approaches have the following advantageous features:

1. *Simple, generic, and versatile framework*: The Bayesian framework based on posterior inference provides a simple and generic way to conduct risk-adjusted monitoring for different types of risk adjustment models. This framework can also be easily adapted to solve different types of monitoring problems.
2. *Intuitive interpretation*: The Bayes factor bears an intuitive interpretation as evidence of the plausibility of one hypothesis against another. Moreover, as will be shown in the case study, the sampled posterior distribution of parameters provides an intuitive graphical representation of possible locations of the change point and associated uncertainty, which will be a valuable tool for medical practitioners in decision making.

3. *Use of priors*: As a defining feature of Bayesian statistics, the use of priors provides a way to incorporate domain knowledge of physicians and other medical professionals in the inference. This fits very well the medical contexts where expert knowledge is very critical.

The main drawback of Bayesian approaches lies in two aspects: the computation load in generating posterior distributions and computing Bayes factors, and the efforts needed for specifying the priors. The former, however, is not a significant challenge with the readily available MCMC algorithms. For the latter, prior setting in medical applications has been studied by many researchers (e.g., Chaloner and Duncan 1983; Chen et al. 2008). Those results need to be adapted to specific applications for RA monitoring.

4 Case Study

This section presents a case study to demonstrate Bayesian approaches for different RA monitoring problems. The data set used in this study is from a UK center for cardiac surgery, part of which is shown in Fig. 2. It contains information on operations during 1992–1998, including time of an operation, surgeon performing the operation, Patient Parsonnet score, and 30-day mortality following the operation. This data set has been used in many studies on RA monitoring (e.g., Steiner et al. 2000; Kamran et al. 2012). In this study, 1701 observations from a single surgeon will be used in the analysis, as displayed in Fig. 3, where the red dots indicate deaths. The logistic model for the data is

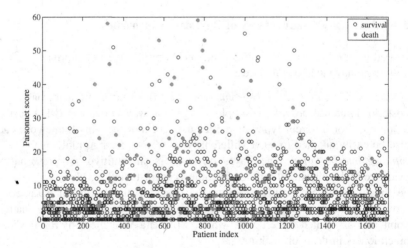

Fig. 3 The data used in the case study

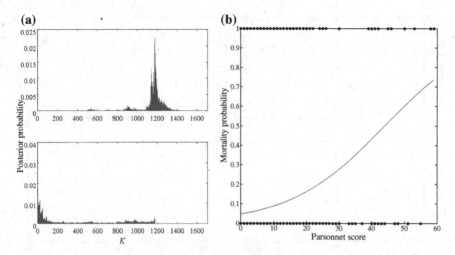

Fig. 4 Results in Phase I monitoring: posterior distribution of K (**a**) and fitted model (**b**)

$$\log\frac{p}{1-p} = \beta_0 + \beta_1 x$$

where x is the Parsonnet score and the parameter of this model is $\theta = [\beta_0, \beta_1]'$.

First, Phase I monitoring is conducted on the data. A vague normal prior $\pi(\beta_1) = N(0, 10^2)$ is used for parameter β_1, and a flat prior $\pi(\beta_0) = \text{Uniform}(0, 0.5)$ is used for β_0. Here β_0 must be positive because a higher risk score tends to lead to a higher mortality probability. The posterior samples of the change point K are obtained through slice sampler, a convenient MCMC algorithm. The empirical distribution of these samples is shown in the upper panel of Fig. 4a. A large number of samples concentrate on a small area around the 1200th patient, a sign that a change point may exist in the data. This is consistent with a rough observation on the raw data in Fig. 3. Since the evidence of change is very strong, the calculated Bayes factor is very large (>100) for any reasonable specification of τ_L and τ_U in (6). Therefore, we conclude that there is a change point in the data. The location of the change point is then estimated using the mode of the posterior of K, which is 1175. To examine if there is any change point before this one, the procedure is applied again to data of the first–1175 patients. The resulting posterior distribution of K is shown in the lower panel of Fig. 4a. The corresponding Bayes factor is very small (<1) for reasonable settings of τ_L and τ_U, meaning that there is no further change point.

Second, since it is determined that there is no change point during the first 1175 observations, these data are used to estimate the baseline parameters of the logistic regression model. Bayesian approach is used to conduct the estimation, which yields point estimates

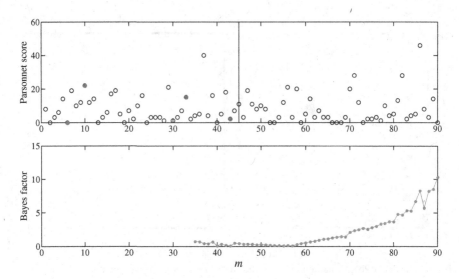

Fig. 5 Results in Phase II monitoring: data (*upper*) and resulting Bayes factors (*lower*)

$$\hat{\beta}_0 = -3.02, \qquad \hat{\beta}_1 = 0.0686$$

The fitted logistic model is shown in Fig. 4b where the dots denote observed outcomes.

Finally, Phase II monitoring is conducted based on the baseline estimated in the Phase I analysis. For simplicity, only a segment of data in Fig. 3 is used. $\tau_L = 15$, $\tau_U = m - 15$, and the monitoring starts when $m = 35$. The data and the resulting Bayes factors are shown in Fig. 5, where the dashed line in the upper panel indicates the location of the change point identified in Phase I analysis. We can see that as m increases, the evidence of change becomes stronger which is manifested clearly by the increasing trend of Bayes factors.

5   Summary and Discussion

Risk-adjusted performance monitoring is a critical research area in healthcare quality control and has received much attention in recent years. Many studies have been done on this topic in different applications and for different purposes. This chapter gives an overview of the existing studies on RA monitoring, encompassing the basic elements of risk adjustment models and popular methods for change detection based on those models. A case study is provided to demonstrate the use of Bayesian approaches for different problems in RA monitoring.

Overall, this topic is an underdeveloped area, and there are many opportunities for future research. One potential direction is variable selection in the construction of risk adjustment models, which is challenged by the existence of large amounts of patient characteristics data, as is often the case in statistical analysis in medical contexts. Powerful statistical/data mining techniques need to be applied to select significant patient risk factors and thus identify a parsimonious risk adjustment model which is the foundation for performance monitoring. Another direction is extending the current RA monitoring methods to more complex types of data than the univariate binary or continuous data. For example, as multiple performance measures are typically needed to characterize quality of care, a simultaneous monitoring scheme on those measures is desired. This can be solved by extending the multivariate control charting methods that have been well studied in industrial contexts to incorporate risk adjustment. Finally, as demonstrated in the case study, Bayesian approaches have great potential to be used for RA performance monitoring in health care. Efforts are needed to develop such approaches for different specific applications.

References

Assareh, H., Smith, I., & Mengersen, K. (2011a). Bayesian change point detection in monitoring cardiac surgery outcomes. *Quality Management in Health Care, 20*(3), 207–222.

Assareh, H., Smith, I., & Mengersen, K. (2011b). Change point detection in risk adjusted control charts. *Statistical Methods for Medical Research* (in press).

Assareh, H., Smith, I., & Mengersen, K. (2011c). Bayesian estimation of the time of a linear trend in risk-adjusted control charts. *International Journal of Computer Science, 38*(4), 409–417.

Assareh, H., & Mengersen, K. (2012). Change point estimation in monitoring survival time. *PLoS One, 7*(3), 1:10.

Benbassat, J., & Taragin, M. (2000). Hospital readmissions as a measure of quality of health care. *Archives of Internal Medicine, 160*, 1074–1081.

Biswas, P., & Kalbfleisch, J. D. (2008). A risk-adjusted CUSUM in continuous time based on cox model. *Statistics in Medicine, 27*, 3382–3406.

Brunelli, A., Morgan-Hughes, N. J., Refai, M., Salati, M., Sabbatini, A., & Rocco, G. (2007). Risk-adjusted morbidity and mortality models to compare the performance of two units after major lung resections. *The Journal of Thoracic and Cardiovascular Surgery, 133*(1), 88–96.

Chaloner, K., & Duncan, G. T. (1983). Assessment of A beta prior distribution: PM elicitation. *The Statistician, 27*, 174–180.

Chen, M., Ibrahim, J. G., & Kim, S. (2008). Properties and implementation of Jeffreys's prior in binomial regression models. *Journal of the American Statistical Association, 103*, 1659–1664.

Cocking, J. G. L., Cook, D. A., & Iqbal, R. K. (2006). Process monitoring in intensive care with the use of cumulative expected minus observed mortality and risk-adjusted p charts. *Critical Care, 10*(1), R28.

Cook, D. A., Coory, M., & Webster, R. A. (2011). Exponentially weighted moving average charts to compare observed and expected values for monitoring risk-adjusted hospital indicators. *BMJ Quality and Safety, 20*, 469–474.

Cook, D. A., Stefan, S. H., Cook, R. J., Farewell, V. T., & Morton, A. P. (2003). Monitoring the evolutionary process of quality: Risk-adjusted charting to track outcomes in intensive care. *Critical Care Medicine, 31*(6), 1676–1682.

Cook, D. A., Duke, G., Hart, G. K., Pilcher, D., & Mullany, D. (2008). Review of the application of risk-adjusted charts to analyse mortality outcomes in critical care. *Critical Care and Resuscitation, 10*(3), 239–251.

Daley, J., Henderson, W. G., & Khuri, S. F. (2001). Risk-adjusted surgical outcomes. *Annual Review of Medicine, 52*, 275–287.

DesHarnais, S., McMahon, L. F., & Wroblewski, R. (1991). Measuring outcomes of hospital care using multiple risk-adjusted indexes. *Health Services Research, 26*(4), 425–445.

Forthman, M. T., Gold, R. S., Dove, H. G., & Henderson, R. D. (2010). Risk-adjusted indices for measuring the quality of inpatient care. *Quality Management in Health Care, 19*(3), 265–277.

Gandy, A., Kvaløy, J. T., Bottle, A., & Zhou, F. (2010). Risk-adjusted monitoring of time to event. *Biometrika, 97*(2), 375–388.

Grigg, O., & Farewell, V. (2004a). An overview of risk-adjusted charts. *Journal of the Royal Statistical Society: Series A, 167*, 523–539.

Grigg, O., & Farewell, V. T. (2004b). A risk-adjusted sets method for monitoring adverse medical outcomes. *Statistics in Medicine, 23*, 1593–1602.

Hendryx, M. S., Dyck, D. G., & Srebnik, D. (1999). Risk-adjusted outcome models for public mental health outpatient programs. *Health Services Research, 34*(1), 171–195.

Hermann, R. C., Rollins, C. K., & Chan, J. A. (2007). Risk-adjusting outcomes of mental health and substance-related care: A review of the literature. *Harvard Review of Psychiatry, 15*(2), 52–69.

Iezzoni, L. I. (ed). 1997. *Risk Adjustment for Measuring Healthcare Outcomes, 3rd edn*. Chicago: Healthcare Administration Press.

Kamran, P., Jin, J., & Yeh, A. B. (2012). Phase I risk-adjusted control charts for monitoring surgical performance by considering categorical covariates. *Journal of Quality Technology, 44*(1), 39–53.

Krumholz, H. M., Chen, J., Wang, Y., Radford, M. J., Chen, Y.-T., & Marciniak, T. A. (1999). Comparing AMI mortality among hospitals in patients 65 years of age and older. *Circulation, 99*(23), 2986–2992.

Leandro, G., Rolando, N., & Gallus, G. (2005). Monitoring surgical and medical outcomes: The Bernoulli cumulative SUM chart. A novel application to assess clinical interventions. *Postgraduate Medical Journal, 81*, 647–652.

Lovegrove, J., Valencia, O., Treasure, T., Sherlaw-Johnson, C., & Gallivan, S. (1997). Monitoring the results of cardiac surgery by variable life-adjusted display. *The Lancet, 350*(18), 1128–1130.

Mukamel, D. B., & Spector, W. D. (2000). Nursing home costs and risk-adjusted outcome measures of quality. *Medical Care, 38*(1), 78–89.

Murtaugh, C. M., Peng, T., Aykan, H., & Maduro, G. (2007). Risk adjustment and public reporting on home health care. *Health Care Financing Review, 28*(3), 77–94.

Novick, R. J., Fox, S. A., Stitt, L. W., Forbes, T. L., & Steiner, S. (2006). Direct comparison of risk-adjusted and non-risk-adjusted CUSUM analyses of coronary artery bypass surgery outcomes. *The Journal of Throracic and Cardiovascular Surgery, 132*(2), 386–391.

Pilcher, D. V., Hoffman, T., Thomas, C., Ernest, D., & Hart, G. K. (2010). Risk-adjusted continuous outcome monitoring with an EWMA chart: Could it have detected excess mortality among intensive care patients at Bundaberg base hospital? *Critical Care and Resuscitation, 12*(1), 36:41.

Pinna-Pintor, P., Bobbio, M., Colangelo, S., Veglia, F., Giammaria, M., Cuni, D., et al. (2002). Inaccuracy of four coronary surgery risk-adjusted models to predict mortality in individual patients. *European Journal of Cardio-Thoracic Surgery, 21*, 199–204.

Poloniecki, J., Valencia, O., & Littlejohns, P. (1998). Cumulative risk adjusted mortality chart for detecting changes in death rate: observational study of heart surgery. *British Medical Journal, 316*, 1697–1700.

Reid, P. P., Compton, D., Grossman, J. H., Fanjiang, G., & Committee on Engineering and the Health Care System. (2005). *Building a better delivery system: A new engineering/health care partnership*. Washington, D.C.: National Academy Press.

Sego, L. H., Reynolds, M. R., & Woodall, W. H. (2009). Risk-adjusted monitoring of survival times. *Statistics in Medicine, 28*, 1386–1401.

Shroyer, A. L. W., Coombs, L. P., Peterson, E. D., Eiken, M. C., Delong, E. R., Chen, A., Ferguson, T. B., Grover, F. L., & Edwards, F. H. (2003). The society of thoracic surgeons: 30-day operative mortality and morbidity risk models. *Annals Thoracic Surgery, 75*, 1856–1865.

Sousa, P., Uva, A. S., & Pinto, F. (2008). Risk-adjustment model in health outcomes evaluation: A contribution to strengthen assessment towards quality improvement in interventional cardiology. *International Journal for Quality in Health Care, 20*(5), 324–330.

Steiner, S. H., Cook, R. J., Farewell, V. T., & Treasure, T. (2000). Monitoring surgical performance using risk-adjusted cumulative sum charts. *Biostatistics, 1*(4), 441–452.

Steiner, S. H., & Jones, M. (2010). Risk-adjusted survival time monitoring with an updated exponentially weighted moving average (EWMA) control chart. *Statistics in Medicine, 29*, 444–454.

Tu, J. V., Jaglal, S. B., & Naylor, C. D. (1995). Multicenter validation of a risk index for mortality, intensive care unit stay, and overall hospital length of stay after cardiac surgery. *Circulation, 91*, 677–684.

Woodall, W. H. (2006). The use of control charts in health-care and public-health surveillance. *Journal of Quality Technology, 38*(2), 89–104.

Zeng, L., & Zhou, S. (2011). A Bayesian approach to risk-adjusted outcome monitoring in healthcare. *Statistics in Medicine, 30*(29), 3431–3446.

Zimmerman, D. R. (2003). Improving nursing home quality of care through outcomes data: The MDS quality indicators. *International Journal of Geriatric Psychiatry, 18*, 250–257.

Univariate and Multivariate Process Capability Analysis for Different Types of Specification Limits

Ashis Kumar Chakraborty and Moutushi Chatterjee

1 Introduction

In the context of statistical quality control, process capability analysis is one of the widely accepted approaches for assessing the ability of a process to produce what it is supposed to produce. Normally, an index known as the process capability index, abbreviated as PCI henceforth, is used to judge the health of the process vis-a-vis the given specification. In this context, the concept of PCI is generally applied in manufacturing industries. PCI mostly gives single valued assessment of the ability of a process to produce items within the pre-assigned specification limits. It is, generally, a higher the better type of index with the 'high' value indicating that the process is capable of producing item that in all likelihood will meet or exceed customers' requirement.

According to Kotz and Johnson (2002), before computing the PCI of a process, one has to ensure that the following two assumptions are satisfied:

1. The quality characteristic under consideration follows normal distribution;
2. The process is under statistical control.

In this context, the assumption of normality is made only due to the fact that such assumption gives some computational advantages. On the other hand, the second assumption is comparatively stronger because, absence of stability in the process makes it unpredictable and hence in that situation, PCI values may not be able to reflect the actual capability level of the process correctly.

A.K. Chakraborty (✉) · M. Chatterjee
SQC & OR Unit, Indian Statistical Institute, Kolkata, India
e-mail: akchakraborty123@rediffmail.com

M. Chatterjee
e-mail: tushi.stats@gmail.com

© Springer-Verlag London 2016
H. Pham (ed.), *Quality and Reliability Management and Its Applications*,
Springer Series in Reliability Engineering, DOI 10.1007/978-1-4471-6778-5_3

Kane (1986), in his seminal paper, first documented some of the process capability indices, which were already being used in industries for quite some times and discussed about the importance of using those PCIs for assessing capability of the process. Due to the unquestionable significance of the concept of PCIs, specially in the context of manufacturing industries, Kane's (1986) paper motivated a huge number of statisticians as well as industrial engineers, working in the field of statistical quality control, to carry out further research work in this field. Kotz and Johnson (2002) have reviewed about 170 of such high quality research papers published within just 15 years of Kane's (1986) paper.

The quality characteristics, which are generally encountered in practice, belong to either of the following three categories viz.,

1. The nominal the best, i.e., processes with both upper specification limit (USL) and lower specification limit (LSL), e.g. height, length;
2. The smaller the better, i.e., processes with only USL, e.g. surface roughness, flatness;
3. The larger the better, i.e., processes with only LSL, e.g. tensile strength, compressive strength.

Moreover, for the quality characteristics of nominal the better type, the corresponding bi-lateral specification limits may be symmetric or asymmetric (with respect to the target) in nature. The consequences of asymmetric bi-lateral specification limits are discussed in detail in Sect. 3.

The four classical PCIs, for symmetric bi-lateral specification limits, which are commonly used, are,

$$\left.\begin{array}{l} C_p = \frac{U-L}{6\sigma} \\ C_{pk} = \frac{d-|\mu-M|}{3\sigma} \\ C_{pm} = \frac{d}{3\sqrt{\sigma^2+(\mu-T)^2}} \\ C_{pmk} = \frac{d-|\mu-M|}{3\sqrt{\sigma^2+(\mu-T)^2}} \end{array}\right\} \tag{1}$$

Here, 'U' and 'L' denote the USL and LSL respectively; $d = (U - L)/2$, $M = (U + L)/2$ and 'T' denotes the targeted value of the quality characteristic under consideration. Also, suppose, 'X' is a random variate corresponding to the measurable quality characteristic under consideration. Then, μ and σ are such that, $X \sim N(\mu, \sigma^2)$.

Vannman (1995) unified these PCIs and proposed the following super-structure of PCIs for symmetric bi-lateral specification limits:

$$C_p(u, v) = \frac{d - u|\mu - M|}{3\sqrt{\sigma^2 + v(\mu - T)^2}}, \quad u, v \geq 0. \tag{2}$$

Note that the PCIs defined in Eq. (1) involve parameters like μ and σ which are often unobservable and consequently, the actual values of these PCIs are also difficult to obtain. To address this problem, the common practice is to compute the values of the plug-in or natural estimators of these PCIs. Such estimators are obtained by replacing the parameters like μ and σ by their corresponding estimators viz., $\overline{X}$ and 's' respectively, based on the random sample(s) drawn from the process. However, such plug-in estimators are subject to sampling fluctuation and hence can not be considered as the substitute of the original PCIs unless their distributional and inferential properties are studied extensively. The properties of the PCIs in Eq. (1) have been studied extensively in literature (refer Kotz and Johnson 2002, Pearn et al. 1992 and the references there in).

Although most of the quality characteristics of nominal the better type have symmetric bi-lateral specification limits, there are some practical situations, where due to some design aspect or to control production cost without compromising with the quality level of the product, asymmetry with respect to the target is solicited in the bi-lateral specification limits. For example, in the context of manufacturing iron rods of specific length, it is easier to cut a longer rod into a smaller one; than to make a shorter rod longer. Accordingly, the specifications should be set such that the distance between USL and T is more than the distance between LSL and T. Similarly, quality characteristics like hole diameter, should have asymmetric specification limits, as it is easier to make a hole with smaller diameter to a larger one through drilling, whereas, turning a larger hole into a smaller one, without compromising with its circularity, requires lot more effort.

A number of remarkable attempts have been made to define PCIs for processes with asymmetric bi-lateral specification limits (see Kane 1986; Boyles 1994; Franklin and Wasserman 1992; Kushlar and Hurley 1992; Vannman 1997 and the references there-in). Chen and Pearn (2001) defined a super-structure of PCIs called $C_p^{''}(u, v)$ for asymmetric specification limits, which is similar to $C_p(u, v)$ of symmetric specification limits. Latter, Chatterjee and Chakraborty (2014) have established exact relationship between the proportion of non-conforming items produced by the process and some member indices of $C_p^{''}(u, v)$, viz., $C_p^{''}$ and $C_{pk}^{''}$. Chatterjee and Chakraborty (2014) have also studied some other interesting properties of $C_p^{''}(u, v)$ including the inter-relationships between the member indices of $C_p^{''}(u, v)$, threshold value of $C_p^{''}(u, v)$ and optimality of $C_p^{''}(u, v)$ on target. These are discussed in more detail in Sect. 3 along with a numerical example.

Apart from the bilateral specification limits, there are also some processes involving larger the better or smaller the better types of quality characteristics which require unilateral or one sided specification limits. In such situations, as the name 'unilateral' suggests, either of USL or LSL exist. For example, quality characteristics like surface roughness and flatness are of smaller the better type in a sense that their values should be as minimum as possible. Hence, only an USL is set for such quality characteristics. On the other hand, tensile strength and compressive strength are the examples of larger the better type of quality characteristics, where, the corresponding quality characteristic values should be as high as possible but should have at least a minimum value, decided by the LSL, for proper functioning of the concerned item.

Among the PCIs defined specifically for unilateral specification limits (see Kane 1986, Vannman 1998, Grau 2009 and the references there-in), the member indices of the super-structures of PCIs called $C_p^U(u, v)$ and $C_p^L(u, v)$, which are defined similar to $C_p(u, v)$, are closer to the practical situations. Chatterjee and Chakraborty (2012) have made an extensive review of the PCIs for unilateral specification region.

Despite the fact that Grau's (2009) super-structure performs better than the other available PCIs for unilateral specification limits, there was some problem in its practical implementation. In fact Grau's (2009) super-structure involves a term 'k' whose purpose is to penalize the deviation of the quality characteristic value from the target towards the opposite side of the existing specification limit. However, no mathematical formulation of 'k' was provided and this left room for favourable manipulation. Chatterjee and Chakraborty (2012) have proposed a formulation of 'k' based on the concept of loss of profit due to the deviation of the quality characteristic value from the target towards the opposite side of the available specification limit. A brief discussion on the PCIs for unilateral specification limits and the formulation of 'k' along with a numerical example are given in Sect. 4.

Although the bi-lateral and unilateral specification limits cover most of the quality characteristics encountered in practice, as has already been discussed earlier in this section, there is another type of quality characteristics which do correspond to neither of these types specification limits. The center of a drilled hole (in case of manufacturing processes) or the case of hitting a target (in ballistics) are some examples of such quality characteristics and the corresponding specification region is circular in nature.

Krishnamoorthi (1990) and Bothe (2006) have defined PCIs for circular specification regions. However, both of them have assumed equal variances (homoscedasticity) and independence of the two axes of the specification region—which may not be practically viable due to several technical reasons. To address these problems, Chatterjee and Chakraborty (2015) have defined a super-structure of PCIs for circular specification region, called $C_{p,c}(u, v)$, which does not require these assumptions. Besides, the authors have studied some important properties of $C_{p,c}(u, v)$, like, inter-relationship among the member indices, the threshold value, relationship with proportion of non-conforming items produced by the process and so on. Moreover, Chatterjee and Chakraborty (2015) have derived the expressions for the expectations and variances of the plug-in estimators of $C_{p,c}(u, v)$ based on the concept of circular normal distribution (see Scheur 1962). Section 5 contains a more detail discussion on the PCIs for circular specification limits.

The PCIs discussed so far, deal with one characteristic of a process at a time. However, with the increasing complexity in the technology, this may not be a valid assumption. In fact, often processes with multiple correlated characteristics are encountered in practice. For example (refer Taam et al. 1993), in an automated paint application process, there are more than one important quality characteristics viz., paint thickness, paint thinner levels, paint lot differences, temperature and so on which are interrelated among themselves. Use of univariate PCIs may not be able to assess the actual capability of the process efficiently, in such situations. One needs to use appropriate multivariate process capability indices (MPCI) in such cases.

Although the literature of statistical quality control is enriched with some mathematically sound MPCIs (see Taam et al. 1993, Chen 1994, Shinde and Khadse 2009, Shahriari et al. 2009 and the references there-in), most of these are difficult to interpret. Moreover, shop-floor people are more conversant with the univariate PCIs C_p, C_{pk}, C_{pm} and C_{pmk} and hence MPCIs which function similar to these PCIs should be easily acceptable to them.

Chakraborty and Das (2007) defined an MPCI called $C_G(u, v)$ which functions similar to $C_p(u, v)$ but takes into account 'p' correlated quality characteristics simultaneously under consideration. Moreover, for $p = 1$, $C_G(u, v)$ boils down to $C_p(u, v)$ which is highly desirable. Later Chatterjee and Chakraborty (2013) have studied some of the properties of $C_G(u, v)$ like interrelationship between member indices and relationship with proportion of non-conforming items produced by the process and observed that these properties are similar to those of $C_p(u, v)$ from multivariate perspective.

Chatterjee and Chakraborty (2011) have also proposed a multivariate analogue of $C_p''(u, v)$, called $C_M(u, v)$, for assessing capability of processes having multiple correlated quality characteristics and asymmetric specification region with respect to the target vector. They have also studied the inter-relationship between the member indices of $C_M(u, v)$. The details of these MPCIs are given in Sect. 6.

Most of the PCIs, available in literature, are based on the common assumption that, the underlying statistical distribution of the concerned quality characteristic is normal. However, this assumption may not always be valid in practice. For example, McCormack et al. (2000) have observed that, in the context of high purity manufacturing, often, the particle count distribution and the distributions of process yield data are found to be non-normal. Some very interesting research work have been carried out in literature, to deal with the impact of such non-normality in the capability assessment of a process. A more detail discussion, in this regard, is made in Sect. 7.

Although, normality is an important, though not indispensable, assumption for process capability assessment, it is often difficult to check the same. However, the situation has somewhat improved in recent times and a number of statistical softwares are now available for testing univariate and multivariate normality. A brief discussion, in this regard, is made in Sect. 8.

Finally, the chapter concludes in Sect. 9 with a brief summarization of the PCIs for different types for specification limits, as have been discussed here.

2   List of Notations

Before going into an elaborate discussion about the univariate and multivariate process capability indices for different types of specification limits, let us first consider the following notations which are used in process capability studies time and again.

1. U: Upper specification limit (USL);
2. L: Lower specification limit (LSL);
3. n: Sample size;
4. $M = \frac{U+L}{2}$;
5. $d = \frac{U-L}{2}$;
6. T: Target;
7. 'X' is a random variable corresponding to the measurable quality characteristic under consideration, such that, $X \sim N(\mu, \sigma^2)$.
8. $D_U = U - T$;
9. $D_L = T - L$
10. $d^* = \min(D_U, D_L)$;
11. $S(x,y) = \frac{1}{3} \times \Phi^{-1}\left\{\frac{\Phi(x) + \Phi(y)}{2}\right\}$;
12. $F^* = \max\left(\frac{d^*(\mu-T)}{D_U}, \frac{d^*(T-\mu)}{D_L}\right)$;
13. $F = \max\left(\frac{d(\mu-T)}{D_U}, \frac{d(T-\mu)}{D_L}\right)$;
14. $k = \max\{\frac{D_U}{D_L}, \frac{D_L}{D_U}\}$;
15. $R_U = \frac{\mu-T}{D_U}$;
16. $R_L = \frac{T-\mu}{D_L}$;
17. $k_1 = \frac{U-T}{\sigma}$;
18. $k_2 = \frac{T-L}{\sigma}$;
19. $A_U^* = \max\{(\mu - T), \frac{T-\mu}{k_U^*}\}$;
20. $A_L^* = \max\{\frac{\mu-T}{k_L^*}, (T - \mu)\}$;
21. D: Diameter of circular specification region;
22. $r_{C,i} = \sqrt{(x_i - \bar{x})^2 + (y_i - \bar{y})^2}$;
23. $\widehat{\mu_C} = \frac{\sum_{i=1}^n r_{C,i}}{n}$;
24. $\overline{\text{MR}} = \frac{\sum_{i=2}^n \text{MR}_i}{n-1}$, where, MR_i's are obtained from moving range chart;
25. $\hat{\sigma}_{\text{ST}} = \frac{\overline{\text{MR}}}{d_2}$. Since for MR chart, information from two samples are considered at a times, we have $d_2 = 1.128$;
26. $r_i = \sqrt{x_i^2 + y_i^2}$;
27. $\widehat{\sigma}_{\text{LT}} = \frac{1}{c_4}\sqrt{\frac{\sum_{i=1}^n (r_i - \bar{r})^2}{n-1}}$;

28. $\widehat{\mu}_r = \frac{\sum_{i=1}^{n} r_i}{n}$;

29. $\widehat{\sigma}_{\text{ST,C}} = \frac{\overline{\text{MR}}_C}{d_2}$, where, MR_C values are obtained from the moving range chart of the data set after the target hole center is shifted to the middle of the cluster of actual hole centers;

30. $\widehat{\sigma}_{\text{LT,C}} = \frac{1}{c_4} \sqrt{\frac{\sum_{i=1}^{n} (r_{C,i} - \overline{r_C})^2}{n-1}}$;

31. c_4, d_2 and d_3 are the common constants of the literature of control chart which are expressed as functions of the sample size 'n';

32. $d_i^{**} = \sqrt{(X_{1_i} - \mu_1)^2 + (X_{2_i} - \mu_2)^2}$;

33. $\mu^* = \frac{1}{n} \sum_{i=1}^{n} d_i^{**}$;

34. 'p' denotes the number of characteristics under consideration;

35. $X = (X_1, X_2, \ldots, X_p)\prime$: Random vector characterizing the 'p' correlated quality characteristics under consideration (Note that now onwards vectors will be denoted by bold-faced letters);

36. $D = (|\mu_1 - M_1|, |\mu_2 - M_2|, \ldots, |\mu_p - M_p|)'$;

37. $d = ((\text{USL}_1 - \text{LSL}_1)/2, ((\text{USL}_2 - \text{LSL}_2)/2, \ldots, ((\text{USL}_p - \text{LSL}_p)/2)'$;

38. $T = (T_1, T_2, \ldots, T_p)'$;

39. $M = (M_1, M_2, \ldots, M_p)'$;

40. T_i is the target value, M_i is the nominal value for the ith characteristic of the item, for $i = 1(1)p$;

41. u and v are the scalar constants that can assume any non-negative integer value;

42. $\mu = (\mu_1, \mu_2, \ldots, \mu_p)'$: Mean vector of a 'p' variate process;

43. $\Sigma = \begin{pmatrix} \sigma_1^2 & \sigma_{12} & \cdots & \sigma_{1p} \\ \sigma_{12} & \sigma_2^2 & \cdots & \sigma_{2p} \\ \vdots & \vdots & \cdots & \vdots \\ \sigma_{1p} & \sigma_{2p} & \cdots & \sigma_{pp} \end{pmatrix}$ is the dispersion matrix of a 'p' variate process;

3 Univariate Process Capability Indices for Asymmetric Specification Limits

Often for the quality characteristics of nominal the best type, the respective upper specification limit (USL) and lower specification limit (LSL) are symmetric with respect to the corresponding target (T). However, this may not always be the case—asymmetry in specification limits with respect to 'T' is also quite common in manufacturing industry. Such asymmetry may generate from a number of very practical situations some of which have been discussed by Boyles (1994). Sometimes, for particular quality characteristic of a product, the customer and or the design engineer is ready to allow more deviation from target towards a particular specification limit than towards the other; generating asymmetry in the specification limits. For example, in the context of drilling holes with hole diameter being the quality characteristic of interest, it is easier to increase the diameter of a hole through repeating the drilling operation than to shorten the existing hole diameter. Therefore, here USL should be closure to target than LSL. Again, it may so happen that, although initially a process starts with symmetric specification limits, after some times, the customer and/ or the manufacturer opts for asymmetric specification limits, to avoid unnecessary increase in production cost or due to some technical or financial issues. Finally, while transforming non-normal data into the normal one, often the symmetric specification limits get converted into asymmetric, owing to the same transformation.

Thus, the quality characteristics having asymmetric specification limits are not rare in industries, though most of the PCIs, available in literature are only applicable to quality characteristics with symmetric specification limits (Sect. 1). To address this problem, Kane (1986) modified C_p and C_{pk} by shifting one of USL and LSL so that the new specification limits are symmetric with respect to the target and defined $C_p^* = \min(\frac{T-\text{LSL}}{3\sigma}, \frac{\text{USL}-T}{3\sigma})$ and $C_{pk}^* = \min(\text{CPL}^*, \text{CPU}^*)$, where, $\text{USL} - T \neq T - \text{LSL}$, $\text{CPL}^* = \frac{T-\text{LSL}}{3\sigma}(1 - \frac{|T-\mu|}{T-\text{LSL}})$ and $\text{CPU}^* = \frac{\text{USL}-T}{3\sigma}(1 - \frac{|T-\mu|}{\text{USL}-T})$. Later, Franklin and Wasserman (1992) and Kushlar and Hurley (1992) proposed shifting both the specification limits $(T - D_L, T + D_U)$ to obtain symmetric ones $(T \pm \frac{D_L+D_U}{2})$, where, $D_U = \text{USL} - T, D_L = T - \text{LSL}$. However, the revised specification limits obtained by such shifting are subsets of the original specification limits and hence assessment of process capability based on these revised limits are often misleading.

Boyles (1994) proposed a new index as $S_{pk} = S(\frac{\text{USL}-\mu}{\sigma}, \frac{\mu-\text{LSL}}{\sigma})$, where, $S(x,y)$ is a smooth function which is defined as $S(x, y) = \frac{1}{3}\Phi^{-1}\{\frac{\Phi(x)+\Phi(y)}{2}\}$. Chen and Pearn (2001) generalized this index as $S_p(v) = S(\frac{\text{USL}-\mu}{\sqrt{\sigma^2 + v(\mu-T)^2}}, \frac{\mu-\text{LSL}}{\sqrt{\sigma^2 + v(\mu-T)^2}})$, where, $v \geq 0$. Although the properties of S_{pk} were studied by Ho (2003), but due to its very complicated nature, it has found very limited application in practice.

Similar to $C_p(u, v)$ of symmetric specification limits, for asymmetric specification limits, Vannman (1997) defined the following two super-structures of PCIs:

$$C_{pv}(u, v) = \frac{d - |T - M| - u|\mu - T|}{3\sqrt{\sigma^2 + v(\mu - T)^2}},$$

and

$$C_{pa}(u, v) = \frac{d - |\mu - M| - u|\mu - T|}{3\sqrt{\sigma^2 + v(\mu - T)^2}}.$$

However, $C_{pv}(u, v)$ fails to capture the asymmetry of the loss function with respect to 'T'; while, $C_{pa}(u, v)$ is not optimum on target.

To address these drawbacks of the PCIs defined so far for asymmetric specification limits, Pearn (1998) proposed a new index analogous to C_{pk} for asymmetric tolerances which is given by

$$C_{pk}^* = \frac{d^* - F^*}{3\sigma} \tag{3}$$

where, $d^* = \min(D_L, D_U)$ and $F^* = \max\{\frac{d^*(\mu - T)}{D_U}, \frac{d^*(T - \mu)}{D_L}\}$. Pearn and Lin (2000) studied some properties of C_{pk}^* and proposed a consistent and asymptotically unbiased estimator which converges to a mixture of two normal distributions. Later Chen and Pearn (2001) generalized C_{pk}^* to a super-structure which is defined as

$$C_p''(u, v) = \frac{d^* - uF^*}{3\sqrt{\sigma^2 + vF^2}} \tag{4}$$

where, $F = \max\{\frac{d(\mu - T)}{D_U}, \frac{d(T - \mu)}{D_L}\}$. $C_p''(u, v)$ is optimum on target and also, high value of $C_p''(u, v)$ indicates high process yield—these are two of the most important properties of any PCI irrespective of the nature of the respective specification limits.

Now, $C_p''(u, v)$ involve parameters of the quality characteristics, viz., μ and σ^2, which are often unobservable. Hence, the plug-in estimator called $\hat{C}_p''(u, v)$ is used for all practical purposes, where $\hat{C}_p''(u, v)$ is obtained by replacing μ and σ^2 in Eq. (4) by the sample mean ($\overline{X}$) and the sample variance s^2 respectively. However, indiscriminate use of such plug-in estimators is not solicited as that may lead to wrong assessment of the process capability. One needs to study the statistical properties of these plug-in estimators. Pearn et al. (2001, 2004) have made thorough studies of some of the distributional and inferential properties of $\hat{C}_{pk}''$ and $\hat{C}_{pmk}''$.

Proportion of non-conformance (PNC) is another measure for assessing the performance of a process apart from PCI. PNC measures the probability of producing items which are non-conforming with respect to the preassigned specification limits. Thus, ability of establishing relationship between these two parallel concepts of process performance analysis, is considered to be an added advantage of using a particular PCI. For symmetric specification limits, PNC is expressed in terms of C_p and C_{pk} as follows:

$$p = 2\Phi(-3C_p) \tag{5}$$

$$p' = \Phi[-3(2C_p - C_{pk})] + \Phi[-3C_{pk}] \tag{6}$$

Note that since C_p measures only the potential capability of a process, 'p' fails to measure the actual PNC unless $\mu = T$; whereas 'p'' measures the observable PNC. In this context, potential capability is the capability a process that can at most be attained given the current dispersion level and specification scenario. Chatterjee and Chakraborty (2014) have explored analogous relationship between C_p'', C_{pk}'' and PNC, where, $C_p'' = \frac{d^*}{3\sigma}$.

3.1 Relationship Between C_p'' and Proportion of Non-conformance

When the process is on target and the distribution of the quality characteristic is normal, the proportion of non-conformance can be defined as

$$P_{NC} = P[X > U|X \sim N(T, \sigma^2)] + P[X < L|X \sim N(T, \sigma^2)] = P_1 + P_2, \text{ say} \quad (7)$$

For establishing relationship between C_p'' and PNC, the following two situations are considered based on the relative position of 'T' with respect to μ, USL and LSL.
Case I: $d^* = D_U = U - T$
Here, $C_p'' = \frac{D_U}{3\sigma}$. From Eq. (7), $P_1 = 1 - \Phi(C_p'')$ and $P_2 = 1 - \Phi[3kC_p'']$, where, $k = \max\{\frac{D_U}{D_L}, \frac{D_L}{D_U}\}$. Hence from Eq. (7), when $\mu = T$, the expression for proportion of non-conformance is,

$$P_{NC} = 2 - \Phi(C_p'') - \Phi[3kC_p''] \quad (8)$$

Case II: $d^* = D_L = T - L$
Here, $C_p'' = \frac{D_L}{3\sigma}$. From Eq. (7), $P_1 = 1 - \Phi(3kC_p'')$, $P_2 = 1 - \Phi[3C_p'']$ and consequently, the expression of P_{NC} is given by Eq. (8).

Thus, when $\mu = T$, the expression for PNC remains same irrespective of the position of 'T' with respect to μ, USL and LSL. Also, for $k = 1$ and $C_p'' = 1$, we have, $P_{NC} = 0.0027$ which is same as the value of 'p' obtained from Eq. (5), when $C_p = 1$. This is due to the fact that, for $k = 1$, the specification limits become symmetric and hence $C_p = C_p''$.

However, P_{NC} measures the proportion of non-conformance only when $\mu = T$ and hence it is required to explore the relationship between PNC, C_p'' and C_{pk}'' (similar to the case of symmetric specification limits) from a more general perspective.

3.2 Relationship Between $C_{pk}^{''}$ and Proportion of Non-conformance

When $\mu \neq T$, PNC can be formulated as

$$
\begin{aligned}
P_{NC}^{E} &= 1 - P[L < X < U | X \sim N(\mu, \sigma^2)] \\
&= 2 - \Phi[\frac{D_U}{\sigma}(1 - R_U)] - \Phi[\frac{D_L}{\sigma}(1 - R_L)] \\
&= 2 - I_1 - I_2, \text{ say}
\end{aligned}
\tag{9}
$$

where, P_{NC}^{E} denotes the expected/ observed PNC, $R_U = \frac{\mu - T}{D_U}$, $R_L = \frac{T - \mu}{D_L}$, $I_1 = \Phi[\frac{D_U}{\sigma}(1 - R_U)]$ and $I_2 = \Phi[\frac{D_L}{\sigma}(1 - R_L)]$. Based on the position of 'T' with respect to μ, USL and LSL, there can be four mutually exclusive and collectively exhaustive situations (see Wu et al. 2009) for each of which Chatterjee and Chakraborty (2014) have established exact relationship between $C_p^{''}$, $C_{pk}^{''}$ and P_{NC}^{E} as follows:

Case I: $d^* = D_U$ and $R_U < R_L$, i.e. $\mu < T$:

$$
P_{NC}^{E} = 2 - \Phi[3kC_{pk}^{''}] - \Phi[3\{C_{pk}^{''} + (k+1)R_L C_p^{''}\}]
\tag{10}
$$

Case II: $d^* = D_U$ and $R_U > R_L$, i.e. $\mu > T$:

$$
P_{NC}^{E} = 2 - \Phi[3C_{pk}^{''}] - \Phi[3\{kC_{pk}^{''} + (k+1)R_U C_p^{''}\}]
\tag{11}
$$

Case III: $d^* = D_L$ and $R_U > R_L$, i.e. $\mu > T$:

$$
P_{NC}^{E} = 2 - \Phi[3kC_{pk}^{''}] - \Phi[3\{C_{pk}^{''} + (k+1)R_U C_p^{''}\}]
\tag{12}
$$

Case IV: $d^* = D_L$ and $R_U < R_L$, i.e. $\mu < T$:

$$
P_{NC}^{E} = 2 - \Phi[3C_{pk}^{''}] - \Phi[3\{kC_{pk}^{''} + (k+1)R_L C_p^{''}\}]
\tag{13}
$$

In this context, $R_U = R_L$ implies $\mu = T$ and hence the specification limits become symmetric about 'T'. Here, one interesting point to note is that, unlike 'p' in Eq. (5), here, P_{NC} does not ensure providing minimum observable proportion of non-conformance; rather, it only measures the observed proportion of non-conformance of the process when $\mu = T$. In particular, the value of $(P_{NC} - P_{NC}^{E})$ increases with the increase in the value of 'k'. Thus, contradicting the usual convention, it may so happen that, a process, with asymmetric specification limits, produces more non-conforming items when it is on target compared to the situation when $\mu = M$ and this is more clearly described in Fig. 1.

Fig. 1 Asymmetric
specification limits with
$P_{\text{NC}} > P_{\text{NC}}^E$

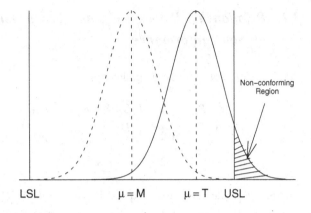

LSL $\mu = M$ $\mu = T$ USL

   Chatterjee and Chakraborty (2014) have extensively studied the interrelationship
between the member indices of $C_p''(u, v)$ and have observed that $C_p'' \geq C_{\text{pk}}'' \geq C_{\text{pmk}}''$
and $C_p'' \geq C_{\text{pm}}'' \geq C_{\text{pmk}}''$, where equality is attained for $\mu = M = T$. Moreover, there is
no clear-cut relationship between C_{pk}'' and C_{pm}''. These are analogous to the
inter-relationship between the member indices of $C_p(u, v)$, as have been observed
by Kotz and Johnson (2002).

   A mathematical expression for the threshold value of C_p'' has also been developed
by Chatterjee and Chakraborty (2014). In this context, threshold value is one of the
most important features of a PCI from the interpretational view point. A process
with a PCI value beyond the threshold value is considered to be capable of producing items within the pre-assigned specification limits; while that with a smaller
value of PCI with respect to the said threshold value is likely to be incapable.
Usually, threshold values are computed for the PCIs like C_p, measuring potential
capability of a process. The common industrial practice is to consider '1' as the
threshold value of a PCI, irrespective of the nature of the corresponding specification limits. Chatterjee and Chakraborty (2014) have formulated the threshold
value of C_p'' as,

$$C_p''^{(T)}(0,0) = C_p''^{(T)} = \begin{cases} \frac{2k_1}{k_1 + k_2} & \text{if } D_U < D_L \\ \frac{2k_2}{k_1 + k_2} & \text{if } D_L < D_U \end{cases} \tag{14}$$

where, k_1 and k_2 are positive real numbers with $k_1 \neq k_2$, such that $k_1 = \frac{U-T}{\sigma}$ and
$k_2 = \frac{T-L}{\sigma}$. From Eq. (13), it is evident that, $C_p''^{(T)}$, the threshold value of C_p'', is a
function of the degree of asymmetry of the specification limits and hence, con-
sidering '1' as the threshold value of C_p'', without properly investigating the nature
of the specification limits leave room for over/under estimation of the actual
capability of a process.

3.3 Example

In order to illustrate the theoretical aspects of the PCIs for asymmetric specification limits discussed so far, we now consider a numerical example based on the data on a high-end audio speaker component called Pulux edge manufactured in Taiwan (Lin and Pearn 2002). For a particular model of Pulux edge, $U = 5.950$, $L = 5.650$ and $T = 5.835$. Lin and Pearn (2002) have collected 90 observations with the corresponding summary statistics found to be as follows:

Sample size $(n) = 90$, sample mean $(\overline{X}) = 5.83$ and sample standard deviation $(s) = 0.023$. Moreover, here $D_U \neq D_L$ indicating asymmetry in the specification limits with respect to T. Based on this data, we compute the values of some of the PCIs and the corresponding PNC values for both the symmetric and asymmetric specification limits to make a comparative study of their performances when the actual specification limits are asymmetric. Thus, $\widehat{C}_p = 2.17, \widehat{p} = 7.515 \times 10^{-11}, \widehat{C}_{pk} = 0.870, \widehat{p}' = 0.004527, \widehat{C}_p'' = 1.6667, \widehat{P}_{NC} = 0.0477, \widehat{C}_{pk}'' = 1.6217$ and $\widehat{P}_{NC}^E = 9.0784 \times 10^{-8}$. Following the standard notations, here 'hat' $((\widehat{\ }))$ is added to the usual PCIs and others to denote their estimated values.

Thus, excluding $\widehat{C}_{pk}$, all the PCIs consider the process to be capable. The threshold value of $\widehat{C}_p''$ is found to be 0.7667 and since both $\widehat{C}_p''$ and $\widehat{C}_{pk}''$ have values higher than $\widehat{C}_p''^{(T)}$, the process is likely to be capable. This assessment of the process is also supported by P_{NC}^E as the $\widehat{P}_{NC}^E$ value is found to be considerably small. Also, since here, $\mu \neq T$, P_{NC} is not applicable here.

Therefore $\widehat{C}_p$ is not applicable here as $\mu \neq T$ while $\widehat{C}_{pk}$ makes an incorrect assessment of the process and hence they are not suitable here. On the other hand, $\widehat{C}_p'', \widehat{C}_{pk}''$ and $\widehat{P}_{NC}^E$ assess the capability of the process correctly. These argue in favour of the selection of appropriate PCIs based on the nature of the specification limits and other related aspects of a process.

4 Process Capability Indices for Unilateral (One Sided) Specification Limits

The PCIs discussed so far, are primarily meant for quality characteristics of nominal the best type and having bi-lateral specification limits. Quality characteristics of smaller the better type (e.g., surface roughness, degree of radiation and so on) and larger the better type (e.g., tensile strength, compressive strength and so on) requiring unilateral (one-sided) specification limits are also common in various manufacturing industries. However, there are only a few PCIs available in literature to assess capability of such processes.

Kane (1986) discussed about two such PCIs viz., $C_{PU} = \frac{U-\mu}{3\sigma}$ and $C_{PL} = \frac{\mu-L}{3\sigma}$. As is the relationship between C_p and 'p', given in Eq. (5), for unilateral specification limits also, analogous relationships, like $p^U = \Phi(3C_{PU})$ and $p^L = \Phi(3C_{PL})$, hold good between PNC, C_{PU} and C_{PL}, where, p^U and p^L denote the proportions of non-conformance generated due to exceeding USL and LSL respectively, when $\mu = T$. Also, 1 is usually considered as the threshold value of C_{PU} and C_{PL}. The distributional as well as inferential properties of these two PCIs, for both the single and multiple sample information, have also been studied extensively (see Lin and Pearn 2002, Pearn and Chen 2002, Shu et al. 2006). In fact, most of the research works on PCIs for unilateral specification limits are based on C_{PU} and C_{PL} only, due to their computational simplicity. Chatterjee and Chakraborty (2012) have made a thorough review of these PCIs for unilateral specification limits.

However, C_{PU} and C_{PL} suffer from the following critical drawbacks:

1. Neither of C_{PU} and C_{PL} incorporate the concept of 'T', the target value for the corresponding variable under consideration, in their respective definitions. As a result, they fail to measure the proximity of the process centering towards the target.
2. Unlike C_p, C_{PU} and C_{PL} can not be considered as the potential PCIs either, due to the presence of the mean μ in their definitions.

Therefore, despite being easy to compute, C_{PU} and C_{PL} are difficult to interpret.

Like $C_p(u, v)$, defined in Eq. (2), Vannman (1998) has defined the following two sets of superstructures of PCIs for unilateral specification limits:

$$\left.\begin{array}{l} C_{pau}(u, v) = \frac{USL-\mu-u|\mu-T|}{3\sqrt{\sigma^2 + v(\mu-T)^2}} \\[3mm] C_{pal}(u, v) = \frac{\mu-LSL-u|\mu-T|}{3\sqrt{\sigma^2 + v(\mu-T)^2}} \end{array}\right\} \quad (15)$$

and

$$\left.\begin{array}{l} C_{pvu}(u, v) = \frac{USL-T-u|\mu-T|}{3\sqrt{\sigma^2 + v(\mu-T)^2}} \\[3mm] C_{pvl}(u, v) = \frac{T-LSL-u|\mu-T|}{3\sqrt{\sigma^2 + v(\mu-T)^2}} \end{array}\right\} \quad (16)$$

Later, Grau (2009) observed some drawbacks in these two superstructures. The values of $C_{pvu}(u, v)$ and $C_{pvl}(u, v)$ are symmetric with respect to T which is not desirable for an ideal PCI for unilateral specification limits. In this context, the basic difference in the nature of asymmetric and unilateral specification limits is that for asymmetric specification limits, deviation from T towards USL and LSL are not of equal importance. However, in the context of the unilateral specification limits, deviation from T towards the existing specification limit (USL or LSL, depending upon the situation) is considered to be serious; while, deviation from T on the opposite side of the said specification limit can not be considered as undesirable, at least from the point of view of the quality of the product. Rather, such products are actually having better quality. Thus for both of these two types of specification limits, the corresponding loss function can by no means be considered as symmetric.

Again, for $0 \le u < 1$, $C_{pau}(u, v)$ and $C_{pal}(u, v)$ are not optimum on target. Also, for $u \ge 1$, $C_{pvu}(u, v)$ and $C_{pvl}(u, v)$ values become negative even before μ reaches U or L, which is highly undesirable. Therefore, neither of the superstructures of PCIs for the processes with unilateral specification limits, defined in Eqs. (15) and (16), are suitable for practical applications.

Grau (2009) has proposed the following superstructure of PCIs for unilateral specification limits, which is free from these drawbacks, and have also studied some of its distributional properties.

$$
\left.
\begin{aligned}
C_p^U(u, v) &= \frac{U - T - uA_U^*}{3\sqrt{\sigma^2 + vA_U^{*2}}} \\[2mm]
C_p^L(u, v) &= \frac{T - L - uA_L^*}{3\sqrt{\sigma^2 + vA_L^{*2}}}
\end{aligned}
\right\}
\tag{17}
$$

where, $A_U^* = \max\{(\mu - T), \frac{T - \mu}{k_U^*}\}$, $A_L^* = \max\{\frac{\mu - T}{k_L^*}, (T - \mu)\}$. Also, $k_I^*(> 1)$ quantifies the amount of loss incurred due to deviation from T towards the opposite side of the existing specification limit, where k_I^* stands for k_U^* or k_L^* depending upon the situation. Note that $C_p^U(u, v)$ and $C_p^L(u, v)$ are defined in such a way that the corresponding PCIs will be free from k_U^* and k_L^* respectively, when the quality characteristic value deviates from T towards the existing specification limit. Since Grau (2009) did not suggest any mathematical formulation of k_I^*, its choice becomes subjective, increasing the scopes for favourable manipulation in the values of $C_p^U(u, v)$ and $C_p^L(u, v)$. In order to eliminate such subjectivity in the definitions of $C_p^U(u, v)$ and $C_p^L(u, v)$, Chatterjee and Chakraborty (2012) have proposed a mathematical formulation of k_I^*.

It is interesting to note that for unilateral specification limits, target is set to maximize profit or to minimize loss. Thus, although deviation of μ from target towards the other side of the existing specification limit will definitely produce items of better quality; the manufacturer is likely to incur a loss of profit per item under constant selling price, since such production will require larger amount of ingredient or higher degree of expertise or more sophisticated machinery. Chatterjee and Chakraborty (2012) have applied this concept of loss of profit to formulate k_I^*.

For the purpose of illustration, suppose the quality characteristic under consideration is of smaller the better type and hence the corresponding process has only USL. Also suppose for this process, there are 'm' stages through which loss of profit can be incurred and let C_i^U is the corresponding loss of profit for the ith stage, where, $i = 1(1)m$. Here, one possible choice for the stages of loss of profit may be per unit or some convenient fraction of the unit of measurement. Moreover, let 'n' denotes the total number of produced items among which n_1 items have the values of the quality characteristic less than 'T', and the remaining n_2 items have the quality characteristic value greater than or equal to the target value such that $n = n_1 + n_2$ with $n_1 > 0$. Also, 'C' is the constant selling price.

Then k_U^* can be formulated as

$$k_U^* = \frac{\text{Selling Price Per Item}}{\text{Average Loss of Profit Per Item}}$$

$$= \frac{C}{\frac{1}{n_1} C_{LP,U}^{Total}} \tag{18}$$

where, $C_{LP,U}^{Total}$, the total loss of profit due to deviation from 'T' towards left, can be defined as

$$C_{LP,U}^{Total} = \sum_{j=1}^{n_1} \sum_{i=1}^{m} C_i^U I_{ij} \tag{19}$$

with

$$I_{ij} = \begin{cases} 1 & \text{if } j\text{th item belongs to the } i\text{th stage of loss of profit}, \forall i = 1(1)m, j = 1(1)n_1, \\ 0 & \text{otherwise}. \end{cases}$$

Similarly, for quality characteristics of the larger-the-better type, k_L^* can be formulated as

$$k_L^* = \frac{C}{\frac{1}{n_1} C_{LP,L}^{Total}} \tag{20}$$

where, $C_{LP,L}^{Total} = \sum_{j=1}^{n_1} \sum_{i=1}^{m} C_i^L I_{ij}$; C_i^L is the loss of profit at the ith stage, when, there exists 'm' such stages through which loss of profit (due to deviation of process mean form 'T' towards right i.e. towards the direction opposite to LSL with respect to 'T') can be incurred and I_{ij} has the same interpretation as before.

4.1 Example

To illustrate the impact of k_I^* on $C_p^U(u, v)$ and $C_L^p(u, v)$, we consider the data set on polarized dependent loss (PDL) of wavelength multiplexer (see Pearn et al. 2009). Here only the data corresponding to supplier I is considered, for which $n = 105$, $\hat{\mu} = 0.061$ decibel (dB), $\hat{\sigma} = 0.0049$ dB and USL $= 0.08$ dB. Moreover, although the original data set did not take into account the values of T, constant selling price per item and stage of loss of profit per item; following Chatterjee and

Chakraborty (2012), we have, $T = 0.064$ dB, C = \$1.00 and loss of profit for per 0.001 dB deviation from T towards left is 0.02 dB. Then, for the present data set, $k_U^* = 8.5337$.

Thus, $\hat{C}_p^U(0,0) = 1.078$, $\hat{C}_p^U(1,0) = 1.053$, $\hat{C}_p^U(0,1) = 1.075$, $\hat{C}_p^U(1,1) = 1.050$, $\hat{C}_{pvu}(0,0) = 0.86$ and $\hat{C}_{pvu}(0,1) = 0.90$. Here, it is easy to observe that out of the total number of 105 observations, 74 have the values of the quality characteristic less than $T = 0.064$. As has already been discussed, these 74 items can not be considered as having inferior quality—the only problem here is in terms of loss of profit. Now, since for $u = 0, 1$ and $v = 0, 1$, all the $\hat{C}_p^U(u, v)$ values are found to be greater than 1 which indicates that the process is performing satisfactorily and the loss of profit is also under control. However, $\hat{C}_{pvu}(u, v)$ does not take into account this aspect of unilateral specification limits. Since, $C_{pvu}(0,0)$ and $C_{pvu}(0, 1)$ merely measure the proximity of μ towards T, irrespective of the direction of such deviation, these PCIs fail to assess actual process performance and consider the process to be incapable which is not actually the case.

5 Process Capability Indices for Circular Specification Region

Apart from the bi-lateral (both symmetric and asymmetric) and unilateral specification limits, there is another type of specification limit which is known as circular specification limit. Such specification limits can be observed in processes like drilling holes (in manufacturing industries) or hitting a target (in ballistics). The uniqueness of circular specification limits is that the so called USL and/or LSL do not exist and consequently, the conventional PCIs are not applicable here.

Krishnamoorthi (1990) first proposed PCIs for processes with circular specification limits which are defined below:

$$\left.\begin{array}{l} PC_p = \frac{\frac{\pi}{4}D^2}{9\pi\sigma^2} = \frac{1}{36} \times \frac{D^2}{\sigma^2} \\[3mm] PC_{pk} = \dfrac{D^2}{4\left[\sqrt{(\overline{X}-a)^2 + (\overline{Y}-b)^2} + 3\sigma\right]^2} \end{array}\right\} \tag{21}$$

where, D is the diameter of the circular specification region, (a, b) is the targeted center of the process and σ is the common standard deviation along the two axes X_1 and X_2, such that, when $\sigma_1 \neq \sigma_2$, $\sigma = \max(\sigma_1, \sigma_2)$. It is assumed that $(X_1, X_2) \sim N_2(\mu_1, \mu_2, \sigma_1^2, \sigma_2^2, \rho = 0)$. Note that, PC_p is defined analogous to C_p and it measures the potential capability of a process; while PC_{pk} measures the actual process capability when the specification region under consideration is circular in nature.

Bothe (2006) has proposed another set of PCIs for circular specification region based on the concept of radial distance. He has considered average radial distance between the centers of various drilled holes or the average radial distance of centers of the drilled holes from the target center as the quality characteristic of interest and has defined the following PCIs analogous to C_p and C_{pk} based on these radial distances:

$$\left.\begin{aligned}
\widehat{C_P} &= \frac{\text{USL} - \widehat{\mu_C}}{3\widehat{\sigma}_{\text{ST},C}} \\
\widehat{P}_P &= \frac{\text{USL} - \widehat{\mu_C}}{3\widehat{\sigma}_{\text{LT},C}} \\
\widehat{C}_{\text{PK}} &= \frac{\text{USL} - \widehat{\mu_r}}{3\widehat{\sigma}_{\text{ST}}} \\
\widehat{P}_{\text{PK}} &= \frac{\text{USL} - \widehat{\mu_r}}{3\widehat{\sigma}_{\text{LT}}}
\end{aligned}\right\} \tag{22}$$

Here, $\widehat{\mu_C} = \frac{\sum_{i=1}^{n} r_{C,i}}{n}$; $r_{C,i} = \sqrt{(x_i - \bar{x})^2 + (y_i - \bar{y})^2}$ and 'n' is the sample size. Also, $\overline{\text{MR}} = \frac{\sum_{i=2}^{n} \text{MR}_i}{n-1}$, MR_i's are obtained from moving range chart; $\widehat{\sigma}_{\text{ST}} = \frac{\overline{\text{MR}}}{1.128}$; $\widehat{\sigma}_{\text{LT}} = \frac{1}{c_4}\sqrt{\frac{\sum_{i=1}^{n}(r_i - \bar{r})^2}{n-1}}$; $r_i = \sqrt{x_i^2 + y_i^2}$; $\widehat{\mu}_r = \frac{\sum_{i=1}^{n} r_i}{n}$; $\widehat{\sigma}_{\text{ST},C} = \frac{\overline{\text{MR}_C}}{d_2}$, where, MR_C values are obtained from the moving range chart of the data set after the target hole location is shifted to the middle of the cluster of actual hole centers and d_2 is a function of the sample size 'n' and $\widehat{\sigma}_{\text{LT},C} = \frac{1}{c_4}\sqrt{\frac{\sum_{i=1}^{n}(r_{C,i} - \overline{r_C})^2}{n-1}}$ and c_4 is a constant based on the sample size 'n'.

Note that, for both these two sets of PCIs defined in Eqs. (21) and (22), it is assumed that the variation in the values of the quality characteristics along the two axes are the same (homoscedastic) and also, these two axes are mutually independent. However, in reality, due to several practical reasons such assumptions of homoscedasticity and independence of X_1 and X_2 are seldom valid. As a result, even if the specification region is circular, the process region is elliptical in nature. Neither of the PCIs defined so far take care of this problem. Moreover, under the assumption of bivariate normality of (X_1, X_2), the distribution of radial distance is no more normal—rather it follows circular normal distribution (Scheur 1962). Thus, PCIs like C_p and C_{pk} are not suitable for assessing capability of such processes. From this view point also, the PCIs defined in Eq. (22) are not suitable for circular specification regions.

Chatterjee and Chakraborty (2015) have addressed these problems by defining a superstructure of PCIs for circular specification region. Suppose $X = (X_1, X_2)' \sim N_2(\mu_1, \mu_2, \sigma_1^2, \sigma_2^2, \rho)$. Also, without loss of generality, suppose the target center of the process is set at $(0,0)$ point of the co-ordinate axes. Then, like $C_p(u, v)$ defined in Eq. (2), Chatterjee and Chakraborty (2015) have defined a superstructure of PCIs for circular specification region as

$$C_{p,c}(u,v) = \frac{\frac{D}{2} - \frac{u}{\sqrt{\pi}}\mu^*}{\sqrt{\chi_{\alpha,2}^2}\sigma_1\sigma_2\sqrt{1-\rho^2}} \times \frac{1}{1+v\mu'\Sigma^{-1}\mu} \tag{23}$$

where, $\mu = (\mu_1, \mu_2)'$ and $\mu^* = \frac{1}{n}\sum_{i=1}^n d_i^{**}$, with $d_i^{**} = \sqrt{(X_{1_i} - \mu_1)^2 + (X_{2_i} - \mu_2)^2} = \sqrt{(X_i - \mu)'(X_i - \mu)}$; $X_i = (X_{1i}, X_{2i})'$ for $i = 1, 2, \ldots, n$ and 'n' is the number of sample observations randomly drawn from a process. Note that bold faced letters are used to denote vector valued variables.

Here $C_{p,c} = C_{p,c}(0,0), C_{pk,c} = C_{p,c}(1,0), C_{pm,c} = C_{p,c}(0,1)$ and $C_{pmk,c} = C_{p,c}(1,1)$ are, by definition, analogous to C_p, C_{pk}, C_{pm} and C_{pmk} respectively.

Note that $C_{p,c}(u,v)$ is defined from a more general perspective compared to the PCIs defined in Eqs. (21) and (22) and hence it does not require the so called assumptions of homoscedasticity and independence along the two axes. $C_{p,c}(u,v)$ is optimum on target as well which is a desirable property of a good PCI. Moreover, for a fixed value of ρ, the values of all the member indices of $C_{p,c}(u,v)$ decrease with the increase in at least one of σ_1^2 and σ_2^2. Similar to the inter-relationships between the member indices of $C_p(u,v)$ and $C_p''(u,v)$ with $u = 0, 1$ and $v = 0, 1$, here also, it is easy to check that

$$C_{p,c}(u,v) \leq C_{p,c}(u,0) \leq C_{p,c}(0,0)$$

$$C_{p,c}(u,v) \leq C_{p,c}(0,v) \leq C_{p,c}(0,0), \quad \forall u \geq 0, v \geq 0$$

and there is no clear-cut relationship between $C_{pk,c}$ and $C_{pmk,c}$.

Since, similar to C_p, $C_{p,c}$ measures the potential capability of a process, Chatterjee and Chakraborty (2015) have derived the expression for the threshold value of $C_{p,c}$ as

$$C_{p,c}^T = \sqrt{\frac{D}{2 \times \chi_{\alpha,2}^2 \times \sigma_{\min} \times \sqrt{1-\rho^2}}} \tag{24}$$

Thus, the threshold value of $C_{p,c}$ is a function of σ_1, σ_2 and ρ and hence is not unique.

However, for $\rho = 0$ and $\sigma_1 = \sigma_2 = \frac{D}{2}$, although the process region becomes circular and coincides with the specification region, $C_{p,c}^T \neq 1$ and for this reason, $C_{p,c}(u,v)$ is not suitable when the correlation between the two axes is very low.

Again, similar to C_{pk}, $C_{pk,c}$ is a yield based PCI and Chatterjee and Chakraborty (2015) have established exact relationships between $C_{p,c}$, $C_{pk,c}$ and PNC as follows.

When the process is on target, PNC can be formulated as

$$
\begin{aligned}
P_{NC} &= P\left[(X - \mathbf{0})'\Sigma^{-1}(X - \mathbf{0}) > \binom{D}{2}0\, I_2\binom{\frac{D}{2}}{0}\right] \\
&= P\left[\chi_{\alpha,2n-4}^2 > 2(n-1)\sqrt{|S|}C_{p,c}^2\right],
\end{aligned}
\tag{25}
$$

where, $A = (n-1)S$, S being the sample variance-covariance matrix.

However, in practice, often the assumption that $\hat{\mu} = (0,0)'$ may not hold and in such cases, P_{NC} measures only the minimum attainable PNC. Considering the more general case, i.e. when $\hat{\mu} \neq (0,0)'$, Chatterjee and Chakraborty (2012) have derived the expression for observed PNC as

$$
\begin{aligned}
P_{NC}^E &= P\left[X'\Sigma^{-1}X > \frac{D^2}{4}\,\middle|\,X \sim N_2(\mu, \Sigma)\right] \\
&= P\left[2(n-1)\sqrt{|S|}C_{pk,c}^2 < \sqrt{F_{\alpha,2,2}(\lambda) \times \chi_{\alpha,2n-4}^2} - \frac{\mu^*}{\sqrt{\pi}} \times \sqrt{F_{\alpha,2n-4,2}}\right]
\end{aligned}
\tag{26}
$$

where, $\chi_{\alpha,2n-4}^2$ and $F_{\alpha,2,2}$ denote respectively the upper $\alpha\%$ point of a χ^2 distribution with $(2n-4)$ degrees of freedom and a F distribution with (2, 2) degrees of freedom.

Moreover, based on the properties of circular normal distribution (refer Scheur 1962), Chatterjee and Chakraborty (2015) have derived the expressions for the expectations and variances of the member indices of $C_{p,c}(u,v)$ for $u = 0, 1$ and $v = 0, 1$.

5.1   Example

To investigate the performance of $C_{p,c}(u,v)$ for assessing capability of processes having circular specification limits, we now consider a manufactured product and we are concerned about the holes drilled subject to some specifications. 20 holes were drilled and for each hole, the values of the corresponding X_1 and X_2 co-ordinates of the centers of the holes were noted. Here, $D = 10$, $\overline{X}_1 = 2.766$, $\overline{X}_2 = 2.776$, $\hat{\sigma}_1^2 = 0.408$, $\hat{\sigma}_2^2 = 0.321$ and $\hat{\rho} = 0.856$. The complete data set is available in Chatterjee and Chakraborty (2015) and Fig. 2 provides a

Fig. 2 Circular specification region and the elliptical process region

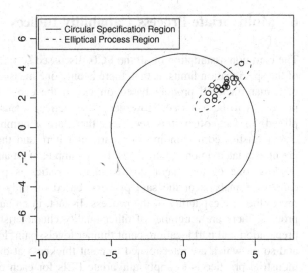

diagrammatic representation of the process region and the corresponding specification region.

Since $\hat{\sigma}_1^2 \neq \hat{\sigma}_2^2$ and the value of $\hat{\rho}$ is also considerably high, the PCIs defined in Eqs. (21) and (22) are not applicable here. Hence values of the member indices of the superstructure $C_{p,c}(u, v)$ are computed as follows:

$\widehat{C}_{p,c} = 3.8097, \widehat{C}_{\text{pk},c} = 3.5184, \widehat{C}_{\text{pm},c} = 0.7605, \widehat{C}_{\text{pmk},c} = 0.7024, \hat{P}_{\text{NC}} = 2 \times 10^{-8}, \hat{P}_{\text{NC}}^E = 0.0598$ and $\hat{C}_{p,c}^T = 1.3613$.

From the above computations it can be observed that $C_{p,c}$ considers the process to be potentially capable and this is also supported by the low value of $\hat{P}_{\text{NC}}$. However, all the other PCIs except $C_{\text{pk},c}$ consider the process to be incapable. Now, from Fig. 2 it is evident that, the data points lie far away from the target center $(0, 0)'$ and this is correctly reflected by the low values of $C_{\text{pm},c}$ and $C_{\text{pmk},c}$. Also, a considerable part of the process region lies outside the circular specification region and this increases the value of $\hat{P}_{\text{NC}}^E$. In fact, $\widehat{C}_{\text{pk},c}$, being a yield based PCI, has rightly reflected the incapability of the process though the high value of $\hat{P}_{\text{NC}}^E$. Thus, although the process is potentially capable, since it is highly off-centered, it is not actually performing satisfactorily. Moreover, the apparent contradiction between the values of $\widehat{C}_{\text{pk},c}$, $\widehat{C}_{\text{pm},c}$ and $\widehat{C}_{\text{pmk},c}$ argue for the judgemental use of PCIs as well as the importance of diagrammatic representations of the process and specification region to have a prima-facie impression about the health of the process.

6 Multivariate Process Capability Indices

The common assumption of all the PCIs discussed so far, irrespective of the nature of the specification limits, is that, there is only one measurable quality characteristic of a manufactured product based on which the capability of the corresponding process is to be assessed. However, the practical scenario is not that much simplified. In fact, often it is seen that there are a number of measurable quality characteristics corresponding to a particular item and these quality characteristics are inter-related among themselves. For example, in an automated paint application process, one of the major quality characteristics is paint thickness. However, capability analysis of the said process, based on only paint thickness, may not reveal the true capability of the process. In fact, in an automated paint application process, there are a number of other quality characteristics like ability of surface preparation and part location, paint thinner levels, paint lot differences, temperature and so on which are inter-related to paint thickness at different degrees. Common industrial practice is to apply univariate PCIs for each of these quality characteristics separately and summarize the process capability as the arithmetic or geometric mean of these individual PCI values. However, this approach may not be able to assess the capability of the process accurately as it ignores the correlation structure among the quality characteristics. This necessitates the application of multivariate process capability indices (MPCI).

Despite having ample scope of industrial applications, there are only a few MPCIs available in literature. Following Shinde and Khadse (2009), the MPCIs, defined so far, are either of the following types:

1. MPCIs defined as the ratio of tolerance region and process region; e.g., Taam et al. (1993), Goethals and Cho (2011) and so on;
2. MPCIs expressed as the probability of non-conforming products; e.g., Chen (1994), Khadse and Shinde (2006), Pearn et al. (2006), Shiau et al. (2013) and so on;
3. MPCIs based on principal component analysis; e.g., Wang and Chen (1998), Wang and Du (2000), Shinde and Khadse (2009), Perakis and Xekalaki (2012), Tano and Vannman (2012) and so on;
4. MPCIs based on the concept of non-parametric statistics; refer Polansky (2001);
5. Other approaches including vector representation of MPCIs; e.g., Kirmani and Polansky (2009), Shahriari et al. (2009); MPCIs based on lowner ordering; refer Kirmani and Polansky (2009) and so on.

6.1 $C_G(u, v)$–A Multivariate Process Capability Index for Symmetric Specification Region

Most of the MPCIs defined so far are difficult to compute and hence are meant for theoreticians. Moreover, since shop-floor people are very much conversant with classical univariate indices, some multivariate analogue of $C_p(u, v)$ would be more palatable to them. Chakraborty and Das (2007) have defined a MPCI called $C_G(u, v)$, analogous to $C_p(u, v)$, to address these problems. For defining the new MPCI, Chakraborty and Das (2007) have made the following realistic assumptions:

1. Underlying process distribution is multivariate normal with mean vector μ and dispersion matrix Σ.
2. The process has hyper-rectangular specification region.
3. For each process variable specification limits are symmetric about its mean.
4. $T = M$ as otherwise the specification region will become asymmetric with respect to the target.

Based on these assumptions, $C_G(u, v)$ can be defined as,

$$C_G(u, v) = \frac{1}{3} \sqrt{\frac{(\mathbf{d} - u\mathbf{D})' \Sigma^{-1} (\mathbf{d} - u\mathbf{D})}{1 + v(\mu - \mathbf{T})' \Sigma^{-1} (\mu - \mathbf{T})}} \tag{27}$$

where, $\mathbf{D} = (|\mu_1 - M_1|, |\mu_2 - M_2|, \ldots, |\mu_p - M_p|)'$;
$\mathbf{d} = (\frac{\text{USL}_1\text{-LSL}_1}{2}, \frac{\text{USL}_2\text{-LSL}_2}{2}, \ldots, \frac{\text{USL}_p\text{-LSL}_p}{2})'$;
$\mathbf{T} = (T_1, T_2, \ldots, T_p)'$;
$\mathbf{M} = (M_1, M_2, \ldots, M_p)'$; $\mu = (\mu_1, \mu_2, \ldots, \mu_p)'$.
Here, T_i is the target value, M_i is the nominal value for the ith characteristic of the item; 'p' denotes the number of characteristics under consideration and
Σ = Variance—covariance matrix of the variable X;
μ_i = Mean of the ith characteristic of the variable X, for $i = 1, 2, \ldots, p$;
μ = The mean vector of the variable X, ·
u and v are the scalar constants that can take any non-negative integer value.
Note that, the member indices of $C_G(u, v)$, viz., $C_G(0, 0)$, $C_G(1, 0)$, $C_G(0, 1)$ and $C_G(1, 1)$ are analogous to the four classical PCIs C_p, C_{pk}, C_{pm} and C_{pmk} of univariate PCIs for symmetric specification limits.
Chatterjee and Chakraborty (2013) have observed that, for $u = 0, 1$ and $v = 0, 1$, the member indices of $C_G(u, v)$ are inter-related among themselves through the following relationships:

$$\left. \begin{array}{l} C_G(0, 0) \geq C_G(1, 0) \geq C_G(1, 1) \\ C_G(0, 0) \geq C_G(0, 1) \geq C_G(1, 1) \end{array} \right\} \tag{28}$$

and there exists no clear-cut relationship between $C_G(1,0)$ and $C_G(0,1)$. Note that such relationships are analogous to those between the member indices of $C_p(u,v)$.

Chatterjee and Chakraborty (2013) have also explored the relationship between the minimum attainable proportion of non-conformance (P_{NC}) and $C_G(0,0)$ and have observed that

$$P_{NC} = 2\left\{1 - P[Y \leq 9C_G^2(0,0)|Y \sim \chi_p^2]\right\} \tag{29}$$

Since by definition, $C_G(0,0)$ is always non-negative, Eq. (26) establishes a one-to-one relationship between $C_G(0,0)$ and P_{NC}. Chatterjee and Chakraborty (2013) have also made an extensive comparative study among the member indices of $C_G(u,v)$ and $C_p(u,v)$ to help these MPCIs gain higher amount of acceptability among the practitioners.

6.2 $C_M(u,v)$ – a Multivariate Process Capability Index for Asymmetric Specification Region

Like in the univariate case, for processes with multiple quality characteristics also, it is common to encounter processes with asymmetric specification regions, i.e., where, $T \neq M$. Although Grau (2007) proposed some MPCIs to assess the capability of such processes, his formulations are complicated in nature and hence are of interest more for theoreticians than the shop-floor people who are ultimately going to use these PCIs.

As have been already discussed in Sect. 3, $C_p''(u,v)$, defined in Eq. (4), is more suitable for measuring capability of the processes with single quality characteristic and asymmetric specification limits with respect to T as compared to the other PCIs available in literature. Chatterjee and Chakraborty (2011) have defined an MPCI called $C_M(u,v)$, which generalizes $C_p''(u,v)$ for processes with multiple quality characteristics. Here, $C_M(u,v)$ is defined as

$$C_M(u,v) = \frac{1}{3}\sqrt{\frac{(\mathbf{d}^* - u\mathbf{G}^*)'\Sigma^{-1}(\mathbf{d}^* - u\mathbf{G}^*)}{1 + v\mathbf{G}'\Sigma^{-1}\mathbf{G}}}, \tag{30}$$

where, $\mathbf{d}^* = \begin{pmatrix} \min(D_{1L}, D_{1U}) \\ \min(D_{2L}, D_{2U}) \\ \vdots \\ \min(D_{pL}, D_{pU}) \end{pmatrix}$ i.e. $d_i^* = \min(D_{iL}, D_{iU})$, for $i = 1(1)p$ with

$D_U = \begin{pmatrix} D_{1U} \\ D_{2U} \\ \vdots \\ D_{pU} \end{pmatrix}$ and $D_L = \begin{pmatrix} D_{1L} \\ D_{2L} \\ \vdots \\ D_{pL} \end{pmatrix}$.

Also $\mathbf{d} = \begin{pmatrix} \dfrac{\text{USL}_1 - \text{LSL}_1}{2} \\ \dfrac{\text{USL}_2 - \text{LSL}_2}{2} \\ \vdots \\ \dfrac{\text{USL}_p - \text{LSL}_p}{2} \end{pmatrix}$ i.e. $d_i = \dfrac{\text{USL}_i - \text{LSL}_i}{2}$, for $i = 1(1)p$

For multivariate case 'G' can be defined as

$\mathbf{G} = \begin{pmatrix} a_1 d_1 \\ a_2 d_2 \\ \vdots \\ a_p d_p \end{pmatrix}$, where, $a_i = [\max\{\frac{\mu_i - T_i}{D_{iU}}, \frac{T_i - \mu_i}{D_{iL}}\}], \forall i = 1(1)p$.

As such $\mathbf{G} = \begin{pmatrix} a_1 & 0 & 0 & \cdots & 0 & 0 \\ 0 & a_2 & 0 & \cdots & 0 & 0 \\ \vdots & \vdots & \vdots & \vdots & \vdots & \vdots \\ 0 & 0 & 0 & \cdots & 0 & a_p \end{pmatrix} \mathbf{d} = \mathbf{Ad}$, say and its univariate

counterpart is given as 'F' in (4).

Similarly, 'F^*' can be generalized as $\mathbf{G}^* = \mathbf{Ad}^*$ for the multivariate case. Also, for $p = 1$, $C_M(u, v) = C_p''(u, v)$. For $u = 0, 1$ and $v = 0, 1$, the authors have observed following relationship between the member indices of $C_M(u, v)$:

$$C_M(0, 0) \geq C_M(1, 0) \geq C_M(1, 1)$$

$$C_M(0, 0) \geq C_M(0, 1) \geq C_M(1, 1)$$

Also, no clear-cut relationship exists between $C_M(1, 0)$ and $C_M(0, 1)$ like in the case of $C_p''(u, v)$.

Note that here $C_M(0, 0)$, which is independent of μ, measures the potential process capability and this is quite justified by the•above relationships as all the other member indices of $C_M(u, v)$ can achieve at most the capability value projected by $C_M(0, 0)$.

6.3 Example

To demonstrate the ability of $C_M(u, v)$, for $u = 0, 1$ and $v = 0, 1$, we consider the data set originally used by Sultan (1986). Here we have two correlated characteristics viz., brinell hardness (H) and tensile strength (S). The USL and LSL for 'H' are 241.3 and 112.7 respectively; while for 'S', these values are 73.30 and 32.70 respectively. Also, the target vector for the said process is $T = (177, 53)'$. Thus, $T = M$ and hence $C_G(u, v)$ will be applicable here.

A random sample of size 25 is drawn from the process and the corresponding summary statistics are as follows:

$$n = 25, \ \overline{X} = \begin{pmatrix} 177.2 \\ 52.316 \end{pmatrix}, \ \hat{\Sigma} = \begin{pmatrix} 338 & 88.8925 \\ 88.895 & 33.6247 \end{pmatrix} \text{ and hence the observed}$$

correlation coefficient between 'H' and 'S' is, $\hat{\rho} = 0.8338$ which is quite high. Thus, $\hat{C}_G(0,0) = 1.2181, \hat{C}_G(1,0) = 1.1971, \hat{C}_G(0,1) = 1.1870$ and $\hat{C}_G(1,1) = 1.1666$. Hence we conclude that the process is capable. Also, the computed MPCI values follow the interrelationship established in Eq. (28). These strongly suggest that the process is performing satisfactorily.

However, before assessing the capability of the process, we need to check the validity of the assumption of multivariate normality of the present data. The p value associated with Shapiro–Wilk test (refer Shapiro and Wilk 1965) is 0.006764 and that with Royston's test (refer Royston 1983) is 0.02586. Since both of these p values are less than 0.05, it is logical to expect that the underlying distribution of the present data set is not multivariate normal (refer Chatterjee and Chakraborty 2013).

In order to assess the capability of the process, the data is transformed using Box–Cox Power Transformation (refer Box and Cox 1964). For the transformed data, p value corresponding to the Shapiro–Wilk multivariate normality test is found to be 0.07627; while that using Royston's test is 0.1103. Therefore, it is logical to expect that the transformed data set indeed follow multivariate normal distribution.

Moreover, since the data set has been transformed to have multivariate normal distribution, it is now required to transform **USL**, **LSL** and **T**, by virtue of the same transformation. The transformed specification limits and targets for H_{new} and S_{new} are as follows:

$$\left. \begin{array}{l} \text{USL}_{H_{\text{new}}} = 240.3 \\ \text{LSL}_{H_{\text{new}}} = 111.7 \\ T_{H_{\text{new}}} = 176 \end{array} \right\} \qquad \left. \begin{array}{l} \text{USL}_{S_{\text{new}}} = 2685.945 \\ \text{LSL}_{S_{\text{new}}} = 534.145 \\ T_{S_{\text{new}}} = 1404.000 \end{array} \right\}$$

Thus, although, apparently the specification region was symmetric with respect to the target vector, the transformed specification region is asymmetric about the transformed target vector, viz., $T_{\text{new}} = (T_{H_{\text{new}}}, T_{S_{\text{new}}}) = (177, 1404)$.

For the transformed data, $d = (64.3, 1075.9)'$, $d^* = (64.3, 869.855)'$, $A = \begin{pmatrix} 0.0031 & 0 \\ 0 & 0.0228 \end{pmatrix}$, $G = Ad = (0.2, 24.5868)$ and $G^* = Ad^* = (0.2, 19.8782)$, $\hat{\mu} = (176.20, 1384.122)'$ and $\hat{\Sigma} = \begin{pmatrix} 338 & 4435.277 \\ 4435.277 & 81311.074 \end{pmatrix}$.

Hence, $\hat{C}_M(0,0) = 1.1672$, $\hat{C}_M(1,0) = 1.1623$, $\hat{C}_M(0,1) = 1.1551$ and $\hat{C}_M(1,1) = 1.1503$. Also, the threshold value of $\hat{C}_M(0,0)$ is computed as 1.1672. Thus, the process is potentially just capable as the threshold value coincides with the value of $\hat{C}_M(0,0)$. However, all of $\hat{C}_M(1,0)$, $\hat{C}_M(0,1)$ and $\hat{C}_M(1,1)$ have values lower than the threshold value. This indicates that the actual capability level of the process is not satisfactory.

Thus, assertion of the underlying distribution of the quality characteristic(s) is utmost necessary before assessing the capability of a process.

7 Process Capability Indices for Non-normal Statistical Distributions

As has been observed by Kotz and Johnson (2002), the assumption of normality of the underlying statistical distribution of the concerned quality characteristic, is one of the basic assumptions for defining process capability indices, irrespective of the nature of the specification limits. Despite of giving some computational advantage, such normality assumption is not valid in many practical situations.

For example, for quality characteristics of smaller the better type (like surface roughness, flatness and so on), for which only USL is available, some times, the quality characteristic has a skewed distribution with a long tail towards the larger values (refer Vannman and Albing 2007).

Clements (Clements 1989) first addressed this problem and suggested replacing 6σ by the length of the interval between the upper and lower 0.135 percentile points of the actual distribution. The author redefined estimators of C_p and C_{pk}, for quality characteristics with non-normal statistical distributions as follows:

$$C'_p = \frac{U - L}{\xi_{1-\alpha} - \xi_\alpha} \tag{31}$$

$$C'_{pk} = \frac{d - |\xi_{0.5} - M|}{(\xi_{1-\alpha} - \xi_\alpha)/2} \tag{32}$$

where, $\xi_{1-\alpha}$ and ξ_α are the upper and lower α percentiles of the distribution of the corresponding random variable X and $\xi_{0.5}$ is the corresponding median. Generally, $\alpha = 0.00135$ is considered for computational purposes.

Following Clements' (1989) approach, Pearn and Kotz (1994), redefined the estimators of C_{pm} and C_{pmk} for non-normal quality characteristics as,

$$C'_{pm} = \frac{U - L}{6\sqrt{\left[\frac{\xi_{1-\alpha} - \xi_\alpha}{6}\right]^2 + (M - T)^2}} \tag{33}$$

$$C'_{pmk} = \min\left\{\frac{U - M}{3\sqrt{\left[\frac{\xi_{1-\alpha} - M}{3}\right]^2 + (M - T)^2}}, \frac{M - L}{3\sqrt{\left[\frac{M - \xi_\alpha}{3}\right]^2 + (M - T)^2}}\right\} \tag{34}$$

Pearn et al. (1999) generalized these indices for asymmetric specification limits. Wright (1995) proposed the following PCI which is sensitive to skewness:

$$C_s = \frac{\frac{d}{\sigma} - \frac{|\mu - M|}{\sigma}}{3\sqrt{1 + \left(\frac{\mu - T}{\sigma}\right)^2 + |\sqrt{\beta_1}|}} \tag{35}$$

where, $\sqrt{\beta_1} = \frac{\mu_3}{\sigma^{3/2}}$ is a widely used measure of skewness and μ_3 is the third order raw moment of the corresponding random variate 'X'.

However, the quantile or percentile based approach of dealing with non-normality, while measuring capability of a process, suffers from a basic problem. Often these PCIs involve extreme percentiles viz., 99.73th or 0.27th percentiles. However, accurate estimation of these extreme percentiles require a huge amount of data, which is often difficult to obtain, especially for processes requiring destructive testing (refer Pearn et al. 1992). Wu et al. (1998) have observed that, PCIs based on Clements' approach fail to measure the capability of a process accurately, especially, when the underlying distribution of the concerned quality characteristic is skewed.

Another approach of dealing with non-normality is to transform the original non-normal data into a normal one through the use of appropriate transformations and then apply the PCIs defined for normal data. Some of the statistical transformations, which are available in literature, are

1. Johnson's (1949) transformation, based on the method of moments;
2. Box–Cox's (1964) power transformation;
3. Somerville and Montgomery's (1996) square-root based transformation for skewed distributions;
4. Hosseinifard et al.'s (2009) root transformation method

Farnum (1996) has extensively discussed the use of Johnson's transformation in the context of non-normal process data. Yang et al. (2010) have carried out a comparative study between the performances of Box–Cox transformation and Johnson's transformation in assessing capability of a process.

One can also choose a process distribution from a smaller family of distributions such as gamma, lognormal or weibull which in turn, simplifies the corresponding inferential problem. Rodriguez (1992) have enlisted the following advantages of using families of distributions for computing PCIs of non-normal processes:

1. Method of maximum likelihood can be used to have stable and straight forward estimation of the concerned parameters.
2. Since the method of maximum likelihood yields asymptotic variances for estimates of the parameters, it can be used to construct confidence intervals for the plug-in estimators of the PCIs.
3. For various families of distributions like gamma, lognormal and weibull, goodness-of-fit tests based on empirical distribution functions are also available.

4. For standard families of distributions, estimated values of the percentiles and proportion of non-conformance, related to the plug-in estimators of the PCIs, can be easily computed using standard results.

It is interesting to note that 'potential capability' means 'possibility of achieving' rather than 'actually achieving' (refer Kotz and Johnson 2002). Veevers (1998) has used the term 'viability' to represent 'capability potential' and has proposed a viability index from a more general perspective, as compared to C_p, in a sense that the viability index is neither restricted to normal distribution of 'X' nor even to univariate situations.

The univariate viability index is defined as

$$V_t = \frac{w}{2d} \tag{36}$$

where, 'w' is the 'window of opportunity' measured by the length of interval of θ for which the distribution of $(X + \theta)$ would generate an expected PNC not greater than the conventional 0.27 %.

Under the assumption of normality of the quality characteristic under consideration,

$$M - (d - 3\sigma) \leq \mu \leq M + (d - 3\sigma) \tag{37}$$

i.e. the window of opportunity for μ can be defined as, $w = 2(d - 3\sigma)$ and the corresponding viability index will be

$$\begin{aligned} V_t &= \frac{2(d - 3\sigma)}{2d} \\ &= 1 - \frac{1}{C_p} \end{aligned} \tag{38}$$

Unlike most of the PCIs, V_t can assume negative values. If V_t is less than zero, there is no possibility of attaining a PNC value of 0.27 % or lower and hence, the process is considered to be 'non-viable'.

For processes with unilateral specification limits also, substantial research work has been done to assess the capability of a process when the underlying statistical distribution is non-normal. Vannman and Albing (2007) modified $C_{pvu}(u, v)$ (see Eq. (16)) for the case, where the quality characteristic has a skewed distribution with a long tail towards large values and a 'USL' with target set at '0', i.e. the quality characteristic has a skewed zero-bound distribution. This superstructure is defined as

$$C_{MA}(\tau, v) = \frac{USL}{\sqrt{q_{1-\tau}^2 + v q_{0.5}^2}} \tag{39}$$

where, $v \geq 0$ and q_τ is the τth quantile of the quality characteristic. The parameter τ should be small and chosen in a suitable way, e.g. $\tau = 0.0027$.

However, Chatterjee and Chakraborty (2012) have observed the following drawbacks in this superstructure:

1. Vannman and Albing (2007) have modified only $C_{pvu}(u, v)$. Neither $C_{pau}(u, v)$ was modified nor any justification for omitting the same was given. However, as has been pointed out by Grau (2009), $C_{pvu}(u, v)$ is not suitable for assessing capability of a process with unilateral specifications.
2. There is room for studying whether considering $\tau = 0.0027$ is justified even if the underlying distribution of the quality characteristic is not normal.
3. Some constants of $C_{pvu}(u, v)$ were omitted just for simplicity without studying the impact of such omission.
4. $C_{MA}(\tau, v)$ fails to perform if the target is other than '0'.
5. The ideal values of v have not been studied.

Albing (2009) has modified the superstructure $C_{MA}(\tau, v)$ which is defined in Eq. (36) for the quality characteristic under Weibull distribution, as follows:

$$C_{MAW}(\tau, v) = \frac{USL}{a\sqrt{(ln(\frac{1}{\tau})^{\frac{2}{b}}) + v(ln2)^{\frac{2}{b}}}} \tag{40}$$

where, 'a' is the scale parameter and 'b' is the shape parameter of a two-parameter Weibull distribution. However, since this super-structure is an extension of $C_{MA}(\tau, v)$, it inherits the drawbacks of $C_{MA}(\tau, v)$ as listed above. Moreover, $C_{MAW}(\tau, v)$ is valid only when the underlying distribution of the quality characteristic is Weibull. It fails to perform in case of all the other types of statistical distributions.

Rodriguez (1992) has also suggested other methods like goodness-of-fit, quantile-quantile plot, kernel density estimation and comparative histograms to assess capabilities of non-normal processes. For a thorough review of the PCIs for non-normal distributions, one can refer to Pearn and Kotz (2007); Tang and Than (1999) and the references there-in.

Finally, the capability assessments for multivariate processes with non-normal process distributions have been studied by Abbasi and Niaki (2010), Ahmad et al. (2009), Polansky (2001) and so on.

The example considered at the end of Sect. 6.2 can be considered here as well. Recall that, there we have transformed the multivariate non-normal data into a multivariate normal one and then applied $C_M(u, v)$. The MPCIs discussed in the present section can also be used for this purpose. In particular, Polansky (2001) used the same data and concluded that the performance of the process is not satisfactory, which supports the observations made by Chatterjee and Chakraborty (2013). Moreover, the approach of transforming the data to multivariate normality and then applying $C_M(u, v)$ is easier to execute as compared to using Polansky's (2001) MPCI.

8   How to Check Univariate and Multivariate Normality of Data

Asserting the underlying distribution of the quality characteristic under consideration plays a major role in capability assessment of a process. Often, in practice, PCIs for univariate and multivariate normal distributions are used to assess the capability of a process, without exploring the actual statistical distribution of the concerned quality characteristic. This may lead to wrong judgement of the actual capability of a process. Hence proper testing of the normality assumption of the available data is utmost solicited.

Now a days, such checking of normality is possible through almost all the statistical softwares available in market, viz., R, SPSS, STATISTICA, MINITAB, SAS, MATLAB and so on. Among these, R is a open source and hence can be freely downloaded from internet. We shall now discuss the procedure of testing normality through the statistical package **R**.

Following are some functions and packages in **R**, which deal with univariate and multivariate normality testing:

 (i) Shapiro–Wilk test (Shapiro and Wilk 1965) for univariate normality can be done using the function **shapiro.test**.
 (ii) qqnorm is a function that produces a normal quantile–quantile (QQ) plot of a data. The corresponding qqline adds a line to a 'theoretical', by default normal, quantile–quantile plot which passes through the first and third quartiles.
 (iii) For testing multivariate normality of a data, one can use the library **MVN** which provides functions for Mardia's multivariate normality test (refer Mardia 1970, 1974) and Royston's multivariate normality test (refer Royston 1983).
 (iv) Generalized Shapiro–Wilk test for multivariate normality (refer Royston 1983 and Villasenor-Alva and Gonzalez-Estrada 2009) can be carried out using libraries like **mvShapiroTest** and **mvnormtest**.

Also, to transform a non-normal data into a normal one, one can use the library **alr3** for Box–Cox transformation (refer Box and Cox 1964) of the data.

9   Concluding Remarks

This chapter deals with measurement of process capability analysis for different situations by mostly suggesting appropriate indices for a given situation. However, there are criticisms for making process capability index as the sole measure of the capability of the process. One can refer to Gunter (1989), Dovich (1991), Carr (1991), Herman (1989), Pignatiello and Ramberg (1993) and many others. Some even suggested that none of the so called PCIs adds any knowledge or understanding about the process beyond that contained in the equivalent basic parameters like μ, σ, target value and the specification limits.

The main problem seems to be that a PCI is taken as a one-time measure or a snap shot of the process and is highly dependent on the chosen sample. This leads to a fear of manipulation which is genuine. We suggest that for a PCI to be calculated, a necessary condition to be fulfilled is that the process should be stable. A sufficient condition could be that the PCI, calculated over a period of time should show stability. This requires an appropriate control charting technique for each PCI depending on the type of distribution a PCI would follow. The authors are now developing these control charts which will settle the issue.

References

Ahmad, S., Abdollahian, M., Zeephongsekul, P., Abbasi, B. (2009). Multivariate non-normal Process capability analysis. *International Journal of Advanced Manufacturing Technology*, *44*, 757–765.

Abbasi, B., & Niaki, S. T. A. (2010). Estimating process capability indices of multivariate non-normal processes. *International Journal of Advanced Manufacturing Technology*, *50*, 823–830.

Albing, M. (2009). Process capability indices for Weibull distributions and upper specification limits. *Quality and Reliability Engineering International*, *25*(3), 317–334.

Bothe, D. R. (2006). Assessing capability for hole location. *Quality Engineering*, *18*, 325–331.

Box, G. E. P., Cox, D. R. (1964). An analysis of tranformation. *Journal of Royal Statistical Society, Series B*, *26*, 211–243.

Boyles, R. A. (1994). Process capability with asymmetric tolerances. *Communications in Statistics —Simulation and Computation*, *23*(3), 615–635.

Carr, W. E. (1991). A new process capability index: parts per million. *Quality Progress*, *24*(2), 152–154.

Chakraborty, A. K., & Das, A. (2007). *Statistical analysis of multivariate process capability indices. Private communication*. Kolkata: Indian Statistical Institute.

Chatterjee, M., & Chakraborty, A. K. (2011). Superstructure of multivariate process capability indices for asymmetric tolerances. *Proceedings of International Congress on Productivity, Quality, Reliability, Optimization and Modelling*, *1*, 635–647.

Chatterjee, M., Chakraborty, A. K. (2012). Univariate process capability indices for unilateral specification region—a review & some modifications. *Internnational Journal of Reliability, Quality and Safety Engineering*, *19*(4), 1250020-1–1250020-18. doi:10.1142/S0218539312500209.

Chatterjee, M., Chakraborty, A. K. (2015). A superstructure of process capability indices for circular specification region. *Communications in Statistics - Theory and Methods*, *44*(6), 1158–1181.

Chatterjee, M., Chakraborty, A. K. (2014). Exact relationship of C_{pk}'' with proportion of non-conformance and some other properties of $C_p''(u, v)$. *Quality and Reliability Engineering International*. *30*(7), 1023–1034 , doi:10.1002/qre.1530.

Chatterjee, M., Chakraborty, A. K. (2013). Some properties of $C_G(u, v)$. *Proceedings of International Conference on Quality and Reliability Engineering* 203–209.

Chatterjee, M., Chakraborty, A. K. (2013). Unification of Some Multivariate Process Capability Indices For Asymmetric Specification Region. *Statistica Neerlandica* (under review).

Chen, H. (1994). A multivariate process capability index over a rectangular solid tolerance zone. *Statistica Sinica*, *4*, 749–758.

Chen, K. S., & Pearn, W. L. (2001). Capability indices for process with asymmetric tolerances. *Journal of the Chinese Institute of Engineers, 24*(5), 559–568.

Clements, J. A. (1989). Process capability calculations for non-normal distributions. *Quality Progress, 22*, 95–100.

Dovich, R. A. (1991). *Statistical terrorists II—it's not safe yet, Cpk is out there, MS.* Rockford, IL: Ingressol Cutting Tools Co.

Farnum, N. R. (1996). Using Johnsoncurves to describe non-normal process data. *Quality Engineering, 9*(2), 329–336.

Franklin, L. A., & Wasserman, G. (1992). Bootstrap lower confidence limits for capability indices. *Journal of Quality Technology, 24*(4), 196–210.

Goethals, P. L., & Cho, B. R. (2011). The development of a target-focused process capability index with multiple characteristics. *Quality and Reliability Engineering International, 27*, 297–311.

Grau, D. (2007). Multivariate capability indices for processes with asymmetric tolerances. *Quality Technology and Quantitative Management, 4*(4), 471–488.

Grau, D. (2009). New process capability indices for one-sided tolerances. *Quality Technology and Quantitative Management, 6*(2), 107–124.

Gunter, B. H. (1989). The use and abuse of Cpk; Parts 1–4. *Quality Progress, 22*(1, 3, 5, 7), 72–73, 108–109, 79–80, 86–87.

Herman, J. T. (1989). Capability index—enough for process industries? In *Transactions on ASQC Congress, Toronto* (pp. 670–675).

Ho, L. L. (2003). Statistical inference from Boyles' capability index. *Economic Quality Control, 18*(1), 43–57.

Hosseinifard, S. Z., Abbasi, B., & Ahmad, S. (2009). A transformation technique to estimate the process capability index for non-normal processes. *International Journal of Advance Manufacturing Technology, 40*, 512–517.

Johnson, N. L. (1949). Systems of frequency curves generated by methods of translation. *Biometrika, 36*, 149–176.

Kane, V. E. (1986). Process capability index. *Journal of Quality Technology, 18*(1), 41–52.

Khadse, K. G., & Shinde, R. L. (2006). Multivariate process capability using relative importance of quality characteristics. *The Indian Association for Productivity, Quality and Reliability (IAPQR) Transactions, 31*(2), 85–97.

Kirmani, S., & Polansky, A. M. (2009). Multivariate process capability via lowner ordering. *Linear Algebra and its Applications, 430*, 2681–2689.

Kotz, S., & Johnson, N. (2002). Process capability indices—a review 1992–2000. *Journal of Quality Technology, 34*(1), 2–19.

Krishnamoorthy, K. S. (1990). Capability indices for processes subject to unilateral and positional tolerances. *Quality Engineering, 2*, 461–471.

Kushler, R. H., & Hurley, P. (1992). Confidence bounds for capability indices. *Journal of Quality Technology, 24*(4), 188–195.

Lin, P. C., & Pearn, W. L. (2002a). Testing process capability for one-sided specification limit with application to the voltage level translator. *Microelectronics Reliability, 42*, 1975–1983.

Lin, G. H., Pearn, W. L. (2002). A note on the interval estimation of C_{pk} with asymmetric tolerances. *Journal of Non-parametric Statistics, 14*(6), 647–654.

Mardia, K. V. (1970). Measures of multivariate skewnees and kurtosis with applications. *Biometrika, 57*(3), 519–530.

Mardia, K. V. (1974). Applications of some measures of multivariate skewness and kurtosis for testing normality and robustness studies. *Sankhy A, 36*, 115–128.

McCormack, D. W, Jr, Harris, I. R., Hurwitz, A. M., & Spagon, P. D. (2000). Capability indices for non-normal data. *Quality Engineering, 12*(4), 489–495.

Pearn, W. L. (1998). New generalization of process capability index C_{pk}. *Journal of Applied Statistics, 25*(6):801–810.

Pearn, W. L., Chen, K. S. (2002). One-sided capability indices C_{PU} and C_{PL}: decision making with sample information. *International Journal of Quality and Reliability Management, 19*(3), 221–245.

Pearn, W. L., Chen, K. S., & Lin, G. H. (1999). A generalization of Clements' method for non-normal pearsonian processes with asymmetric tolerances. *International Journal of Quality and Reliability Management, 16*(5), 507–521.

Pearn, W. L., Hung, H. N., & Cheng, Y. C. (2009). Supplier selection for one-sided processes with unequal sample size. *European Journal of Operational Research, 195*, 381–393.

Pearn, W. L., & Kotz, S. (1994). Application of Clements' method for calculating second and third generation process capability indices for non-normal Pearsonian populations. *Quality Engineering, 7*, 139–145.

Pearn, W. L., & Kotz, S. (2007). *Encyclopedia and handbook of process capability indices—series on quality, reliability and engineering statistics*. Singapore: World Scientific.

Pearn, W. L., Kotz, S., & Johnson, N. L. (1992). Distributional and inferential properties of process capability indices. *Journal of Quality Technology, 24*(4), 216–231.

Pearn, W. L., Lin, P. C. (2000). Estimating capability index C_{pk} for processes with asymmetric tolerances. *Communications in Statistics: Theory and Methods, 29*(11), 2593–2604.

Pearn, W. L., Lin, P. C., Chen, K. S. (2001). Estimating process capability index C''_{pmk} for asymmetric tolerances: distributional properties. *Metrika, 54*:261–279.

Pearn, W. L., Lin, P. C., Chen, K. S. (2004). The C''_{pk} index for asymmetric tolerances: implications and inferences. *Metrika, 60*, 119–136.

Pearn, W. L., Wang, F. K., & Yen, C. H. (2006). Measuring production yield for processes with multiple quality characteristics. *International Journal of Production Research, 44*(21), 4649–4661.

Perakis, M., & Xekalaki, E. (2012). On the implimentation of the principal component analysis—based approach in measuring process capability. *Quality and Reliability Engineering International, 28*, 467–480.

Pignatiello, J. J. Jr., Ramberg, J. S. (1993). Process capability indices: just say "no". In *Transactions on ASQC Congress* (pp. 92–104).

Polansky, A. (2001). A smooth nonparametric approach to multivariate process capability. *Technometrics, 53*(2), 199–211.

Rodriguez, R. N. (1992). Recent developments in process capability analysis. *Journal of Quality Technology, 24*(4), 176–187.

Royston, J. P. (1983). Some Techniques for Assessing Multivariate Normality Based on the Shapiro-Wilk W. *Applied Statistics, 32*(2), 121–133.

Scheur, E. M. (1962). Moments of the radial error. *Journal of the American Statistical Association, 57*, 187–190.

Shahriari, H., & Abdollahzadeh, M. (2009). A new multivariate process capability vector. *Quality Engineering, 21*(3), 290–299.

Shapiro, S. S., & Wilk, M. B. (1965). An analysis of variance test for normality (complete samples). *Biometrika, 52*, 591–611.

Shinde, R. L., & Khadse, K. G. (2009). Multivariate process capability using principal component analysis. *Quality and Reliability Engineering International, 25*, 69–77.

Shiau, J. J. H., Yen, C. L., Pearn, W. L., & Lee, W. T. (2013). Yield related process capability indices for processes of multiple quality characteristics. *Quality and Reliability Engineering International, 29*, 487–507.

Shu, M. H., Lu, K. H., Hsu, B. M., Lou, K. R. (2006). Testing quality assurance using process capability indices C_{PU} and C_{PL} based on several groups of samples with unequal sizes. *Information and Management Sciences, 17*(1), 47–65.

Somerville, S. E., & Montgomery, D. C. (1996). Process capability indices and non-normal distributions. *Quality Enginerring, 9*, 305–316.

Sultan, T. L. (1986). An acceptance chart for raw materials of two correlated properties. *Quality Assurance, 12*, 70–72.

Taam, W., Subbaiah, P., & Liddy, W. (1993). A note on multivariate capability indices. *Journal of Applied Statistics, 20*, 339–351.

Tang, L. C., & Than, S. E. (1999). Computing process capability indices for non-normal data: A review and comparative study. *Quality and Reliability Engineering International, 35*(5), 339–353.

Tano, I., & Vannman, K. (2012). A multivariate process capability index based on the first principal component only. *Quality and Reliability Engineering International, 29*(7), 987–1003.

Vannman, K. (1995). A unified approach to capability indices. *Statistica Sinica, 5*, 805–820.

Vannman, K. (1997). A general class of capability indices in the case of asymmetric tolerances. *Communications in Statistics—Theory and Methods, 26*, 2049–2072.

Vannmam, K. (1998). Families of process capability indices for one—sided specification limits. *Statistics, 31*(1), 43–66.

Vanmannman, K., & Albing, M. (2007). Process capability indices for one-sided specification intervals and skewed distributions. *Quality and Reliability Engineering International, 23*, 755–765.

Veevers, A. (1998). Viability and capability indices for multi-response processes. *Journal of Applied Statistics, 25*(4), 545–558.

Villasenor-Alva, J. A., & Gonzalez-Estrada, E. (2009). A generalization of Shapiro-Wilk's test for multivariate normality. *Communications in Statistics: Theory and Methods, 38*(11), 1870–1883.

Wang, F. K., & Chen, J. C. (1998). Capability index using principal component analysis. *Quality Engineering, 11*, 21–27.

Wang, F. K., & Du, T. C. T. (2000). Using principal component analysis in process performance for multivariate data. *Omega, 28*, 185–194.

Wright, P. A. (1995). A process capability index sensitive to skewness. *Journal of Statistical Computation and Simulation, 52*, 195–203.

Wu, C.-W., Pearn, W. L., & Kotz, S. (2009). An overview of theory and practice on process capability indices for quality assurance. *International Journal of Production Economics, 117*(2), 338–359.

Wu, H. H., Wang, J. S., Liu, T. L. (1998). Discussions of the Clements' based process capability indices. In *Proceedings of the CIIE National Conference, Taiwan* (pp. 561–566).

Yang, J. R., Song, X. D., Ming, Z. (2010). Comparison between non-normal process capability study based on Two kinds of transformations. In *Proceedings of the First ACIS International Symposium on Cryptography and Network Security, data Mining and Knowledge Discovery, E-Commerce and Its Applications and Embedded Systems* (pp. 201–205). doi:10.1109/CDEE. 2010.97.

Part II
Reliability Management

Modeling and Analyzing System Failure Behavior for Reliability Analysis Using Soft Computing-Based Techniques

Harish Garg

List of Symbols

$\tilde{A}$	Fuzzy set
$\mu_{\tilde{A}}$	Membership functions of fuzzy set $\tilde{A}$
$\upsilon_{\tilde{A}}$	Nonmembership functions of fuzzy set $\tilde{A}$
$\tilde{\lambda}_i$	Fuzzy failure rate of ith component
$\tilde{T}_i$	Fuzzy repair time of ith component
$A^{(\alpha)}$	Alpha-cut of the fuzzy set
MTBF_i	Mean time between failures of the ith components
MTTR_i	Mean time to repair of the ith components
CMTBF_i	Cost of mean time between failures of the ith components
CMTTR_i	Cost of mean time to repair of the ith components
LbMTBF_i	Lower limit of the mean time between failures of the ith components
UbMTBF_i	Upper limit of the mean time between failures of the ith components
LbMTTR_i	Lower limit of mean time to repair of the ith components
UbMTTR_i	Upper limit of mean time to repair of the ith components
R_s	System reliability
A_s	System availability
M_s	System maintainability
iter	Current iteration number
$\text{iter}_{\max}$	Maximum iteration number
T_ω	Weakest t-norm
TFN	Triangular fuzzy number
IFN	Intuitionistic fuzzy number
$\alpha_i, \beta_i, \gamma_i$	Physical feature of each component
c_1	Individual intelligence coefficient
c_2	Social intelligence coefficient

H. Garg (✉)
School of Mathematics, Thapar University Patiala, Patiala, Punjab 147004, India
e-mail: harishg58iitr@gmail.com

© Springer-Verlag London 2016
H. Pham (ed.), *Quality and Reliability Management and Its Applications*,
Springer Series in Reliability Engineering, DOI 10.1007/978-1-4471-6778-5_4

1 Introduction

With modern technology and higher reliability requirements, systems are getting more complicated day-by-day, and hence job of the system analyst or plant personnel becomes so difficult to run the system under failure-free pattern. In the competitive market scenario, reliability and maintainability are the most important parameters that determine the quality of the product with their aim of estimating and predicting the probability of the failure, and optimizing the operation management. Therefore, the primary objective of any industrial system is to acquire quality products/systems that satisfy user needs with measurable improvements to mission capability and operational support in a timely manner, and at a fair and reasonable price. These features address reliability, availability, and maintainability (RAM) as essential elements of mission capability. Generally, system performance can be improved either by incremental improvements of component reliability/availability or by provision of redundant components in parallel; both methods result in an increase in system cost. Therefore, optimization methods are necessary to obtain allowable costs at the same time as high availability levels. Extensive reliability design techniques have been introduced by the researchers during the past two decades for solving the optimization problem on the specific applications. Comprehensive overviews of these models have been addressed in Kuo et al. (2001) and Gen and Yun (2006). However, the heuristic techniques require derivatives for all nonlinear constraint functions that are not derived easily because of the high computational complexity. To overcome this difficulty, several methods have been proposed based on the so-called computational intelligence or meta-heuristic search methods which proved itself to be able to approach the optimal solution against these problems. These heuristics include genetic algorithm (GA) (Holland 1975; Goldberg 1989), differential evolution (DE) (Storn and Price 1995, 1997; Brest et al. 2006), particle swarm optimization (PSO) (Kennedy and Eberhart 1995; Eberhart and Kennedy 1995), and artificial bee colony (ABC) (Karaboga 2005; Karaboga and Basturk 2007).

In that direction, Bris et al. (2003) attempted to optimize the maintenance policy, for each component of the system, minimizing the cost function, with respect to the availability constraints using genetic constraints. Lapa et al. (2006) presented a methodology for preventive maintenance policy evaluation based upon a cost-reliability model using a genetic algorithm. Leou (2006) proposed a formulation considering both reliability and cost reduction for maintenance scheduling. The genetic algorithm combined with the simulated annealing method was adopted as a solution method. Juang et al. (2008) proposed a genetic algorithm-based optimization model to optimize the availability of a series parallel system where the objective is to determine the most economical policy of component's MTBF and MTTR. Saraswat and Yadava (2008) reviewed the literature from 1988 to 2005 on RAMS engineering. Coelho (2009) presented an efficient PSO algorithm based on Gaussian distribution and chaotic sequence to solve the reliability–redundancy optimization problems. Rajpal et al. (2006) explored the application of artificial neural networks to model the behavior of a complex, repairable system. A composite measure of RAM parameters called as the RAM index has been proposed for measuring the system performance by simultaneously considers all the three key indices which influence the system performance directly. Their index was static in nature, while Komal et al. (2010) introduced RAM index which was time dependent and used historical uncertain data for its evolution. Garg and Sharma (2013) have investigated the multi-objective reliability–redundancy allocation problem of a repairable industrial system with PSO and GA. The solution of series–parallel reliability redundancy allocation problem has been solved by Yeh and Hsieh (2011), Hsieh and Yeh (2012) with ABC and found the supremacy over the other techniques. Garg and Sharma (2012) had discussed the two-phase approach for analyzing the reliability and maintainability analysis of the industrial system using the PSO algorithm. Recently, Garg et al. (2012, 2013) have solved the reliability optimization problem with ABC algorithm and compared their performance with other evolutionary algorithm.

Conventionally, it was assumed that all the parameters and goals are precisely known. However, this is not occurring in the real-life world, as we often encounter the situation that we have to make a decision under uncertainty due to the presence of incomplete or imprecise or vagueness in information. Thus, quantification of uncertainty in reliability analysis is very important as it helps for effective decision making. For this, fuzzy theoretic approach (Zadeh 1965) has been used to handle the subjective information or uncertainties during the evaluation of the reliability of a system. After their successful applications, a lot of progress has been made in both theory and application, and hence several researches were conducted on the extensions of the notion of fuzzy sets. Among these extensions, the one that have drawn the attention of many researches during the last decades is the theory of intuitionistic fuzzy sets (IFS) introduced by Attanassov (1986, 1989). The concepts of IFS can be viewed as an appropriate/alternative approach to define a fuzzy set in the case where available information is not sufficient for the definition of an imprecise concept by means of a conventional fuzzy set. IFS add an extra degree to the usual fuzzy sets in order to model hesitation and uncertainty about the membership degree of belonging. In fuzzy sets, the degree of acceptance is only considered but IFS is characterized by a membership function and a nonmember function so that the sum of both values is less than or equal to one. Gau and Buehrer (1993) extended the idea of fuzzy sets by vague sets. Bustince and Burillo (1996) showed that the notion of vague sets coincides with that of IFSs. Therefore, it is expected that IFSs could be used to simulate any activities and processes requiring human expertise and knowledge, which are inevitably imprecise or not totally reliable. As far as reliability field is concerned, IFSs has been proven to be highly useful to deal with uncertainty and vagueness, and a lot of work has been done to develop and enrich the IFS theory given in Chen (2003), Chang et al. (2006), Garg and Rani (2013), Garg et al. (2013, 2014), Taheri and Zarei (2011), and their corresponding references.

The entire above researchers have analyzed the reliability index only for measuring the performance of the system. But it is commonly known that other reliability parameters such as failure rate, mean time between failures, etc. also affect the system performance and consequently their behavior. Therefore, it is necessary that all these reliability parameters are analyzed simultaneously for assessing the behavior of the system deeply. For this, Garg (2013) presented a new methodology named as vague Lambda-Tau methodology (VLTM) for analyzing various reliability parameters in terms of intuitionistic fuzzy membership functions. As it has been observed from their study, the computed reliability indices contain a wide range of spreads, in the form of support, and hence do not give the accurate results or not giving the exact behavior of the system. Thus, it is necessary that the spread in these indices must be reduced during the analysis so that plant personnel may use these for increasing the production as well as productivity of the system.

In recent years, research implications of reliability, availability, and maintainability (RAM) aspects of reliability engineering systems have increased substantially due to rising operating and maintenance costs. For industrial systems, the cost is considered to be the most significant factor and RAM is an increasingly important issue for determining the performance of the system. On the other hand, the information available from the collected databases or records is most of the time imprecise, limited, and uncertain, and the management decisions are based on experience. Thus it is difficult for job analysts to analyze the performance of the system by utilizing these uncertain data. Therefore, the objective of this chapter is to quantify the uncertainties that make the decisions realistic, generic, and extensible for the application domain. For this, an optimization model has been constructed by taking composite measure of RAM parameters called RAM index and system cost as an objective function and solved with evolutionary techniques algorithm. The obtained failure rates and repair times of all constituent components are used for measuring the performance of the system in terms of various reliability parameters using intuitionistic fuzzy set theory and weakest t-norm based arithmetic operations. Performance analysis on system RAM index has also been analyzed to show the effect of taking wrong combinations of their reliability parameters on its performance. The suggested framework has been illustrated with the help of a case.

2 Critical Comments on Reviewing Literature

2.1 Shortcoming of the Existing Literature

The following shortcomings are observed after critically reviewing the literature.

- The conventional/empirical methods (dynamic programming, integer/mixed integer programming, etc.) do not provide a globally optimal solution to the problem, and hence the design cost increases.
- The cost associated with the system design such as manufacturing and repairing cost are not well taken into account.

- Probability theory does not always provide useful information to the practitioners due to the limitation of being able to handle only quantitative information.
- The subjective information is not captured during reliability analysis.
- It is unable to assess and predict the critical component of the system, as per preferential order,

2.2 Objective of the Work

The objective of this work is to analyze the performance of the complex repairable industrial systems, in terms of various reliability parameters using weakest t-norm (T_ω) based arithmetic operations on intuitionistic fuzzy set theory, while improving upon the above-mentioned critical shortcomings. The following tools are adopted for this purpose, which may give better results (close to real condition):

- An optimization model is constructed from the system by considering RAM index, manufacturing cost, and repairing cost as an objective function for obtaining the design parameters such as MTBF and MTTR corresponding to each of its associated components.
- As compared to the traditional conventional optimization technique, ABC has been used for finding optimal (or near to) values as it always gives a global solution.
- Weakest t-norm (T_ω) based arithmetic operations over vague Lambda-Tau methodology have been adopted for assessing the analysis of the system behavior.
- Performance analysis of the system index has been addressed for ranking the critical component of the system as per preferential order.

3 Intuitionistic Fuzzy Set (IFS) Theory

The fuzzy set theory (Zadeh 1965) has been successfully applied in various disciplines for handling the uncertainties in the data in terms of their membership functions. After their successful applications, several researches were conducted on the extensions of the notion of fuzzy sets. Among these extensions to the theory of intuitionistic fuzzy sets (IFS), first proposed by Attanassov (1986), is the most widely used by defining two characteristic functions expressing the degree of membership and nonmembership of element in the universe. Mathematically, if we consider X be a universe of discourse, then $\tilde{A} = \{ <x, \mu_{\tilde{A}}(x), v_{\tilde{A}}(x) > |x \in X\}$ is called an IFS where the function $\mu_{\tilde{A}}, v_{\tilde{A}} : X \to [0, 1]$ be the degree of membership and nonmembership of the element x in the fuzzy set $\tilde{A}$, respectively, such that $\mu_{\tilde{A}}(x) + v_{\tilde{A}}(x) \leq 1$ for every $x \in X$. In addition, $\pi_{\tilde{A}}(x) = 1 - \mu_{\tilde{A}}(x) - v_{\tilde{A}}(x)$ is called the degree of hesitation or uncertainty level of the element x in the set $\tilde{A}$. If $\pi_{\tilde{A}}(x) = 0$ for all $x \in X$, then the IFS is reduced to a fuzzy set.

3.1 (α, β)-cuts

(α, β)-cut of the IFS set is defined as the values of x, when the membership value corresponding to x is greater than or equal to the specified cut level. Mathematically, it is expressed as

$$A_{(\alpha,\beta)} = \{x \in X | \mu_{\tilde{A}}(x) \geq \alpha \quad \text{and} \quad v_{\tilde{A}}(x) \leq \beta\} \tag{1}$$

In other words, $A_{(\alpha,\beta)} = A_\alpha \cap A^\beta$ where $A_\alpha = \{x \in X | \mu_{\tilde{A}}(x) \geq \alpha\}$ and $A^\beta = \{x \in X | v_{\tilde{A}}(x) \leq \beta\}$; $\alpha, \beta \in [0, 1]$.

3.2 t-Norm and Weakest t-Norm

A triangular norm (t-norm) T is a binary operation on $[0, 1]$, i.e., a function $T : [0, 1]^2 \rightarrow [0, 1]$ such that (i) T is associative, (ii) T is commutative, (iii) T is nondecreasing, and (iv) T has 1 as a neutral element such that $T(x, 1) = x$ for each $x \in [0, 1]$.

A t-norm is called the weakest t-norm, denoted by T_ω, iff

$$T(x, y) = \begin{cases} 0; & \max(x, y) < 1 \\ \min(x, y); & \text{otherwise} \end{cases} \tag{2}$$

3.3 Triangular Intuitionistic Fuzzy Numbers (TIFNs)

Let $\tilde{A}_i = \langle (a_{i1}, a_{i2}, a_{i3}); \mu_i, v_i \rangle$, $i = 1, 2, \ldots, n$ be the 'n' IFS on $\mathbb{R}$ where $a_{i1}, a_{i2}, a_{i3} \in \mathbb{R}$ representing the lower, middle, and upper values of a triangular intuitionistic fuzzy membership functions, then the set $\tilde{A}_i$ is said to be triangular intuitionistic fuzzy number (TIFN) if its membership and nonmembership functions are defined as

$$\mu_{\tilde{A}_i}(x) = \begin{cases} \mu \times \frac{x - a_{i1}}{a_{i2} - a_{i1}}; & i1 \leq x \leq a_{i2} \\ \mu; & x = a_{i2} \\ \mu \times \frac{a_{i3} - x}{a_{i3} - a_{i2}}; & a_{i2} \leq x \leq a_{i3} \\ 0; & \text{otherwise} \end{cases};$$

$$1 - v_{\tilde{A}_i}(x) = \begin{cases} (1 - v) \times \frac{x - a_{i1}}{a_{i2} - a_{i1}}; & a_{i1} \leq x \leq a_{i2} \\ 1 - v; & x = a_{i2} \\ (1 - v) \times \frac{a_{i3} - x}{a_{i3} - a_{i2}}; & a_{i2} \leq x \leq a_{i3} \\ 0; & \text{otherwise} \end{cases}$$

The α-cut of IFS defined for the above set is $A_i^{(\alpha)} = [a_{i1}^{(\alpha)}, a_{i3}^{(\alpha)}]$ and $B_i^{(\alpha)} = [b_{i1}^{(\alpha)}, b_{i3}^{(\alpha)}]$ corresponding to $\mu_{\tilde{A}_i}$ and $1 - v_{\tilde{A}_i}$, respectively, where $a_{i1}^{(\alpha)}, b_{i1}^{(\alpha)}$ are the increasing functions, and $a_{i3}^{(\alpha)}, b_{i3}^{(\alpha)}$ are decreasing functions of α and are defined as

$$a_{i1}^{(\alpha)} = a_{i1} + \frac{\alpha}{\mu_i}(a_{i2} - a_{i1}); \qquad b_{i1}^{(\alpha)} = a_{i1} + \frac{\alpha}{1 - v_i}(a_{i2} - a_{i1})$$

$$a_{i3}^{(\alpha)} = a_{i3} - \frac{\alpha}{\mu_i}(a_{i3} - a_{i2}); \qquad b_{i3}^{(\alpha)} = a_{i3} - \frac{\alpha}{1 - v_i}(a_{i3} - a_{i2})$$

The basic arithmetic operations, i.e., addition, subtraction, multiplication, and division, of IFNs depend upon the arithmetic of the interval of confidence. Therefore, these operations using weakest T_ω-based operations with and $\mu = \min(\mu_i)$ and $v = \min(v_i)$ are defined as follows:

1. Addition of $T_w(\oplus)$:

$$\tilde{A}_1 \oplus_{T_w}^{\alpha} \cdots \oplus_{T_w}^{\alpha} \tilde{A}_n = \begin{cases} \left[\sum_{i=1}^{n} a_{i1}^{(\alpha)}, \sum_{i=1}^{n} a_{i3}^{(\alpha)} \right] & \text{if } \tilde{A}_i \in \text{TFNs} \\ \sum_{i=1}^{n} a_{i2} - \max_{1 \leq i \leq n}\left(a_{i2} - a_{i1}^{(\alpha)}\right), \\ \sum_{i=1}^{n} a_{i2} - \max_{1 \leq i \leq n}\left(a_{i3}^{(\alpha)} - a_{i1}^{(\alpha)}\right), & \text{otherwise} \end{cases} \qquad (3)$$

2. Subtraction of $T_w(\ominus)$:

$$\tilde{A}_1 \ominus_{T_w}^{\alpha} \cdots \ominus_{T_w}^{\alpha} \tilde{A}_n = \begin{cases} \left[a_{11}^{(\alpha)} - \sum_{i=2}^{n} a_{i3}^{(\alpha)}, a_{13}^{(\alpha)} - \sum_{i=2}^{n} a_{i1}^{(\alpha)} \right] & \text{if } \tilde{A}_i \in \text{TFNs} \\ \left[a_{12} - \sum_{i=2}^{n} a_{i2} - \max_{1 \leq i \leq n}\left(a_{i2} - a_{i1}^{(\alpha)}\right), \right. \\ \left. a_{12} - \sum_{i=2}^{n} a_{i2} + \max_{1 \leq i \leq n}\left(a_{i3}^{(\alpha)} - a_{i2}\right) \right] & \text{otherwise} \end{cases} \qquad (4)$$

3. Multiplication of $T_w(\otimes)$: Here, multiplication of the approximate fuzzy operations is shown for $\tilde{A}_i \in \mathbb{R}^+$ and others can easily be derived with similar ways.

$$\tilde{A}_1 \otimes_{T_w}^{\alpha} \cdots \otimes_{T_w}^{\alpha} \tilde{A}_n = \begin{cases} \left[\prod_{i=1}^{n} a_{i1}^{(\alpha)}, \prod_{i=1}^{n} a_{i3}^{(\alpha)} \right] & \text{if} \tilde{A}_j \in \text{TFNs} \\[2em] \left[\prod_{i=1}^{n} a_{i2} - \max_{1 \leq i \leq n} \left((a_{i2} - a_{i1}^{(\alpha)}) \prod_{\substack{j=1 \\ j \neq i}}^{n} a_{j2} \right), \right. \\[2em] \left. \prod_{i=1}^{n} a_{i2} + \max_{1 \leq i \leq n} \left((a_{i3}^{(\alpha)} - a_{i2}) \prod_{\substack{j=1 \\ j \neq i}}^{n} a_{j2} \right) \right] & \text{otherwise} \end{cases} \tag{5}$$

4. Division of $T_w(\emptyset)$: Here, division of the approximate fuzzy operations is shown for $\tilde{A}_i \in \mathbb{R}^+$

$$\tilde{A}_1 \emptyset_{T_w}^{\alpha} \cdots \emptyset_{T_w}^{\alpha} \tilde{A}_n = \tilde{A}_1 \otimes_{T_w}^{\alpha} \frac{1}{\tilde{A}_2} \cdots \otimes_{T_w}^{\alpha} \frac{1}{\tilde{A}_n} \quad ; \quad \text{if} \quad 0 \notin \tilde{A}_i, i \geq 2 \tag{6}$$

4 Evolutionary Algorithms: GA, DE, PSO, ABC

The brief overview of the evolutionary algorithm, namely, GA, DE, PSO, and ABC, is discussed here.

4.1 Genetic Algorithm (GA)

Genetic algorithms (GAs) are a part of evolutionary algorithms, which is a rapidly growing area of artificial intelligence. Holland (1975) is considered to the father of GA. GA is a model or concept of biological evolution based on Charles Darwin's theory of natural selection. The essence of GAs involves the encoding of an optimization function as arrays of bits or character strings to represent the solutions (represented by chromosomes). Start from possible solutions termed as the population, evolution cycle, or iterations by evaluating the fitness of all the individuals in the population, creating a new population by performing crossover, mutation, etc., and replacing the old population and then iteratively again using the new population. The above process is repeated until some stopping condition is satisfied. A more detailed implementation of a genetic algorithm can be found in Gen and

Yun (2006), Goldberg (1989), etc. The pseudocode of the GA algorithm is described in Algorithm 1:

Algorithm 1 Pseudo code of Genetic algorithm (GA)

1: Objective function: $f(\mathbf{x})$
2: Define Fitness F (eg. $F \propto f(x)$ for maximization)
3: Initialize population
4: Initial probabilities of crossover (p_c) and mutation (p_m)
5: **repeat**
6:    Generate new solution by crossover and mutation
7:    if p_c >rand, Crossover; end if
8:    if p_m >rand, Mutate; end if
9:    Accept the new solution if its fitness increases.
10:   Select the current best for the next generation.
11: **until** requirements are met

4.2 Differential Evolution (DE)

DE is a relatively recent heuristic proposed by Storn and Price (1995, 1997), Brest et al. (2006), which was designed to optimize the problems over continuous domains. DE shares similarities with evolutionary algorithms but it does not use binary encoding as in GA and does not use a probability density function to self-adapt its parameters as an evolution strategy. Instead of this, DE performs mutation based on the distribution of the solutions in the current population. The main difference between GA and DE is that, in GA, mutation is the result of small perturbations to the genes of an individual (potential solution), while in DE, mutation is the result of arithmetic combinations of individuals. Basically, there are three important factors: the population size, mutation constant factor, and crossover rate to be necessarily considered when the DE algorithm is utilized. In this, mutation is carried out by the mutation scheme. For each vector, $x_i(t)$ at any time or generation t for the ith individual of a population, then the donor vector v_i is generated by randomly chosen three distinct vectors r_1, r_2 and r_3 at t by the mutation scheme as given in Eq. (7):

$$v_i(t+1) = x_{i,r_1}(t) + F \cdot \left(x_{i,r_2}(t) - x_{i,r_3}(t)\right) \tag{7}$$

where mutation or scaling factor $F \in [0, 1]$ is a real parameter, which controls the amplification of the differences between two individuals with indexes r_2 and r_3.

The crossover is controlled by a crossover probability $C_r \in [0, 1]$ and by generating a uniformly distributed random number rand $\in [0, 1]$, and the jth component of v_i is manipulated as

$$y_{ij}(t+1) = \begin{cases} v_{ij}(t+1) & \text{if rand}(j) \leq C_r \text{ or } j=k \\ x_{ij}(t) & \text{if rand}(j) > C_r \text{ or } j \neq k \end{cases} \qquad (8)$$

where $j,k \in \{1,2,\ldots,D\}$ is the random parameter index and D is the dimensional vector. In this way, each component is updated randomly and can be decided to be acceptable if it improves on the fitness of the parent individual, i.e., if the cost of the trial vector is better than that of the target, the target vector is allowed to advance to the next generation. Otherwise, the target vector is replaced by a trial vector in the next generation. Set the generation number for $t = t + 1$ and repeat the process until the termination criterion is met.

The pseudocode of the algorithm is described in Algorithm 2:

Algorithm 2 Pseudo code of Differential Evolutio (DE)

1: Initialize population x with randomly generated solutions
2: Set the scaling factor F and the crossover probability $C_r \in [0,1]$.
3: **repeat**
4:     Generate a new vector v by DE scheme (7)
5:     Generate a randomly distributed number $rand \in [0,1]$
6:     for each dimension, if $rand \leqslant C_r$, Crossover it as defined in eq. (8)
7:     Accept the new solution if its fitness is better than previous.
8:     Select the current best for next generation
9:     Set counter $t = t + 1$
10: **until** requirements are met

4.3 Particle Swarm Optimization (PSO)

Particle swarm optimization (PSO) (Kennedy and Eberhart 1995; Eberhart and Kennedy 1995) is a population-based optimization technique of swarm intelligence field inspired by social behavior of bird flocking or fish schooling in which each solution called "particle" flies around in a multidimensional problem search space. Unlike the genetic algorithm, PSO algorithm has no crossover and mutation operators. In this algorithm, the particle follows the piecewise paths formed by positional vectors in a quasi-stochastic manner. During movement, every particle adjusts its position according to its own experience of neighboring particles, using the best position encountered by itself and its neighbors. The former one is known as personal best (pbest, p_i) and the latter one is globally best (gbest, p_g). Acceleration is weighted by random terms, with the separate random number being generated for acceleration toward pbest and gbest locations, respectively. Based on the pbest and gbest information of the each particle's, the velocity (v_i) and position of the particle (x_i) are updated according to Eqs. (9) and (10), respectively, as

$$v_i(t+1) = w \cdot v_i(t) + c_1 \cdot \text{ud} \cdot (p_i(t) - x_i(t)) + c_2 \cdot \text{Ud} \cdot (p_g(t) - x_i(t)) \quad (9)$$

$$x_i(t+1) = x_i(t) + v_i(t+1) \quad (10)$$

where w is the inertia weight; $i = 1, 2, \ldots, N$ indicates the number of the particles of the population (swarm), $t = 1, 2, \ldots, t_{\max}$ indicates the iterations. Positive constants c_1 and c_2 are the cognitive and social components, respectively, which are the acceleration constants responsible for varying the particle velocity toward pbest and gbest, respectively. Variables ud and Ud are two random functions in the range [0, 1]. Equation (10) represents the position update, according to its previous position and its velocity.

The essential steps of the particle swarm optimization can be summarized as the pseudocode given in Algorithm 3.

Algorithm 3 Pseudo code of Particle swarm optimization (PSO)

1: Objective function: $f(\mathbf{x})$, $\quad \mathbf{x} = (x_1, x_2, \ldots, x_D)$;
2: Initialize particle position and velocity for each particle and set $t = 1$.
3: Initialize the particle's best known position to its initial position
4: **repeat**
5: Update the best known position (p_i) for each particle
6: Update the swarm's best known position (p_g)
7: Calculate particle velocity according to the velocity equation (9).
8: Update particle position according to the position equation (10).
9: **until** requirements are met.

4.4 Artificial Bee Colony (ABC) Algorithm

The artificial bee colony (ABC) optimization algorithm was first developed by Karaboga (2005). Since then, Karaboga and Basturk and their colleagues have symmetrically studied the performance of the ABC algorithm concerning unconstrained optimization problems and its extension (Karaboga and Basturk 2007; Karaboga and Akay 2009; Karaboga and Ozturk 2011). In the ABC algorithm, the bees in a colony are divided into three groups: employed bees, onlooker bees, and scouts. The employed bee shares information with the onlooker bees in a hive so that onlooker bees can choose a food source to the forager. The employed bee of a discarded food site is forced to become a scout for searching new food source randomly. At the initialization, the ABC generates a randomly distributed population of n employed bees' solutions representing the food source positions. Based on these solutions x, nectar (fitness) amount corresponding to each position is evaluated by a fitness function, and thus the probability of an onlooker bee chooses to go the preferred food source which can be defined by $p_i = f(x_i) / \sum_{j=1}^{N} f(x_j)$, where N is the number of food sources. After a solution is generated, that solution is improved using a local search process called greedy selection process carried out by an onlooker and employed bees and is given by Eq. (11):

$$Z_{hj} = x_{kj} + \phi(x_{hj} - x_{kj}) \qquad\qquad (11)$$

where $k \in \{1, 2, \ldots, N\}$ and $j \in \{1, 2, \ldots, D\}$ are randomly chosen indexes. Here, D is number of solution parameters and k is different from h, ϕ is a random number between -1 and 1, and Z_k is the solution in the neighborhood of X_k. If a particular food source solution does not improve for a predetermined iteration number, then a new food source will be searched out by its associated bee and it becomes a scout which discovers a new food source to be replaced with the randomly generated food source within its domain. So this randomly generated food source is equally assigned to this scout and changing its status from scout to employ and hence other iteration/cycle of the algorithm begins until the termination condition, maximum cycle number (MCN) or relative error, is not satisfied. The pseudocode of the algorithm is described in Algorithm 4.

Algorithm 4 Pseudo code of Artificial Bee Colony (ABC) optimization

---
1: Objective function: $f(\mathbf{x})$, $\mathbf{x} = (x_1, x_2, \ldots, x_D)$;
2: Initialization Phase
3: **repeat**
4:     Employed Bee Phase
5:     Onlooker Bee Phase
6:     Scout Bee Phase
7:     Memorize the best position achieved so far.
8: **until** requirements are met.

---

5 Methodology

In the first fold, the optimal values of MTBF and MTTR for each of its constituent components are computed. For this, an optimization model has been proposed by considering system RAM index, manufacturing cost, and repairing cost, and ABC is used for finding an optimal solution. Obtained results are shown to be statistically significant by means of pooled t test with other evolutionary algorithms. The computed optimal results, MTBF and MTTR, are used for analyzing the behavior of the system in order to increase the efficiency of the system in the second fold. In nutshells, the objective of this study is threefold as given below

(i) develop an optimization model for the considered system,
(ii) obtain the optimal values of MTBF and MTTR, and

(iii) comput various reliability parameters, which affects the system performance, using weakest t-norm based arithmetic operations based on vague Lambda-Tau methodology.

The detailed description of the methodology is described as follows:

5.1 Formulation of an Optimization Model for Obtaining Values of MTBF and MTTR

To improve the quality and quantity of a manufacturing-related curriculum, there is need to emphasize more on operational management. To achieve this end, availability and reliability of equipment in the process must be maintained at the highest order. But unfortunately, failure is an unavoidable phenomenon associated with technological products and systems. Over time, however, a given system suffers failures and even though it can be minimized by proper maintenance, inspection, proper training to the operators, motivation, and by inculcating positive attitude in the workmen. Thus, maintainability is also to be a key index to enhance the performance of these systems. These features are interrelated in such a way that it is necessary to have both a high reliability and a good maintainability in order to achieve a high availability. Implementation of these for improving the system availability or reliability will normally consume resources such as cost, weight, volume, and so forth. Thus, it is very important for decision makers to fully consider both the actual business and the quality requirements. Thus keeping in view the competitive environment, the behavior of such systems can be studied in terms of their reliability, availability, and maintainability (RAM). To this, a composite measure of RAM parameter named as the RAM index has been used for accessing the impacts of system parameters on its performance. For this, the expressions of the index using system reliability, availability, and maintainability of the given industrial system, from its constituent components and based on a reliability block diagram (RBD), can be written as

$$\text{RAM}(t) = w_1 \times R_s(t) + w_2 \times A_s(t) + w_3 \times M_s(t) \qquad (12)$$

where $w_i \in (0, 1)$ are the weights such that $\sum_{i=1}^{3} w_i = 1$. The values of weight vectors, $w_1 = 0.36$, $w_2 = 0.30$, $w_3 = 0.34$, are used here for the analysis corresponding to reliability, availability, and maintainability of the system, respectively.

On the other hand, the system analyst wants to minimize the total cost manufacturing and repairing of the system. The manufacturing cost varies with different product specifications. For instance, the component will be higher reliable if their corresponding failure rate (MTBF) will be lower (higher), and hence it leads to a sharp increase in the manufacturing cost. The mathematical relationship between the MTBF and manufacturing cost of the system can be expressed as Juang et al. (2008) and Garg et al. (2012).

$$\text{CMTBF}_i = \alpha_i(\text{MTBF})^{\beta_i} + \gamma_i \qquad (13)$$

where CMTBF_i represents the component's manufacturing cost to the MTBF of the ith component, while $\alpha_i, \beta_i, \gamma_i$ are constants representing the physical property of the component.

Additionally, the failure of the component will reduce the efficiency of the system, and hence the system analyst is always intended for recovery as soon as possible. For this, the analyst has to maintain an experienced staff which may reduce the repair time, money, and manpower for repairing the component within a reasonable time. Thus, assuming a linear relationship between mean time to repair MTTR and the repairing cost of the individual components (CMTTR) with the relation represented mathematically as Juang et al. (2008) and Garg et al. (2012)

$$\text{CMTTR}_i = a_i \cdot \text{MTTR}_i - b_i \qquad (14)$$

where a_i and b_i are the constants depending upon the ith component.

Thus, the total cost of the system can be written as

$$\text{Tc} = \sum(\alpha_i(\text{MTBF})^{\beta_i} + \gamma_i) + \sum(a_i \cdot \text{MTTR}_i - b_i) \qquad (15)$$

Using these achieved cost (15) and RAM index of the system (12), the optimization model is formulated for maximizing the system performance per unit capital, i.e.,

$$
\begin{aligned}
&\text{Maximize} \quad \frac{\text{RAM}}{\text{Tc}} \\
&\text{subject to} \quad \text{LbMTBF}_i \leq \text{MTBF}_i \leq \text{UbMTBF}_i \\
&\qquad\qquad\;\; \text{LbMTTR}_i \leq \text{MTTR}_i \leq \text{UbMTTR}_i \\
&\qquad\qquad\;\; i = 1, 2 \ldots n \quad \text{All variables} \geq 0
\end{aligned} \qquad (16)
$$

where n is the number of components in the system and LbMTBF_i, UbMTBF_i, LbMTTR_i, and UbMTTR_i are, respectively, the lower and upper bound of MTBF and MTTR for ith component of the system. The optimization model thus formulated, for obtaining the systems parameters, i.e., MTBF and MTTR of each component, is solved using evolutionary algorithms as described in Sect. 5.

5.2 Analyzing the System Behavior Using Weakest t-Norm Based Operations

The main seek of this phase is to reduce the uncertainty level in the reliability parameters up to a desired degree of accuracy so that plant personnel or decision makers may use these indices for increasing the performance as well as analyzing the behavior of the system in a more sensitive zone. For this, the optimal values as obtained during the above phase are used in it. For increasing the significance of the study, the obtained data related to basic events of the components are represented in the form of triangular intuitionistic fuzzy numbers for handling the uncertainties, on both sides of the data. The weakest t-norm based arithmetic operations have been used, instead of ordinary fuzzy or intuitionistic fuzzy arithmetic operations, for analyzing the system reliability parameters in the form of membership and non-membership functions on vague set theory. The detail of the procedure for analyzing the system behavior has been explained below under the assumptions:

- After repairs, the repaired component is considered as good as new.
- The standby units are of the same nature and capacity as that of active units.
- Product of failure rate and repair time is very small.
- Separate maintenance facility is available for each component.

Step 1 The technique starts from the information extraction phase, in which data related to the main component of the systems in the form of failure rate and repair times extract from the various resources such as historical records/logbooks/databases, etc., and are integrated with the help of plant personnel. In the present study, data related to failure rate and repair time are obtained using phase-I of the proposed technique.

Step 2 In order to increase the relevance of the study, the obtained data are converted into the intuitionistic fuzzy numbers. Triangular intuitionistic fuzzy numbers (TIFNs) are used here for representing these data as fuzzy numbers allow expert opinion, linguistic variables, operating conditions, uncertainties, and imprecision in reliability information. Another feature that adds to the decision of selecting TIFN lies in their ease to represent the membership function effectively and to incorporate the judgment distribution of multiple experts. For instance, imprecise or incomplete information such as low/high failure rate, i.e., about 4 or between 5 and 7, is well represented by TIFN as compared to other numbers. Thus, corresponding to each component data, the fuzzified input data are obtained. For instance, the intuitionistic triangular fuzzy numbers corresponding to the ith component of the system with ±15 % spreads on both sides of the data for failure rate and repair times are depicted graphically in Fig. 1, where $\tilde{\lambda}_{ij}$ is the fuzzy failure rate and $\tilde{\tau}_{ij}$ be fuzzy repair time, of component i, with $j = 1, 2, 3$, being lower, mean (crisp), and upper limit of the membership functions, respectively.

Step 3 As soon as fuzzified input data are obtained for all the main component of the system, the resultant expression of the fuzzy numbers for the system failure rate and repair time can be obtained using the extension principle, coupled with α-cuts and the weakest t-norm based arithmetic operations given in Eqs. (3) to (6) on conventional AND/OR expressions as listed in Table 1. To analyze system behavior quantitatively, various reliability parameters of interest, given in Table 2, such as system failure rate, repair time, MTBF, etc., with left and right spreads can be obtained and shown graphically.

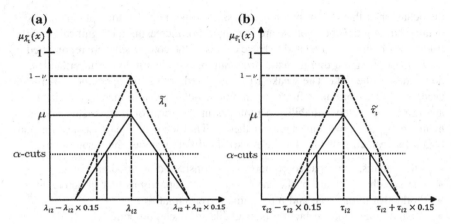

Fig. 1 Input intuitionistic triangular fuzzy numbers

Table 1 Basic expressions of Lambda-Tau methodology

Gate	λ_{AND}	τ_{AND}	λ_{OR}	τ_{OR}
Expression	$\prod_{j=1}^{n} \lambda_j \left[\sum_{i=1}^{n} \prod_{\substack{j=1 \\ i \neq j}}^{n} \tau_j \right]$	$\dfrac{\prod_{i=1}^{n} \tau_i}{\sum_{j=1}^{n} \prod_{\substack{i=1 \\ i \neq j}}^{n} \tau_i}$	$\sum_{i=1}^{n} \lambda_i$	$\dfrac{\sum_{i=1}^{n} \lambda_i \tau_i}{\sum_{i=1}^{n} \lambda_i}$

Table 2 Some reliability parameters

Parameters	Expressions
Failure rate	$MTTF_s = \frac{1}{\lambda_s}$
Repair time	$MTTR_s = \frac{1}{\mu_s} = \tau_s$
ENOF	$W_s(0,t) = \frac{\lambda_s \mu_s t}{\lambda_s + \mu_s} + \frac{\lambda_s^2}{(\lambda_s + \mu_s)^2}[1 - e^{-(\lambda_s + \mu_s)t}]$
MTBF	$MTBF_s = MTTF_s + MTTR_s$
Reliability	$R_s = e^{-\lambda_s t}$
Availability	$A_s = \frac{\mu_s}{\lambda_s + \mu_s} + \frac{\lambda_s}{\lambda_s + \mu_s} e^{-(\lambda_s + \mu_s)t}$

Step 4 In fuzzification process, the crisp quantities are converted into fuzzy quantities; however, in several applications as well as most of actions or decisions implemented by human or machines are binary or crisp in nature. So it is necessary to defuzzify the fuzzy results that have generated through fuzzy analysis. The process of converting the fuzzy output to a crisp value is said to be defuzzification. Out of existence of variation of defuzzification technique in literature such as center of sum, center of gravity, max-membership principle, center of largest area, etc., center of gravity (COG) method is selected for defuzzification as it is equivalent to the mean of data and so it is very appropriate for reliability calculations (Ross 2004).

6 Illustrative Example: Pulping Unit

To demonstrate the application of the proposed methodology, a case from a paper mill situated in the northern part of India is taken. The mill produces approximately 200 tons of paper per day. The paper mills are large capital-oriented engineering systems, comprising units/subsystems, namely feeding, pulping, washing, screening, bleaching, and paper formulation system, arranged in a predefined configuration (Garg and Sharma 2012; Garg et al. 2013, 2014). The present analysis is based on the study of one of the important units, i.e., pulping unit whose brief description is described as follows.

6.1 System Description

The pulping unit is one of the important functioning parts of the paper mill, which are carried out four major operations in the unit: (i) cooking of chips, (ii) separation of knots, (iii) washing of pulp, and (iv) opening of fibers. The wooden chips (wood composition varies with the quality of paper) from storage are fed into the digester for cooking through feeding system. After mixing with white liquor (NaOH), cooking is done for several hours in digester using dry and saturated steam. The cooked pulp contains knots which preclude the production of paper and are removed by passing the pulp through the knotters. This knot free pulp is then flowed over a series of the large size drums (known as deckers) to remove the used liquor (called black liquor) from the pulp to the maximum extent. The liquor and knot free pulp is then washed through several stages. Finally, the washed pulp is passed through openers (rotating at high speed) for segregating the fibers through combing action. Thus, the prepared pulp (called pulp with fine fibers) is then sent to washing system for further treatment. In brief, the pulping system consists of the following four subsystems (Garg et al. 2013).

- Digester (A): It consists of a single unit, used for cooking the chips. Here, a mixture of wooden chips and NaOH + Na_2S (1:3.5 ratio) is heated by steam at 175 °C. Failure of digester stops the cooking process, and hence leads to system failure.
- Knotter (B): It consists of two units, one working and other standby, used to remove the knots from the cooked chips because the knots preclude the production of paper. Knotter subsystem's complete failure occurs only if both of its units fail.
- Decker (C): It has three units, arranged in series configuration and is used to remove liquor from the cooked chips. Failure of any one causes the complete failure of the pulping system. Although production is possible even with two or single decker, but it will reduce the quality of paper, which is less requirement and consequently lead to lesser profit.

- Opener (D): This subsystem possesses two units, one working and other standby, and is used to break the walls of the fibers into ribbons ensuring the availability of large surface area for bonding. Complete failure of this subsystem occurs when both the units fail.

The system diagram and the interaction between the main components of the system are modeled using reliability block diagram (RBD) which is shown in Fig. 2a, b, respectively.

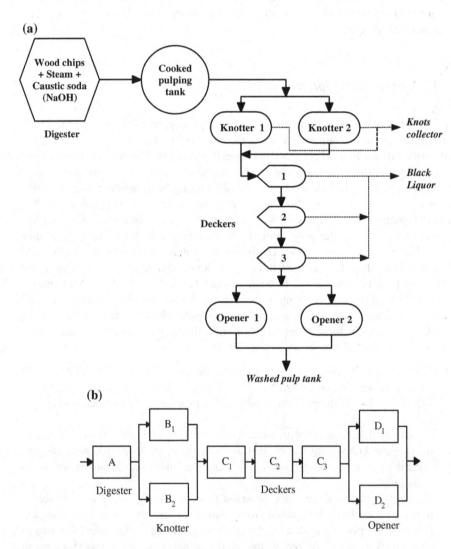

Fig. 2 Pulping unit: **a** systematic diagram and **b** RBD model

Table 3 Range of MTBF, MTTR, and their associated cost

Components	MTBF in hours		MTTR in hours		CMTBF in $		CMTTR in $	
	Lb	Ub	Lb	Ub	Lb	Ub	Lb	Ub
Digester	2995	3150	2.0	4.0	46,93,002	51,74,171	19,329	19,192
Knotter	1850	1950	2.0	5.0	20,92,624	23,18,692	18,655	18,490
Decker	1880	1920	2.0	4.0	22,56,196	23,50,687	18,765	18,675
Opener	1860	1910	2.0	5.0	25,13,331	26,48,081	18,734	18,587

6.2 Formulation of Optimization Model for the System

The system performance per capital optimization model is formulated for the considered system based on its system RAM index and cost. The approximate expression of the system RAM index is derived based on the assumption that the components are operated independently of each other, failure rate ($\lambda_i \sim 1/\text{MTBF}_i$), and repair rate ($\mu_i = 1/\text{MTTR}_i$) are constants such that $\lambda_i \ll \mu_i$ and each component is supported by a maintenance team. Based on the interaction among the working components of the system, the optimization model (16) is formulated for the system. The corresponding lower and upper bound of MTBF and MTTR with their manufacturing and repairing cost are tabulated in Table 3.

6.2.1 Parameters Setting

In the experiments by GA, DE, PSO, and ABC, the values of the common parameters used in each algorithm such as population size and total evaluation number are chosen to be randomly as $20 \times D$ and 1000, respectively, where D is the dimension of the problem. The method has been implemented in Matlab (MathWorks) and the program has been run on a T6400 @ 2 GHz Intel Core (TM) 2 Duo processor with 2 GB of random access memory (RAM). In order to eliminate stochastic discrepancy, 30 independent runs have been made that involves 30 different initial trial solutions. The termination criterion has been set either limited to a maximum number of generations or to the order of relative error equal to 10^{-6}, whichever is achieved first. The other specific parameters of algorithms are given below:

GA setting In our experiment, real coded genetic algorithm is utilized to find optimal values. Roulette wheel selection criterion is employed to choose better fitted chromosomes. One-point crossover with the rate of 0.9 and random point mutation with the rate of 0.01 are used in the present analysis for the reproduction of new solutions.

DE setting In DE, F is a real constant which affects the differential variation between two solutions and set to 0.5 in our experiments. Value of crossover rate, which controls the change of the diversity of the population, was chosen to be 0.9.

PSO setting In the experiment, cognitive and social components, c_1 and c_2, in Eq. (9) are both set to be 1.49, while the inertia weight (w) was defined as the linear decreases from initial weight w_1 to final weight w_2 with the relation $w = w_2 + (w_1 - w_2)(\text{iter}_{max} - \text{iter})/\text{iter}_{max}$. Here, iter_{max} represents the maximum generation number and 'iter' is used a generation number.

ABC setting Except common parameters (population number and maximum evaluation number), the basic ABC used in this study employs only one control parameter, which is called limit which are defined as limit $= SN \times D$ where SN is the number of food sources or employed bees.

6.2.2 Computational Results

Using these settings, the optimal design parameters for the system performance optimization are obtained by solving the optimization problem and their corresponding results are tabulated in Table 4. Using these optimal design—MTBF and MTTR—results, the plant personnel may change their initial goals so as to reduce the operational and maintenance cost by adopting suitable maintenance strategies from their design results. The best, mean, worst, median, and standard deviation (SD) values of the objective functions are summarized in Table 5. It has been noticed here that the worst value obtained by ABC is far better than the best solutions found by other algorithms.

6.2.3 Statistical Analysis

In order to analyze whether the results as obtained in the above tables are statistically significantly with each other or not, we performed t test on pair of algorithms. For this, first equality of variances will be tested, since the t test assumes equality of variances, using an F test on the pair of algorithms. For this, one tail F test has been performed with significant level of $\alpha = 0.01$. The calculated values of F-statistics (=2.278326, 1.965759 and 1.962456, respectively, for GA, DE, and PSO when pair with ABC) are less than the F-critical value (=2.423438) at (29, 29) degree of freedom. Hence, the null hypothesis of equal variances, i.e., of equal variances, may be accepted. Now a single-tail t test has been performed with the null hypothesis that their mean difference is zero at 1 % significance level in the case of ABC results with other results. The results computed are tabulated in Table 6 and it indicates that the value of their t stat is much greater than the t critical values. Also, the p value obtained during the test is less than the significance level α. Thus, it is highly significant and null hypothesis, i.e., mean of the two algorithms is identical which is rejected. Hence, the two types of means differ significantly. Furthermore, since mean of the performance function value of the system with ABC is greater than others, we conclude that ABC is definitely better than others' results and this difference is statistically significant.

Table 4 Optimal design parameter for the system

Components	GA		DE		PSO		ABC	
	MTBF	MTTR	MTBF	MTTR	MTBF	MTTR	MTBF	MTTR
Digester	3021.337646	2.124445	3001.378004	2.240216	2998.921009	3.008118	2995.000000	2.000000
Knotter	1852.765677	3.633385	1851.228695	4.390708	1851.759748	4.267327	1850.590735	2.465347
Decker	1880.616785	2.401734	1881.797006	2.253575	1880.828655	2.093784	1880.015217	2.015455
Opener	1861.380676	3.194290	1861.097750	2.246486	1863.653044	4.986936	1860.211903	3.849707
Objective function	$4.71649730 \times 10^{-8}$		$4.73456701 \times 10^{-8}$		$4.73412895 \times 10^{-8}$		$4.75194028 \times 10^{-8}$	

Table 5 Statistical simulation results of the objective function

Method	Best ($\times 10^{-8}$)	Mean ($\times 10^{-8}$)	Worst ($\times 10^{-8}$)	Median ($\times 10^{-8}$)	SD ($\times 10^{-8}$)
GA	4.71649730	4.68618455	4.67235774	4.68340019	0.89303101
DE	4.73456701	4.71761563	4.69839448	4.71813194	0.82951411
PSO	4.73412895	4.71860403	4.70278758	4.71960230	0.82881691
ABC	4.75194028	4.74333998	4.73988872	4.74854536	0.59164143

Table 6 t test for statistical analysis

	GA	DE	PSO	ABC
Mean ($\times 10^{-8}$)	4.68618455	4.71761563	4.71860403	4.74333998
SD ($\times 10^{-10}$)	0.89303101	0.82951411	0.82881691	0.59164143
Variance ($\times 10^{-20}$)	0.79750437	0.68809366	0.68693748	0.35003959
Observation	30	30	30	30
Pooled variance ($\times 10^{-20}$)	0.59355722	0.53696547	0.53636745	
Hypothesized mean difference	0	0	0	
Degree of freedom	58	58	58	
t stat	28.73239506	13.59617848	13.08106236	
$P\ (T \leq t)$ one tail	0	0	0	
T critical one tail	2.39237747	2.39237747	2.39237747	

6.3 Analyze the Behavior of the System

In order to analyze the behavior of the considered system using weakest t-norm based arithmetic operations on vague set theory, the obtained optimal values of system MTBF and MTTR are used here for increasing the relevance of the study. Uncertainty always exits in the data. Thus for handling of this, the collected data are converted into triangular intuitionistic fuzzy numbers (TIFNs) with ±15 % spread on both sides of the data. Based on their RBD, the minimal cut set of the system is $\{A\}$, $\{B_1 B_2\}$, $\{C_1\}$, $\{C_2\}$, $\{C_3\}$, and $\{D_1 D_2\}$. Using these cut sets and the input data in the form of intuitionistic fuzzy numbers with a degree of acceptance level is 0.6 and rejection level is 0.2, the top events of the system parameters are calculated using weakest t-norm based arithmetic operation on vague Lambda-Tau methodology in the form of membership and nonmembership functions and are plotted in Fig. 3 along with the existing technique results (Garg 2013; Knezevic and Odoom 2001). The results are explained as follows:

(i) The results computed by traditional or crisp methodology are independent of the confidence level and thus their corresponding results are remaining constant. Hence, this technique is beneficial for a system whose data are precise, i.e., those which do not contain the uncertainty/impreciseness in the data.

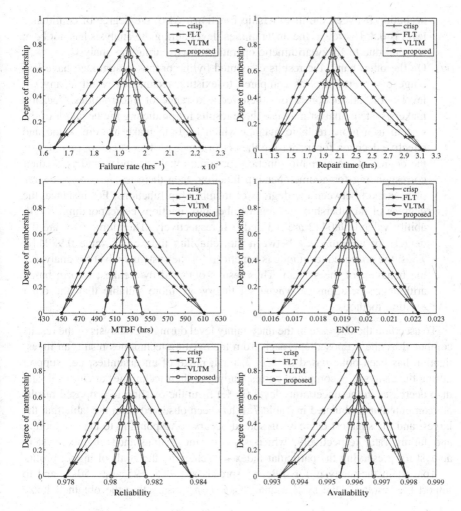

Fig. 3 Reliability plots for the system at ±15 % spreads

(ii) The results computed by Knezevic and Odoom (2001) using fuzzy arithmetic operations have a wide range of uncertainties in the form of spread. In their approach, the degree of nonmembership functions is simply one minus the degree of membership functions, and hence there is zero degree of hesitation between the membership functions. Also, the highest level of domain of confidence is taken as 1. Thus results computed by their approach are not so much ideal with the real-life conditions as it does not consider the degree of interminancy between the membership functions.

(iii) The results computed by Garg (2013) approach using intuitionistic fuzzy arithmetic operations on vague set theory rather than on fuzzy set theory are shown in figures with VLTM legend. In their approach, there is 0.2 degree of hesitation between the membership functions. Also, the degree of confidence level is 0.8. However, the uncertainties' level during the analysis has not been reduced due to fuzzy arithmetic operations used during the analysis.

(iv) On the other hand, the results computed by the proposed approach have less range of uncertainties as compared to existing techniques result at any cut level α of satisfaction. From their reduced range of uncertainties, decision makers/system analyst may use these results for predicting the behavior of the system in a more realistic manner which leads to make a more sound and effective decision for increasing the performance of the system. In this, if γ_1 is the degree of membership function for some reliability index and γ_2 be their corresponding for nonmembership function, then there is $1 - \gamma_1 - \gamma_2$ degree of hesitation between the degrees of membership functions. For instance, the degrees of membership and nonmembership functions corresponding to reliability value 0.980442 are 0.3 and 0.4, respectively. Therefore, there are 0.3 degrees of interminancy between the reliability indexes of value 0.980442. Thus, the proposed technique is beneficial for the system analyst for analyzing the behavior of the system. Thus results obtained by weakest t-norm based arithmetic operations on vague set theory are more suitable than the other existing methods.

To ascertain the decrease in the uncertainty level during the analysis of the results computed by the proposed approach from the existing techniques result, an investigation has been done based on Fig. 3. First, range of uncertainties, i.e., support during the analysis, is computed corresponding to their techniques at cut level, and then decreases in the uncertainty level (in %) from the existing to proposed results are computed and tabulated in Table 7. It has been observed from the table that the largest and the smallest decrease in spread occurs corresponding to the availability and failure rate, respectively, which suggests the maintenance engineer/system analyst for preserving the particular index for achieving the goals of higher profit.

To sustain the analysis for different spreads say ±15, ±25 and ±50 % and to import the results to the system analysts, it is necessary that the obtained fuzzy

Table 7 Decrease in spread corresponding to reliability parameters

Methods	Spread of the reliability parameters					
	Failure rate	Repair time	MTBF	ENOF	Reliability	Availability
I or II	0.0005280	1.878033	161.06219	0.005878	0.005716	0.004667
III	0.0001597	0.500704	43.227339	0.001611	0.001566	0.001244
Decrease in spread of the reliability parameters from						
I or II to III	69.753787	73.338913	73.161088	72.592718	72.603219	73.344761

I Knezevic and Odoom (2001) approach, *II* Garg (2013) approach, *III* proposed approach

output is converted into crisp value so that decision maker/system analyst may implement these results into the system. For this, defuzzification has been done using the center of gravity method and their corresponding values at different levels of uncertainties ±15, ±25 and ±50 % along with their crisp results are tabulated in Table 8. It has been concluded from the table that the defuzzified values of reliability indices are much wider when computed with the existing techniques results as compared to proposed technique. This is mainly due to the reason that the existing techniques, Knezevic and Odoom (2001) and Garg (2013), used the fuzzy arithmetic operations and hence a wide range of spread. It has also been evident from the table that defuzzified values change with change of spread, whereas a crisp value remains constant. It shows that when uncertainty level are varies from ±15 to ±25 % and further to ±50 %, the variations in their defuzzified values by the proposed approach for almost all the indices are quite less as compared to other results. Thus due to their reduced range of prediction, the values obtained are beneficial for the system analyst for future course of actions.

6.4 Performance Analysis Using RAM index

In order to maintain the system performance satisfactory, it is necessary that proper maintenance actions should be adopted at a regular interval of time. Also, it is necessary that the current condition of the equipment should be changed according to the effective maintenance program. But it is very difficult for the system analyst to predict the component on which more attention should be given for saving money, manpower, and time. This is mainly due to various inherent factors that affect the system performance. Moreover, failure of one component will reduce the efficiency of the system, and hence consequently the performance. To handle this problem, an analysis has been carried out on system RAM index which shows the effects of the component parameters on its performance. Failure is an inevitable phenomenon in the system, and hence it is necessary that uncertainty levels in the analysis should be reduced up to a desired degree of accuracy. For this, RAM index has been computed in terms of membership functions by proposed approach and compared their results with the other existing techniques in Fig. 4a, while the variations of RAM index at different levels of uncertainties (spreads ranging from 0 to 100 %) are plotted in Fig. 4b. On the other hand, the variation of this index at 15 % spread for a long-run period is shown in Fig. 4c and it is concluded that in order to increase the performance of the system, a necessary action should be taken after $t = 12$ h, since after $t = 12$ h, system performance is decreasing rapidly.

Since the performance of the system directly depends on its component parameters and hence investigation has been done on the system index by varying the system parameters, failure rates, and repair times, of the main component of the

Table 8 Crisp and defuzzified values of reliability parameters

Spread	Technique	Failure rate (×10⁻³) h⁻¹	Repair time (h)	MTBF (h)	ENOF	Reliability	Availability
	Crisp	1.93523506	2.01344562	518.746537	0.01929219	0.980833703	0.99614515
Defuzzified values for reliability indices							
±15 %	I	1.93535961	2.10768701	524.797326	0.01929171	0.98083457	0.99586949
	II	A: 1.93536034	2.10634228	524.832890	0.01929175	0.98083457	0.99587144
		B: 1.93535984	2.10594575	524.807269	0.01929176	0.98083457	0.99587258
	III	A: 1.93523506	2.02360992	519.201529	0.01929199	0.98083386	0.99611878
		B: 1.93523506	2.02356357	519.199691	0.01929200	0.98083386	0.99611889
±25 %	I	1.93558102	2.28560303	536.155284	0.01929075	0.98083611	0.99535541
	II	A: 1.93558306	2.28421131	536.269840	0.01929080	0.98083612	0.99535640
		B: 1.93558166	2.28296213	536.189389	0.01929081	0.98083611	0.99535994
	III	A: 1.93523506	2.04417751	520.016098	0.01929158	0.98083414	0.99606735
		B: 1.93523506	2.04404761	520.010946	0.01929159	0.98083414	0.99606767
±50 %	I	1.93661896	3.35320443	602.738637	0.01928453	0.98084335	0.99227442
	II	A: 1.93662714	3.35048271	603.625814	0.01928473	0.98084341	0.99226644
		B: 1.93662155	3.34067112	603.010083	0.01928479	0.98084337	0.99229627
	III	A: 1.93523506	2.14236217	523.891671	0.01928965	0.98083545	0.99582370
		B: 1.93523506	2.14182030	523.870177	0.01928966	0.98083545	0.99582501

I FLT, *II* VLTM, *III* proposed approach
A Membership function, *B* nonmembership function

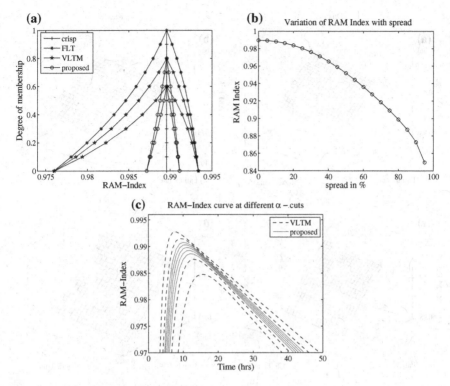

Fig. 4 Variation of the RAM index plots

system simultaneously. The results corresponding to this analysis has been plotted through the surface plot in Fig. 5, which contain four subplots corresponding to the four main components of the system. The ranges of their indexes have been notified during the analysis and have been tabulated in Table 9. From the analysis, it has been observed that variation of the opener component parameters will affect the system performance significantly, while digester component has less significance. For instance, a variation of the failure rate from 0.457884×10^{-3} to 0.619491×10^{-3} and repair time from 3.272251 to 4.427163 for opener component will reduce the RAM index by $2.411\,\%$. Similar effect has been observed for other component also. On the basis of results shown in tabular form, it can be analyzed that for improving the performance of the system, decision maker/plant personnel pay more attention to the components as per the preferential order; opener, knotter, decker, and digester.

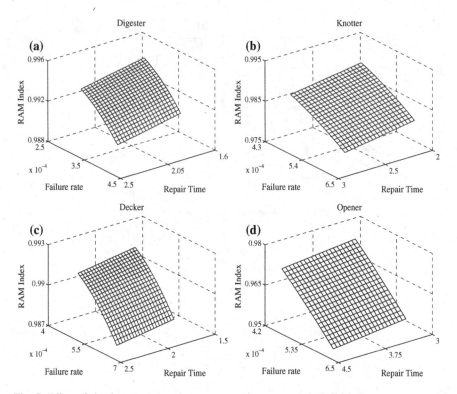

Fig. 5 Effect of simultaneously varying components parameters on RAM index

Table 9 Effect of simultaneously variations of system's components' failure and repair times on its RAM index for washing system

Component	Range of failure rate $\lambda \times 10^{-3}$ (h^{-1})	Range of repair time τ (h)	RAM index
Digester	0.283995–0.384229	1.700000–2.300000	Min: 0.98949531
			Max: 0.99440821
Knotter	0.459925–0.622252	2.095544–2.835149	Min: 0.97912669
			Max: 0.98941737
Decker	0.452609–0.612353	1.713136–2.317773	Min: 0.98708771
			Max: 0.99162133
Opener	0.457884–0.619491	3.272251–4.427163	Min: 0.95280552
			Max: 0.97578153

7 Conclusion

The present chapter investigates the failure behavior analysis of a repairable industrial system. Industrial system is growing in nature and complexities are increasing due to various interconnected components working in the system. Thus it is difficult, if not impossible, for the system analyst to predict the behavior of the system, and hence increasing the performance and productivity of the system. For this, a structural framework has been developed to model, analyze, and predict the failure pattern of the system behavior in both quantitative as well as qualitative manner. Applying soft computing techniques to analyze and to optimize the design problems of repairable series–parallel systems, it appears to be very helpful in the decision making of system parameter design. For this, design parameters of the system are obtained from their system performance optimization model by considering the system RAM index and cost (manufacturing as well as repairing) as an objective. The estimation of optimal design parameter (MTBF and MTTR) will generally help the maintenance engineers to understand the behavioral dynamics of the system. Due to complexity in the system configuration, the data obtained from historical records are imprecise and inaccurate. Keeping this point in view, efficiency for analyzing the behavior of the system is increased using computed design parameters. For strengthening the analysis, various reliability parameters such as system failure rate, MTBF, etc. are analyzed in the form of membership and nonmembership functions using weakest t-norm based arithmetic operations on vague Lambda-Tau methodology. These computed parameters may help the concerned managers to plan and adapt suitable maintenance practices/strategies for improving system performance and thereby reduce operational and maintenance costs. The major advantage of the proposed technique over the existing technique is that it gives compressed search space for each computed reliability index by utilizing available information and uncertain data. This suggests that decision maker/ system analyst has more sensitive region to make a more sound and effective decision to improve the system performance in lesser time. In order to take a decision for improving the performance of the system, as per preferential order, an analysis has been done on the system performance RAM index which helps the plant personnel to rank the system components. Components of the system which has excessive failure rates, long repair times, or high degree of uncertainty associated with these values are identified and reported in preferential order as opener, knotter, decker, and digester. The methodology will assist the managers (i) to carry out design modifications, if any, required to achieve minimum failures; and (ii) to help in maintenance (repair and replacement) decision making. Computed result will facilitate the management in reallocating the resources, making maintenance decisions, achieving long-run availability of the system, and enhancing the overall productivity of the system.

References

Attanassov, K. T. (1986). Intuitionistic fuzzy sets. *Fuzzy Sets and Systems, 20,* 87–96.

Attanassov, K. T. (1989). More on intuitionistic fuzzy sets. *Fuzzy Sets and Systems, 33*(1), 37–46.

Brest, J., Greiner, S., Boskovic, B., Mernik, M., & Zumer, V. (2006). Self- adapting control parameters in differential evolution: A comparative study on numerical benchmark problems. *Transactions on Evolutionary Computation, 10*(6), 646–657.

Bris, R., Chatelet, E., & Yalaoui, F. (2003). New method to minimize the preventive maintenance cost of series-parallel systems. *Reliability Engineering and System Safety, 82,* 247–255.

Bustince, H., & Burillo, P. (1996). Vague sets are intuitionistic fuzzy sets. *Fuzzy Sets and Systems, 79*(3), 403–405.

Chang, J. R., Chang, K. H., Liao, S. H., & Cheng, C. H. (2006). The reliability of general vague fault tree analysis on weapon systems fault diagnosis. *Soft Computing, 10,* 531–542.

Chen, S. M. (2003). Analyzing fuzzy system reliability using vague set theory. *International Journal of Applied Science and Engineering, 1*(1), 82–88.

Coelho, L. S. (2009). An efficient particle swarm approach for mixed-integer programming in reliability redundancy optimization applications. *Reliability Engineering and System Safety, 94*(4), 830–837.

Eberhart, R., & Kennedy, J. (1995). A new optimizer using particle swarm theory. In *Proceedings of the Sixth International Symposium on Micro Machine and Human Science* (pp. 39–43).

Garg, H. (2013). Reliability analysis of repairable systems using Petri nets and Vague Lambda-Tau methodology. *ISA Transactions, 52*(1), 6–18.

Garg, H., & Rani, M. (2013). An approach for reliability analysis of industrial systems using PSO and IFS technique. *ISA Transactions, 52*(6), 701–710.

Garg, H., & Sharma, S. P. (2012). A two-phase approach for reliability and maintainability analysis of an industrial system. *International Journal of Reliability, Quality and Safety Engineering, 19*(3), 1250013 (19 pages).

Garg, H., & Sharma, S. P. (2012). Stochastic behavior analysis of industrial systems utilizing uncertain data. *ISA Transactions, 51*(6), 752–762.

Garg, H., & Sharma, S. P. (2013). Multi-objective reliability-redundancy allocation problem using particle swarm optimization. *Computers & Industrial Engineering, 64*(1), 247–255.

Garg, H., Sharma, S. P., & Rani, M. (2012). Cost minimization of washing unit in a paper mill using artificial bee colony technique. *International Journal of System Assurance Engineering and Management, 3*(4), 371–381.

Garg, H., Rani, M., & Sharma, S. P. (2013a). Predicting uncertain behavior of press unit in a paper industry using artificial bee colony and fuzzy Lambda-Tau methodology. *Applied Soft Computing, 13*(4), 1869–1881.

Garg, H., Rani, M., & Sharma, S. P. (2013). Reliability analysis of the engineering systems using intuitionistic fuzzy set theory. *Journal of Quality and Reliability Engineering,* Article ID 943972, 10 pages.

Garg, H., Rani, M., & Sharma, S. P. (2013). Preventive maintenance scheduling of the pulping unit in a paper plant, Japan. *Journal of Industrial and Applied Mathematics, 30*(2), 397–414.

Garg, H., Rani, M., & Sharma, S. P. (2013). Predicting uncertain behavior and performance analysis of the pulping system in a paper industry using PSO and Fuzzy methodology. In P. Vasant (Ed.), *Handbook of Research on Novel Soft Computing Intelligent Algorithms: Theory and Practical Applications* (pp. 414–449). IGI Global, USA.

Garg, H., Rani, M., Sharma, S. P., & Vishwakarma, Y. (2014a). Intuitionistic fuzzy optimization technique for solving multi-objective reliability optimization problems in interval environment. *Expert Systems with Applications, 41,* 3157–3167.

Garg, H., Rani, M., & Sharma, S. P. (2014b). An approach for analyzing the reliability of industrial systems using soft computing based technique. *Expert Systems with Applications, 41,* 489–501.

Gau, W. L., & Buehrer, D. J. (1993). Vague sets. *IEEE Transaction on Systems, Man, and Cybernetics, 23*, 610–613.

Gen, M., & Yun, Y. S. (2006). Soft computing approach for reliability optimization: State-of-the-art survey. *Reliability Engineering and System Safety, 91*(9), 1008–1026.

Goldberg, D. E. (1989). *Genetic algorithm in search, optimization and machine learning.* MA: Addison-Wesley.

Holland, J. H. (1975). *Adaptation in natural and artificial systems, Ann Arbor.* MI: The University of Michigan Press.

Hsieh, T.-J., & Yeh, W.-C. (2012). Penalty guided bees search for redundancy allocation problems with a mix of components in series parallel systems. *Computers & Operations Research, 39*(11), 2688–2704.

Juang, Y. S., Lin, S. S., & Kao, H. P. (2008). A knowledge management system for series-parallel availability optimization and design. *Expert Systems with Applications, 34*, 181–193.

Karaboga, D. (2005). An idea based on honey bee swarm for numerical optimization. Tech. rep., TR06, Erciyes University, Engineering Faculty, Computer Engineering Department.

Karaboga, D., & Akay, B. (2009). A comparative study of artificial bee colony algorithm. *Applied Mathematics and Computation, 214*(1), 108–132.

Karaboga, D., & Basturk, B. (2007). A powerful and efficient algorithm for numerical function optimization: Artificial bee colony (ABC) algorithm. *Journal of Global Optimization, 39*, 459–471.

Karaboga, D., & Ozturk, C. (2011). A novel clustering approach: Artificial bee colony (ABC) algorithm. *Applied Soft Computing, 11*(1), 652–657.

Kennedy, J., & Eberhart, R. C. (1995) Particle swarm optimization. In *IEEE International Conference on Neural Networks, Vol. IV* (pp. 1942–1948). Piscataway.

Knezevic, J., & Odoom, E. R. (2001). Reliability modeling of repairable systems using Petri nets and Fuzzy Lambda-Tau Methodology. *Reliability Engineering and System Safety, 73*(1), 1–17.

Komal, Sharma, S. P., & Kumar, D. (2010). RAM analysis of repairable industrial systems utilizing uncertain data. *Applied Soft Computing, 10*, 1208–1221.

Kuo, W., Prasad, V. R., Tillman, F. A., & Hwang, C. (2001). *Optimal reliability design: Fundamentals and applications.* Cambridge: Cambridge University Press.

Lapa, C. M. F., Pereira, C. M., & Barros, M. P. D. (2006). A model for preventive maintenance planning by genetic algorithms based on cost and reliability. *Reliability Engineering and System Safety, 91*, 233–240.

Leou, R. (2006). A method for unit maintenance scheduling considering reliability and operation expense. *Electrical Power and Energy Systems, 28*, 471–481.

Rajpal, P. S., Shishodia, K. S., & Sekhon, G. S. (2006). An artificial neural network for modeling reliability, availability and maintainability of a repairable system. *Reliability Engineering and System Safety, 91*(7), 809–819.

Ross, T. J. (2004). Fuzzy logic with engineering applications, 2nd edn. New York: Wiley.

Saraswat, S., & Yadava, G. (2008). An overview on reliability, availability, maintainability and supportability (RAMS) engineering. *International Journal of Quality and Reliability Management, 25*(3), 330–344.

Storn, R., & Price, K. V. (1995). Differential evolution: A simple and efficient adaptive scheme for global optimization over continuous spaces. Tech. Rep. Technical Report TR-95-012, International Computer Science Institute, Berkley.

Storn, R., & Price, K.V. (1997). Differential evolution—a simple and efficient heuristic for global optimization over continuous spaces. *Journal of Global Optimization, 11*(4), 341–359.

Taheri, S., & Zarei, R. (2011). Bayesian system reliability assessment under the vague environment. *Applied Soft Computing, 11*(2), 1614–1622.

Yeh, W. C., & Hsieh, T. J. (2011). Solving reliability redundancy allocation problems using an artificial bee colony algorithm. *Computer and operational research, 38*(11), 1465–1473.

Zadeh, L. A. (1965). Fuzzy sets. *Information and Control, 8*, 338–353.

System Reliability Evaluation
of a Multistate Manufacturing Network

Yi-Kuei Lin, Ping-Chen Chang and Cheng-Fu Huang

1 Introduction

This chapter studies the system reliability as a performance indicator to measure the demand satisfaction of the manufacturing system. Three characteristics are considered in this chapter: (i) multiple production lines in parallel, (ii) multiple reworking actions, and (iii) distinct defect rate of each station. To evaluate performance of a manufacturing system, network analysis is an applicable methodology to be adopted. A great deal of studies has been devoted to performance evaluation for a manufacturing system by adopting network analysis (Lee and Garcia-Diaz 1993, 1996; Chen and Lan 2001; Lan 2007; Listeş 2007; Paquet et al. 2008; Kemmoe et al. 2013; Li and Li 2013; Mourtzis et al. 2013). Previous literatures, however, did not emphasize stochastic capacities of each station while evaluating the performance of a manufacturing system. Those works mainly focused on costs, profits, and sales. The capacity of stations in a manufacturing network reflects the demand satisfaction ability directly. Nevertheless, evaluations of costs, profits, and sales mainly provide the financial information rather than the production performance.

From the perspective of network analysis, a manufacturing system in which each station consists of a group of machines or workers indicates that such a station has stochastic capacity levels (i.e., multistate). For instance, a station with K machines implies that $(K + 1)$ capacity levels are available. The lowest level zero corresponds to all machines complete malfunction, while K is the highest level of operation. Therefore, the manufacturing system is also multistate and it can be treated as the so-called multistate network (Aven 1985; Hudson and Kapur 1985; Xue 1985; Alexopoulos 1995; Zuo et al. 2007; Lin 2009a, b; Lin and Chang 2012a, b, c,

Y.-K. Lin (✉) · P.-C. Chang · C.-F. Huang
Department of Industrial Management, National Taiwan University of Science &
Technology, Taipei 106, Taiwan
e-mail: yklin@mail.ntust.edu.tw

© Springer-Verlag London 2016

H. Pham (ed.), *Quality and Reliability Management and Its Applications*,
Springer Series in Reliability Engineering, DOI 10.1007/978-1-4471-6778-5_5

2013). For the manufacturing network with stochastic-flow capacities, we name it as the multistate manufacturing network (MMN) herein. To measure the capability that an MMN satisfies the customers' requirements, Lin (2009a) focused on two-commodity reliability evaluation of an MMN in terms of minimal path (MP), in which an MP is a path the proper subsets of which are no longer paths. In Lin's work, the system reliability is defined as the probability that the MMN satisfies two-commodity demand. A significant amount of research (Zuo et al. 2007; Lin 2009a, b; Lin and Chang 2012a, b, c) has also been devoted to studying the system reliability of a multistate network (such as manufacturing, computer, and communication systems, etc.) in terms of MP. In those works, the demand transmitted through a network must obey the flow conservation law (Ford and Fulkerson 1962), meaning that flow will not increase or decrease during transmission.

However, some attributes related to the MMN such as reworking and scraps were not considered in the above literatures. For a practical MMN, the input flow processed by each station would not be the same as output flow due to the defect rate of each station. The defect rate of each station influences the capability of a manufacturing system and leads to defective WIP (work-in-process) or products, in which defective WIP/products might be reworked or scrapped (Pillai and Chandrasekharan 2008; Liu et al. 2009). Thus, an important issue should be concerned is that how reworking actions affect the amount of output in a manufacturing system, especially for those production lines with multiple reworking actions. Furthermore, reworking actions may be implemented at the same stations, implying that a manufacturing system would produce products by the general production path(s) and the reworking path(s), for satisfying demand (Buscher and Lindner 2007; Teunter et al. 2008). That is, the output products of the MMN might be less than the input raw materials. Hence, the traditional methodology for the multistate network problem could not be applied in such a case due to the violation of the flow conservation. Moreover, based on the MP concept, an arc (station) would not appear on the same path more than once; otherwise it is not an MP. However, defective WIP from a station would be reworked starting from a previous station(s) or the same station(s) (Buscher and Lindner 2007; Teunter et al. 2008), which is breaking the basic concept of MP.

To overcome the limitations of flow conservation and MP, this chapter proposes a graphical model for the MMN with multiple production lines in parallel and multiple reworking actions. Furthermore, a novel technique of "prior-set" is proposed to deal with multiple reworking actions. The prior-set is utilized to record all stations prior to a specific station, which is beneficial for input flow determination. Once the input flows of stations are derived, we may obtain the minimal capacity vectors that stations should provide. Subsequently, we evaluate the probability that the MMN could produce a given demand d in terms of such vectors. Such a probability is referred to as the system reliability, which is a performance indicator of MMN.

1.1 Notation

N	set of nodes
L_j	jth production line, $j = 1, 2$
n_j	number of stations in L_j, $j = 1, 2$
a_i	ith arc (station) where $i = 1, 2, \ldots, n_1 + n_2$
A	$\{a_i \mid i = 1, 2, \ldots, n_1 + n_2\}$: set of arcs (stations)
p_i	success rate of a_i, where $0 \leq p_i \leq 1$
q_i	defect rate of a_i, where $q_i = 1 - p_i$
M_i	maximal capacity of a_i, $i = 1, 2, \ldots, n_1 + n_2$
w_i	loading of a_i, $i = 1, 2, \ldots, n_1 + n_2$
W	$(w_1, w_2, \ldots, w_{n_1 + n_2})$: the loading vector
x_i	current capacity of a_i, $i = 1, 2, \ldots, n_1 + n_2$
X	$(x_1, x_2, \ldots, x_{n_1 + n_2})$: the capacity vector
c_i	number of possible capacities of a_i, $i = 1, 2, \ldots, n_1 + n_2$
x_{ij}	jth possible capacity of a_i, $j = 1, 2, \ldots, c_i$. Thus, x_i takes possible values $0 = x_{i1} < x_{i2} < \cdots < x_{ic_i} = M_i$, $i = 1, 2, \ldots, n_1 + n_2$
d	demand
d_j	demand assigned to L_j and $\sum_{j=1}^{2} d_j = d$
I	units of input raw materials
I_j	units of input raw materials for L_j and $\sum_{j=1}^{2} I_j = I$
O	units of output products
O_j	units of output products from L_j and $\sum_{j=1}^{2} O_j = O$
a_i'	ith station doing the reworking action
$L_j^{(G)}$	jth general production path (without reworking action)
$L_j^{(R\mid r_1 \to k_1)}$	jth one-pass reworking path with only the first reworking action, in which the first reworking action indicates that defective WIP output fromstation r_1 is reworked starting from station k_1
$L_j^{(R\mid r_2 \to k_2)}$	jth one-pass reworking path with only the second reworking action, in which the second reworking action indicates that defective WIP output from station r_2 is reworked starting from station k_2
$L_j^{(R\mid r_1 \to k_1, r_2 \to k_2)}$	two-pass reworking path with both the first and the second reworking actions
$f_{j,i}^{(G)}$	input flow for $a_i \in L_j^{(G)}$
$f_{j,i}^{(R\mid r_1 \to k_1)}$	input flow for $a_i \in L_j^{(R\mid r_1 \to k_1)}$
$f_{j,i}^{(R\mid r_2 \to k_2)}$	input flow for $a_i \in L_j^{(R\mid r_2 \to k_2)}$
$f_{j,i}^{(R\mid r_1 \to k_1, r_2 \to k_2)}$	input flow for $a_i \in L_j^{(R\mid r_1 \to k_1, r_2 \to k_2)}$
$M(L_j)$	maximum capacity of L_j
Ψ_i^-	prior-set to store the stations prior to the station a_i

$\Phi(\Psi)$ product of success rates of stations in the set Ψ
Λ_j overall success rate of L_j
$V(X)$ maximum output under X

1.2 Definition

$Y \geq X$ $(y_1, y_2, \ldots, y_{n_1 + n_2}) \geq (x_1, x_2, \ldots, x_{n_1 + n_2})$: $y_i \geq x_i$ for each $i = 1, 2, \ldots, n_1 + n_2$
$Y > X$ $(y_1, y_2, \ldots, y_{n_1 + n_2}) > (x_1, x_2, \ldots, x_{n_1 + n_2})$: $Y \geq X$ and $y_i > x_i$ for at least one i

2 Problem Description and Assumptions

This chapter evaluates the system reliability of the MMN, which is defined as the probability of demand satisfaction. Once the order is requested by customers, the production manager may determine the integration of instantaneous production rate over a period of time. Such an instantaneous production rate is defined as demand, which is measured in d units of product that should be produced per unit time.

To analyze the MMN, we emphasize the input flow of each station, in which the input flow is defined as the input amount that each station processes per unit time. For those defective WIP, multiple reworking actions may be taken to repair them. In the MMN, products are produced by multiple production lines in parallel. That is, the sequence of stations and their functions in all production lines are the same. For convenience, we concentrate on two production lines case first. The proposed methodology and algorithm can then be easily extended to more than two production lines case. To evaluate the system reliability of a manufacturing system, some assumptions are addressed as follows.

   I. Each inspection point (node) is perfectly reliable. It indicates that inspection would not damage any WIP/products.
   II. The capacity of each station (arc) is a random variable according to a given probability distribution.
   III. The capacities of different stations (arcs) are statistically independent.
   IV. Each defective WIP is reworked at most once by the same station. This implies that such a defective WIP is repaired until a usable state. If the defective WIP after reworking is still defective, it means that such a defective WIP is non-repairable. Then it is scrapped.

3 Graphical Methodology

An AOA (activity-on-arrow) diagram is applied to represent a manufacturing system. Each arc (arrow) is regarded as a station consisting of the identical functional machines and each node denotes an inspection point following the station. Based on the AOA diagram, a graphical transformation is utilized to model the manufacturing system as an MMN. Let $(\mathbf{N}, \mathbf{A})$ be a manufacturing system with two production lines in parallel, where $\mathbf{N}$ represents the set of nodes and $\mathbf{A} = \{a_i | i = 1, 2, \ldots, n_1 + n_2\}$ represents the set of arcs with n_1 (resp. n_2) is the number of stations in the first production line L_1 (resp. the second production line L_2). For instance, Fig. 1 represents a manufacturing system with two production lines in the form of AOA diagram, in which L_1 (consisting of $a_1, a_2, \ldots, a_8$) and L_2 (consisting of $a_9, a_{10}, \ldots, a_{16}$) producing the same type of product. Each production line consists of eight stations ($n_1 = n_2 = 8$) and the success rate of ith station is p_i, where $0 \le p_i \le 1$ (i.e., the defect rate is $q_i = 1 - p_i$). In Fig. 1, the two black nodes in each production line represent the input inspection point (check for raw materials) and output inspection point (check for final products), respectively. White nodes between two stations denote the inspection points for WIP to check if it can enter the next process (success) or should be scrapped (failure). The meshed nodes indicates that defective WIP (output from a_3 and a_6 or a_{11} and a_{14}) inspected by this inspection point can be reworked.

However, it is difficult to determine the input flows from the regular process (without reworking) or the reworking process from Fig. 1 because it only shows the direction which defective WIP should enter. Thus, we transform the manufacturing system into an MMN as shown in Fig. 2, in which a dummy-station a_i' is set to denote the station a_i doing the reworking action. The transformed network is beneficial for further analysis because it is easily to distinguish the input amount from regular process and reworking process of each station.

To analyze the MMN, each production line is decomposed into the form of paths. Consider the jth production line L_j with two reworking actions, there are four combinations of these paths: (i) general production path $L_j^{(G)}$ without reworking action; (ii) reworking path with only the first (resp. the second) reworking action, in which the first (resp. the second) reworking action indicates that defective WIP output from station r_1 (resp. r_2) is reworked starting from station k_1 (resp. k_2); and

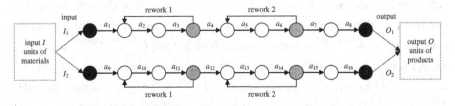

Fig. 1 A manufacturing system with two identical production lines in parallel

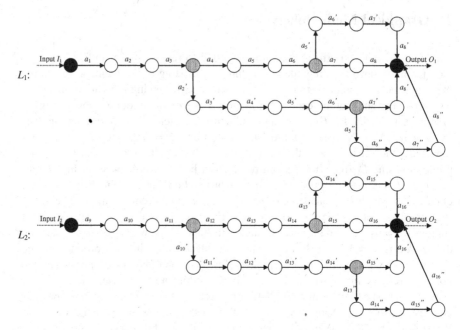

Fig. 2 The transformed MMN for Fig. 1

(iii) reworking path with both the first and the second reworking actions. We further named combination (ii) as one-pass reworking paths $L_j^{(R|r_1 \to k_1)}$ or $L_j^{(R|r_2 \to k_2)}$ while combination (iii) is named two-pass reworking path $L_j^{(R|(r_1 \to k_1, r_2 \to k_2)}$. For the special case that r_1 is the same as k_1, it implies that the defective WIP is reworked at the same station.

Take the first production line L_1 in Fig. 2 for instance, the set $\{a_1, a_2, a_3, a_4, a_5, a_6, a_7, a_8\}$ is a general production path $L_1^{(G)}$ (Fig. 3a). On the other side, the one-pass reworking path with the first reworking action is $L_j^{(R|r_1 \to k_1)} = L_j^{(R|a_3 \to a_2)} = \{a_1, a_2, a_3, a_2', a_3', a_4', a_5', a_6', a_7', a_8'\}$, the notation (R| $a_3 \to a_2$) indicates that defective WIP output from a_3 can be reworked starting from a_2 (note that, a_2' is a dummy-station of a_2). In fact, no defective WIP is processed by a_1, a_2, and a_3 and the input flow would be zero for these stations. Thus, a_1, a_2, and a_3 can be ignored and merely the stations a_2', a_3', ..., a_8' doing reworking action are retained. Since the station a_i and the dummy-station a_i' are the same, the one-pass reworking path $L_1^{(R|a_3 \to a_2)} = \{a_2, a_3, a_4, a_5, a_6, a_7, a_8\}$ (Fig. 3b). The other paths for L_1 are $L_1^{(R|a_6 \to a_5)} = \{a_5, a_6, a_7, a_8\}$ (Fig. 3c) and $L_1^{(R|a_3 \to a_2, a_6 \to a_5)} = \{a_5, a_6, a_7, a_8\}$ (Fig. 3d). Similarly, we have $L_2^{(G)} = \{a_9, a_{10}, a_{11}, a_{12}, a_{13}, a_{14}, a_{15}, a_{16}\}$, $L_2^{(R|a_{11} \to a_{10})} = \{a_{10}, a_{11}, a_{12}, a_{13}, a_{14}, a_{15}, a_{16}\}$, $L_2^{(R|a_{14} \to a_{13})} = \{a_{13}, a_{14}, a_{15}, a_{16}\}$, and $L_2^{(R|a_{11} \to a_{10}, a_{14} \to a_{13})} = \{a_{13}, a_{14}, a_{15}, a_{16}\}$ for the second production line.

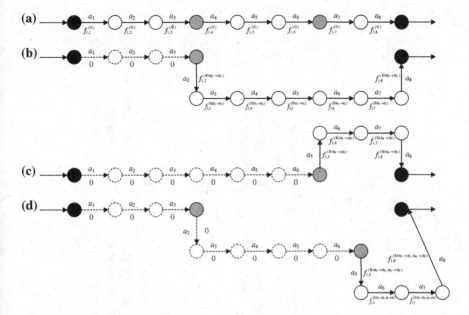

Fig. 3 Decomposed paths of L_1

We utilize $f_{j,i}^{(G)}$, $f_{j,i}^{(R|r_1 \to k_1)}$, $f_{j,i}^{(R|r_2 \to k_2)}$ and $f_{j,i}^{(R|r_1 \to k_1, r_2 \to k_2)}$ to represent the input amount of the ith station in the jth production line for $L_j^{(G)}$, $L_j^{(R|r_1 \to k_1)}$, $L_j^{(R|r_2 \to k_2)}$ and $L_j^{(R|r_1 \to k_1, r_2 \to k_2)}$, respectively. Table 1 provides the input flow of each station. Thus, we obtain the input flows (from the regular process and the reworking process) for each station clearly.

4 Model Construction

In order to produce sufficient products for satisfying demand, the input amount of raw materials should be predetermined. Once the input amount of raw materials is determined, the input flow of each station can be derived. The determination of input flows in the case of multiple reworking actions should be calculated in terms of so-called "prior-set." Subsequently, capacity of stations and system reliability of MMN are evaluated.

4.1 Prior-Set and Input Determination

For each production line L_j, suppose that I_j units of raw materials are able to produce O_j units of product; we intend to obtain the relationship between I_j and O_j

Table 1 Input flow of stations for Fig. 3

I	$f_{1,i}^{(G)}$	$f_{1,i}^{(R|a_3 \to a_2)}$	$f_{1,i}^{(R|a_6 \to a_5)}$	$f_{1,i}^{(R|a_3 \to a_2,\, a_6 \to a_5)}$
1	I_1	0	0	0
2	$I_1 P_1$	$I_1 P_1 q_2 q_3$	0	0
3	$I_1 P_1 P_2$	$I_1 P_1 q_2 q_3 P_2$	0	0
4	$I_1 P_1 P_2 P_3$	$I_1 P_1 q_2 q_3 P_2 P_3$	0	0
5	$I_1 P_1 P_2 P_3 P_4$	$I_1 P_1 q_2 q_3 P_2 P_3 P_4$	$I_1 P_1 P_2 P_3 P_4 P_5 q_6$	$I_1 P_1 q_2 q_3 P_2 P_3 P_4 P_5 q_6$
6	$I_1 P_1 P_2 P_3 P_4 P_5$	$I_1 P_1 q_2 q_3 P_2 P_3 P_4 P_5$	$I_1 P_1 P_2 P_3 P_4 P_5 q_6 P_5$	$I_1 P_1 q_2 q_3 P_2 P_3 P_4 P_5 q_6 P_5$
7	$I_1 P_1 P_2 P_3 P_4 P_5 P_6$	$I_1 P_1 q_2 q_3 P_2 P_3 P_4 P_5 P_6$	$I_1 P_1 P_2 P_3 P_4 P_5 q_6 P_5 P_6$	$I_1 P_1 q_2 q_3 P_2 P_3 P_4 P_5 q_6 P_5 P_6$
8	$I_1 P_1 P_2 P_3 P_4 P_5 P_6 P_7$	$I_1 P_1 q_2 q_3 P_2 P_3 P_4 P_5 P_6 P_7$	$I_1 P_1 P_2 P_3 P_4 P_5 q_6 P_5 P_6 P_7$	$I_1 P_1 q_2 q_3 P_2 P_3 P_4 P_5 q_6 P_5 P_6 P_7$

i	$f_{2,i}^{(G)}$	$f_{2,i}^{(R|a_{11} \to a_{10})}$	$f_{2,i}^{(R|a_{14} \to a_{13})}$	$f_{2,i}^{(R|a_{11} \to a_{10},\, a_{14} \to a_{13})}$
9	I_2	0	0	0
10	$I_2 P_9$	$I_2 P_9 P_{10} q_{11}$	0	0
11	$I_2 P_9 P_{10}$	$I_2 P_9 P_{10} q_{11} P_{10}$	0	0
12	$I_2 P_9 P_{10} P_{11}$	$I_2 P_9 P_{10} q_{11} P_{10} P_{11}$	0	0
13	$I_2 P_9 P_{10} P_{11} P_{12}$	$I_2 P_9 P_{10} q_{11} P_{10} P_{11} P_{12}$	$I_2 P_9 P_{10} P_{11} P_{12} P_{13} q_{14}$	$I_2 P_9 P_{10} q_{11} P_{10} P_{11} P_{12} P_{13} q_{14}$
14	$I_2 P_9 P_{10} P_{11} P_{12} P_{13}$	$I_2 P_9 P_{10} q_{11} P_{10} P_{11} P_{12} P_{13}$	$I_2 P_9 P_{10} P_{11} P_{12} P_{13} q_{14} P_{13}$	$I_2 P_9 P_{10} q_{11} P_{10} P_{11} P_{12} P_{13} q_{14} P_{13}$
15	$I_2 P_9 P_{10} P_{11} P_{12} P_{13} P_{14}$	$I_2 P_9 P_{10} q_{11} P_{10} P_{11} P_{12} P_{13} P_{14}$	$I_2 P_9 P_{10} P_{11} P_{12} P_{13} q_{14} P_{13} P_{14}$	$I_2 P_9 P_{10} q_{11} P_{10} P_{11} P_{12} P_{13} q_{14} P_{13} P_{14}$
16	$I_2 P_9 P_{10} P_{11} P_{12} P_{13} P_{14} P_{15}$	$I_2 P_9 P_{10} q_{11} P_{10} P_{11} P_{12} P_{13} P_{14} P_{15}$	$I_2 P_9 P_{10} P_{11} P_{12} P_{13} q_{14} P_{13} P_{14} P_{15}$	$I_2 P_9 P_{10} q_{11} P_{10} P_{11} P_{12} P_{13} q_{14} P_{13} P_{14} P_{15}$

fulfilling $O_j \geq d_j$ and $\sum_{j=1}^{2} d_j = d$, where d_j is the assigned demand for L_j. Each I_j cannot exceed the maximum capacity of L_j, where the maximum capacity of L_j is $M(L_j) = \min\{M_i \mid i: a_i \in L_j\}$. That is, the maximum capacity of L_j is determined by the bottleneck station at that production line. Thus, we have the following constraint.

$$I_j \leq M(L_j) = \min\{M_i \mid i: a_i \in L_j\}. \tag{1}$$

A prior-set Ψ_i^- is proposed to storage the stations in $L_j^{(G)}$ prior to a_i (excluding a_i itself). This implies that the prior-set records the processes operated before entering a_i. Hence, the input flow of a station can be determined by its corresponding prior-set in terms of success rates. Besides, the output amount can be derived in a similar manner. A function $\Phi(\Psi)$ is further defined as the product of success rates of stations in the set Ψ by

$$\Phi(\Psi) = \prod_{t:a_t \in \Psi} p_t \tag{2}$$

For instance, $\Phi(\Psi_6^- \cap L_1^{(R|a_3 \to a_2)}) = \prod_{t:a_t \in \Psi_6^- \cap L_1^{(R|a_3 \to a_2)}} p_t = p_2 p_3 p_4 p_5$, in which prior-set of station a_6 is $\Psi_6^- = \{a_1, a_2, a_3, a_4, a_5\}$ and the set of first production line is $L_1^{(R|a_3 \to a_2)} = \{a_2, a_3, a_4, a_5, a_6, a_7, a_8\}$ with intersection set $\{a_2, a_3, a_4, a_5\}$. For the special case $\Psi = \varnothing$, it is defined $\Phi(\varnothing) = 1$. Let $a_{\lambda_{j,u}}$ denote the station whose output defective WIP can be reworked by the uth reworking action in L_j. For the case in Fig. 3, $a_{\lambda_{1,1}} = a_3$ indicates that the defective WIP output from a_3 in L_1 can be reworked by the first reworking action (starting from a_2). The amount of output products for the jth production line is calculated as follows,

$$
\begin{aligned}
O_j = \{ & [I_j \Psi(L_j^{(G)})] \\
& + [I_j \Phi(\Psi_{\lambda_{j,1}}^-)(1 - p_{\lambda_{j,1}})\Phi(L_j^{(R|r_1 \to k_1)})] \\
& + [I_j \Phi(\Psi_{\lambda_{j,2}}^-)(1 - p_{\lambda_{j,2}})\Phi(L_j^{(R|r_2 \to k_2)})] \\
& + [I_j \Phi(\Psi_{\lambda_{j,1}}^-)(1 - p_{\lambda_{j,1}})\Phi(\Psi_{\lambda_{j,2}}^- \cap L_j^{(R|r_1 \to k_1)})(1 - p_{\lambda_{j,2}})\Phi(L_j^{(R|r_1 \to k_1, r_2 \to k_2)})] \},
\end{aligned} \tag{3}
$$

where the first term $[I_j \Phi(L_j^{(G)})]$ is produced by the general production path $L_j^{(G)}$. The second term $[I_j \Phi(\Psi_{\lambda_{j,1}}^-)(1 - p_{\lambda_{j,1}})\Phi(L_j^{(R|r_1 \to k_1)})]$ and the third term $[I_j \Phi(\Psi_{\lambda_{j,2}}^-)(1 - p_{\lambda_{j,2}})\Phi(L_j^{(R|r_2 \to k_2)})]$ are processed by the one-pass reworking paths $L_j^{(R|r_1 \to k_1)}$ and $L_j^{(R|r_2 \to k_2)}$, respectively. The last term $[I_j \Phi(\Psi_{\lambda_{j,1}}^-)(1 - p_{\lambda_{j,1}})\Phi(\Psi_{\lambda_{j,2}}^- \cap L_j^{(R|r_1 \to k_1)})(1 - p_{\lambda_{j,2}})\Phi(L_j^{(R|r_1 \to k_1, r_2 \to k_2)})]$ is processed by the two-pass reworking paths $L_j^{(R|r_1 \to k_1, r_2 \to k_2)}$. Note that $[I_j \Phi(\Psi_{\lambda_{j,1}}^-)p_{\lambda_{j,1}}]$ is the amount of (success) WIP output from $a_{\lambda_{j,1}}$ and thus $[I_j \Phi(\Psi_{\lambda_{j,1}}^-)(1 - p_{\lambda_{j,1}})]$ in the second term is the amount of defective WIP. Those defective WIP is input to the reworking path $L_j^{(R|r_1 \to k_1)}$ afterwards. Therefore, the output amount of products from $L_j^{(R|r_1 \to k_1)}$ is $[I_j \Phi(\Psi_{\lambda_{j,1}}^-)(1 - p_{\lambda_{j,1}})\Phi(L_j^{(R|r_1 \to k_1)})]$. Similar procedure can be applied to calculate the output amount of $L_j^{(R|r_2 \to k_2)}$ and $L_j^{(R|r_1 \to k_1, r_2 \to k_2)}$. For convenience, let $\Gamma_{j,1} = \Phi(\Psi_{\lambda_{j,1}}^-)(1 - p_{\lambda_{j,1}})$, $\Gamma_{j,2} = \Phi(\Psi_{\lambda_{j,2}}^-)(1 - p_{\lambda_{j,2}})$, and $\Gamma_{j,1\&2} = \Phi(\Psi_{\lambda_{j,1}}^-)(1 - p_{\lambda_{j,1}})\Phi(\Psi_{\lambda_{j,2}}^- \cap L_j^{(R|r_1 \to k_1)})(1 - p_{\lambda_{j,2}})$ for the following derivation. Thus, given the input I_j, the output amount of L_j is simplified as

$$O_j = I_j\{\Phi(L_j^{(G)}) + \Gamma_{j,1}\Phi(L_j^{(R|r_1 \to k_1)}) + \Gamma_{j,2}\Phi(L_j^{(R|r_2 \to k_2)}) + \Gamma_{j,1\&2}\Phi(L_j^{(R|r_1 \to k_1, r_2 \to k_2)})\}. \quad (4)$$

Let $\Sigma_j = \Phi(L_j^{(G)}) + \Gamma_{j,1}\Phi(L_j^{(R|r_1 \to k_1)}) + \Gamma_{j,2}\Phi(L_j^{(R|r_2 \to k_2)}) + \Gamma_{j,1\&2}\Phi(L_j^{(R|r_1 \to k_1, r_2 \to k_2)})$. That is, Σ_j is the overall success rate of L_j. Thus, Eq. (4) is further simplified as follows.

$$O_j = I_j \Sigma_j. \quad (5)$$

It is necessary that $O_j \geq d_j$ for obtaining sufficient output that satisfies the demand of L_j. Thus, we have following constrain,

$$I_j \geq d_j / \Sigma_j. \quad (6)$$

The following equation guarantees the MMN can produce exact sufficient output O_j that satisfies demand d_j,

$$I_j = d_j / \Sigma_j. \quad (7)$$

Once the input amount of raw materials I_j is determined, the input flow of each station in Table 1 can be represented in terms of prior-sets as shown in Table 2. Such a representation may possess a general form for denoting the input flow of each station. In a special case of the same station success rate, the prior-set can be reduced to denote "the number of prior stations."

4.2 Determination of Input Flow for Each Station

The input raw materials/WIP processed by the ith station a_i should satisfy the following constraint,

$$\sum_{j=1}^{2} \left(f_{j,i}^{(G)} + f_{j,i}^{(R|r_1 \to k_1)} + f_{j,i}^{(R|r_2 \to k_2)} + f_{j,i}^{(R|r_1 \to k_1, r_2 \to k_2)} \right) \leq M_i, \quad i = 1, 2, \ldots, n_1 + n_2.$$

$$(8)$$

Table 2 Revised input flow representation by prior-set

l	$f_{1,j}^{(G)}$	$f_{1,j}^{R(a_3\to a_2)}$	$f_{1,j}^{R(a_4\to a_5)}$	$f_{1,j}^{R(a_3\to a_2,\,a_4\to a_5)}$
1	$I_1\Phi(\Psi_1^-)$	0	0	0
2	$I_1\Phi(\Psi_2^-)$	$I_1\Gamma_{1,1}\Phi(\Psi_2^-\cap L_1^{R(a_3\to a_2)})$	0	0
3	$I_1\Phi(\Psi_3^-)$	$I_1\Gamma_{1,1}\Phi(\Psi_3^-\cap L_1^{R(a_3\to a_2)})$	0	0
4	$I_1\Phi(\Psi_4^-)$	$I_1\Gamma_{1,1}\Phi(\Psi_4^-\cap L_1^{R(a_3\to a_2)})$	0	0
5	$I_1\Phi(\Psi_5^-)$	$I_1\Gamma_{1,1}\Phi(\Psi_5^-\cap L_1^{R(a_3\to a_2)})$	$I_1\Gamma_{1,2}\Phi(\Psi_5^-\cap L_1^{R(a_4\to a_5)})$	$I_1\Gamma_{1,1\&2}\Phi(\Psi_5^-\cap L_1^{R(a_3\to a_2,\,a_4\to a_5)})$
6	$I_1\Phi(\Psi_6^-)$	$I_1\Gamma_{1,1}\Phi(\Psi_6^-\cap L_1^{R(a_3\to a_2)})$	$I_1\Gamma_{1,2}\Phi(\Psi_6^-\cap L_1^{R(a_4\to a_5)})$	$I_1\Gamma_{1,1\&2}\Phi(\Psi_6^-\cap L_1^{R(a_3\to a_2,\,a_4\to a_5)})$
7	$I_1\Phi(\Psi_7^-)$	$I_1\Gamma_{1,1}\Phi(\Psi_7^-\cap L_1^{R(a_3\to a_2)})$	$I_1\Gamma_{1,2}\Phi(\Psi_7^-\cap L_1^{R(a_4\to a_5)})$	$I_1\Gamma_{1,1\&2}\Phi(\Psi_7^-\cap L_1^{R(a_3\to a_2,\,a_4\to a_5)})$
8	$I_1\Phi(\Psi_8^-)$	$I_1\Gamma_{1,1}\Phi(\Psi_8^-\cap L_1^{R(a_3\to a_2)})$	$I_1\Gamma_{1,2}\Phi(\Psi_8^-\cap L_1^{R(a_4\to a_5)})$	$I_1\Gamma_{1,1\&2}\Phi(\Psi_8^-\cap L_1^{R(a_3\to a_2,\,a_4\to a_5)})$
i	$f_{2,i}^{(G)}$	$f_{2,i}^{R(a_{11}\to a_{10})}$	$f_{2,i}^{R(a_{14}\to a_{13})}$	$f_{2,i}^{R(a_{11}\to a_{10},\,a_{14}\to a_{13})}$
9	$I_2\Phi(\Psi_9^-)$	0	0	0
10	$I_2\Phi(\Psi_{10}^-)$	$I_2\Gamma_{2,1}\Phi(\Psi_{10}^-\cap L_2^{R(a_{11}\to a_{10})})$	0	0
11	$I_2\Phi(\Psi_{11}^-)$	$I_2\Gamma_{2,1}\Phi(\Psi_{11}^-\cap L_2^{R(a_{11}\to a_{10})})$	0	0
12	$I_2\Phi(\Psi_{12}^-)$	$I_2\Gamma_{2,1}\Phi(\Psi_{12}^-\cap L_2^{R(a_{11}\to a_{10})})$	0	0
13	$I_2\Phi(\Psi_{13}^-)$	$I_2\Gamma_{2,1}\Phi(\Psi_{13}^-\cap L_2^{R(a_{11}\to a_{10})})$	$I_2\Gamma_{2,2}\Phi(\Psi_{13}^-\cap L_2^{R(a_{14}\to a_{13})})$	$I_2\Gamma_{2,1\&2}\Phi(\Psi_{13}^-\cap L_2^{R(a_{11}\to a_{10},\,a_{14}\to a_{13})})$
14	$I_2\Phi(\Psi_{14}^-)$	$I_2\Gamma_{2,1}\Phi(\Psi_{14}^-\cap L_2^{R(a_{11}\to a_{10})})$	$I_2\Gamma_{2,2}\Phi(\Psi_{14}^-\cap L_2^{R(a_{14}\to a_{13})})$	$I_2\Gamma_{2,1\&2}\Phi(\Psi_{14}^-\cap L_2^{R(a_{11}\to a_{10},\,a_{14}\to a_{13})})$
15	$I_2\Phi(\Psi_{15}^-)$	$I_2\Gamma_{2,1}\Phi(\Psi_{15}^-\cap L_2^{R(a_{11}\to a_{10})})$	$I_2\Gamma_{2,2}\Phi(\Psi_{15}^-\cap L_2^{R(a_{14}\to a_{13})})$	$I_2\Gamma_{2,1\&2}\Phi(\Psi_{15}^-\cap L_2^{R(a_{11}\to a_{10},\,a_{14}\to a_{13})})$
16	$I_2\Phi(\Psi_{16}^-)$	$I_2\Gamma_{2,1}\Phi(\Psi_{16}^-\cap L_2^{R(a_{11}\to a_{10})})$	$I_2\Gamma_{2,2}\Phi(\Psi_{16}^-\cap L_2^{R(a_{14}\to a_{13})})$	$I_2\Gamma_{2,1\&2}\Phi(\Psi_{16}^-\cap L_2^{R(a_{11}\to a_{10},\,a_{14}\to a_{13})})$

Constraint (8) ensures that the total amount of input flow entering station a_i does not exceed the maximal capacity M_i. The term $\sum_{j=1}^{2} \left(f_{j,i}^{(G)} + f_{j,i}^{(R|r_1 \to k_1)} + f_{j,i}^{(R|r_2 \to k_2)} + f_{j,i}^{(R|r_1 \to k_1, r_2 \to k_2)} \right)$ is further defined as the loading of each station, say w_i, and we have the following equation,

$$w_i = \sum_{j=1}^{2} \left(f_{j,i}^{(G)} + f_{j,i}^{(R|r_1 \to k_1)} + f_{j,i}^{(R|r_2 \to k_2)} + f_{j,i}^{(R|r_1 \to k_1, r_2 \to k_2)} \right), \quad i = 1, 2, \ldots, n_1 + n_2.$$

(9)

Let x_i denote the capacity of each station a_i. According to assumption II, the capacity x_i of each station a_i is a random variable and thus the manufacturing network is multistate, where x_i takes possible values $0 = x_{i1} < x_{i2} < \cdots < x_{ic_i} = M_i$ for $i = 1, 2, \ldots, n_1 + n_2$ with c_i denoting number of possible capacities of a_i. Under the state $X = (x_1, x_2, \ldots, x_{n_1+n_2})$, constraint (10) is necessary to guarantee that a_i can process the input raw materials/WIP,

$$x_i \geq w_i, \text{ for } i = 1, 2, \ldots, n_1 + n_2.$$

(10)

4.3   Evaluation of System Reliability

Given the demand d, the system reliability R_d is the probability that the output product from the MMN is no less than d. Thus, the system reliability is $\Pr\{X|V(X) \geq d\}$, where $V(X)$ is defined as the maximum output under X. It implies that each station should provide sufficient capacity to process the input raw materials/WIP and finally produce enough units of output products. However, it is not a wise way to find all X such that $V(X) \geq d$ and then cumulate their probabilities to derive R_d. Any minimal vector Y in the set $\{X|V(X) \geq d\}$ is claimed to be a minimal capacity vector for d. That is, Y is a minimal capacity vector for d if and only if (i) $V(Y) \geq d$ and (ii) $V(X) < d$ for any capacity vectors X such that $X < Y$. Given $Y_1, Y_2, \ldots, Y_h$, the set of minimal capacity vectors satisfying demand d, the system reliability R_d is

$$R_d = \Pr\{\bigcup_{v=1}^{h} B_v\},$$

(11)

where $B_v = \{X|X \geq Y_v\}$, $v = 1, 2, \ldots, h$. Several methods such as the recursive sum of disjoint products (RSDP) algorithm (Zuo et al. 2007; Lin and Chang 2012b, 2013; Lin et al. 2012, 2013), inclusion-exclusion method (Hudson and Kapur 1985; Xue 1985; Lin 2009a, b), disjoint-event method (Hudson and Kapur 1985; Yarlagadda and Hershey 1991), and state-space decomposition (Alexopoulos 1995; Aven 1985) may be applied to compute $\Pr\{\bigcup_{v=1}^{h} B_v\}$. Jane and Laih (2008) proved that the state-space decomposition performs a better efficiency in computation and storage space than inclusion-exclusion principle and disjoint-event method. In addition, Zuo et al. (2007) pointed out that the RSDP is more efficient than the state-space decomposition, especially for larger networks. Hence, the RSDP algorithm is beneficial to be applied for the system reliability evaluation in terms of MP.

5 Algorithm to Generate the Minimal Capacity Vectors

Given two identical production lines in parallel, $L_1 = \{a_1, a_2, \ldots, a_{n_1}\}$ and $L_2 = \{a_{n_1+1}, a_{n_1+2}, \ldots, a_{n_1+n_2}\}$. Defective WIP output from station r_1 (resp. r_2) is reworked starting from station k_1 (resp. k_2). The success rate of station a_i is denoted as p_i, the minimal capacity vectors for d are derived with the following steps.

Step 1: Find the maximum output for each path.

$$
\begin{aligned}
O_{1,\max} &= \min\{M_i|i\colon a_i \in L_1\}\Sigma_1 \text{ and} \\
O_{2,\max} &= \min\{M_i|i\colon a_i \in L_2\}\Sigma_2.
\end{aligned}
\tag{12}
$$

Step 2: Find the demand assignment (d_1, d_2) satisfying $d_1 + d_2 = d$ under constraints $d_1 \leq O_{1,\max}$ and $d_2 \leq O_{2,\max}$.

Step 3: For each demand pair (d_1, d_2), do the following steps.

3.1 Determine the amount of input materials for each production line by

$$
\begin{aligned}
I_1 &= d_1/\Sigma_1 \text{ and} \\
I_2 &= d_2/\Sigma_2.
\end{aligned}
\tag{13}
$$

3.2 Determine the input flows in terms of prior-set for each station a_i. For $a_i \in L_j$, the input flows are derived as follows,

$$
\begin{aligned}
f_{j,i}^{(G)} &= I_j\Phi(\Psi_i^-) \text{ for } i \text{ such that } a_i \in L_j^{(G)}, \\
f_{j,i}^{(R|r_1 \to k_1)} &= I_j\Gamma_{j,1}\Phi(\Psi_i^- \cap L_j^{(R|r_1 \to k_1)}) \text{ for } i \text{ such that } a_i \in L_j^{(R|r_1 \to k_1)}, \\
f_{j,i}^{(R|r_2 \to k_2)} &= I_j\Gamma_{j,2}\Phi(\Psi_i^- \cap L_j^{(R|r_2 \to k_2)}) \text{ for } i \text{ such that } a_i \in L_j^{(R|r_2 \to k_2)}, \text{ and} \\
f_{j,i}^{(R|r_1 \to k_1, r_2 \to k_2)} &= I_j\Gamma_{j,1\&2}\Phi(\Psi_i^- \cap L_j^{(R|r_1 \to k_1, r_2 \to k_2)}) \text{ for } i \text{ such that } a_i \in L_j^{(R|r_1 \to k_1, r_2 \to k_2)}.
\end{aligned}
\tag{14}
$$

3.3 Transform input flows from general production paths and reworking paths into stations' loading vector $W = (w_1, w_2, \ldots, w_{n_1+n_2})$ via

$$
w_i = \sum_{j=1}^{2}\left(f_{j,i}^{(G)} + f_{j,i}^{(R|r_1 \to k_1)} + f_{j,i}^{(R|r_2 \to k_2)} + f_{j,i}^{(R|r_1 \to k_1, r_2 \to k_2)}\right).
\tag{15}
$$

3.4 For each station, find the smallest possible capacity x_{ic} such that $x_{ic} \geq w_i > x_{i(c-1)}$. Then $Y = (y_1, y_2, \ldots, y_{n_1+n_2})$ is a minimal capacity vector for d where $y_i = x_{ic}$ for all i.

Step 4: Those Y obtained in Step 3 are the minimal capacity vectors for d.

The following theorem guarantees that all Y generated from the proposed algorithm are the minimal capacity vectors for d.

Theorem *The capacity vector Y generated from the algorithm is a minimal capacity vector for d.*

Proof Suppose that Y is not a minimal capacity vector for d, then there exists a minimal capacity vector Z for d such that $Z < Y$. Without loss of generality, we set $Z = (z_1, z_2, \ldots, z_i, \ldots, z_n)$ and there exists at least one $z_i < y_i$. The situation $z_i < y_i$ implies that $z_i < \sum_{j=1}^{2} \left(f_{j,i}^{(G)} + f_{j,i}^{(R|r_1 \to k_1)} + f_{j,i}^{(R|r_2 \to k_2)} + f_{j,i}^{(R|r_1 \to k_1, r_2 \to k_2)} \right)$ and cannot provide sufficient capacity for the input units of WIP which contradicts the assumption that Z is the minimal capacity vector for d (note that $w_i = \sum_{j=1}^{2} \left(f_{j,i}^{(G)} + f_{j,i}^{(R|r_1 \to k_1)} + f_{j,i}^{(R|r_2 \to k_2)} + f_{j,i}^{(R|r_1 \to k_1, r_2 \to k_2)} \right)$ and y_i is the minimal capacity satisfying l_i). Thus, we conclude that Y generated from the algorithm is a minimal capacity vector for d. □

6 Case-Based Examples

This section addresses two case studies, including a typical PCB and a touch panel manufacturing systems, to demonstrate the system reliability evaluation procedure. Based on the derived system reliability, a decision-making issue is addressed to determine a reliable production strategy.

6.1 Case 1: PCB Manufacturing System

Printed circuit boards (PCBs) are widely used in cell phones, laptop/desktop computers, 3C products, etc. Extensive application in numerous modern electronic products has placed high demand for PCBs. Focusing on capacity analysis, this example studies demand satisfaction and decision making for a PCB manufacturing system. The PCB manufacturing system with 10 stations ($n_1 = n_2 = 10$) in two production lines is shown in the form of AOA diagram in Fig. 4. For the single-sided board manufacturing, the input raw material is a board with a thin layer of copper foil. For different product types, the manufacturing processes and sequences may be different. Generally, the regular manufacturing process of PCB is

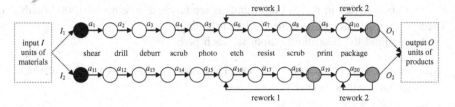

Fig. 4 A PCB manufacturing system

starting from shearing (a_1, a_{11}) in which the board is cut as a specific size. Subsequently, automated drilling machines (a_2, a_{12}) drill holes through the board for mounting electronic components on it. After drilling, the deburring machines (a_3, a_{13}) remove copper particles from the board then the scrubbing machines (a_4, a_{14}) are for cleaning the board. Following cleaning, the photo imaging machines (a_5, a_{15}) create the circuit pattern on the board. By chemical etching (a_6, a_{16}) and resist stripping (a_7, a_{17}), the copper not part of the circuit pattern is removed. Once again, scrubbing machines (a_8, a_{18}) are for cleaning the chemicals and resist on the board. After scrubbing, the legend printing stations (a_9, a_{19}) put the required logos or letters on the boards. Finally, the regular manufacturing process is finished by packaging stations (a_{10}, a_{20}). We assume that the capacities of stations are independent each other. That is, the performance of a station will not affect another because each station is a single individual in physic.

In the PCB manufacturing network, some defective WIP output from stations a_8 and a_{18} are reworked starting from stations a_6 and a_{16}, respectively. The defective products output from a_{10} and a_{20} can be reworked by the same station. To analyze the PCB manufacturing network in terms of paths, the first production line L_1 is divided into one general production path $L_1^{(G)} = \{a_1, a_2, a_3, a_4, a_5, a_6, a_7, a_8, a_9, a_{10}\}$ and three reworking paths $L_1^{(R|r_1 \rightarrow k_1)} = L_1^{(R|a_8 \rightarrow a_6)} = \{a_6, a_7, a_8, a_9, a_{10}\}$, $L_1^{(R|a_{10} \rightarrow a_{10})} = \{a_{10}\}$, and $L_1^{(R|a_8 \rightarrow a_6, a_{10} \rightarrow a_{10})} = \{a_{10}\}$. Similarly, the second production line L_2 is divided into one general production path $L_2^{(G)} = \{a_{11}, a_{12}, a_{13}, a_{14}, a_{15}, a_{16}, a_{17}, a_{18}, a_{19}, a_{20}\}$ and three reworking paths $L_2^{(R|a_{18} \rightarrow a_{16})} = \{a_{16}, a_{17}, a_{18}, a_{19}, a_{20}\}$, $L_2^{(R|a_{20} \rightarrow a_{20})} = \{a_{20}\}$, and $L_2^{(R|a_{18} \rightarrow a_{16}, a_{20} \rightarrow a_{20})} = \{a_{20}\}$.

The capacity of each station is measured in terms of number of boards can be processed per day. For instance, the specification of a deburring machine is estimated as 300,000 ft^2/month. That is, for the board size of 24″ × 24″, the capacity of the deburring machine is 75,000 boards/month (i.e., 2500 boards/day). A deburring station comprising four machines has five capacity levels, say {0, 2500, 5000, 7500, 10000}, evaluated in terms of boards/day. The lowest level 0 corresponds to complete malfunction of all machines, while 10,000 will be the highest level when all machines operate successfully. The success rate and capacity data of each station is provided in Table 3. The PCB manufacturing network has to satisfy demand

d = 8000 boards/day in which output boards are batched in terms of 1000 boards. The system reliability R_{8000} is derived as follows.

Step 1: Find the maximum output for each path.

$$O_{1,\max} = \min\{13500, 10000, 10000, 10500, 9000, 12000, 10000, 10500, 13500, 12000\} \times 0.76082$$
$$= 9000 \times 0.76082 = 6847.38 \text{ and}$$
$$O_{2,\max} = 9000 \times 0.75877 = 6828.93.$$

Step 2: Find the demand assignment (d_1, d_2) satisfying $d_1 + d_2 = 8000$ under constraints $d_1 \leq 6847.38$ and $d_2 \leq 6828.93$. Since the output boards are batched in terms of 1000 boards, the feasible demand pairs are $D_1 = (6000, 2000)$, $D_2 = (5000, 3000)$, $D_3 = (4000, 4000)$, $D_4 = (3000, 5000)$, and $D_5 = (2000, 6000)$.

Step 3: For each demand pair (d_1, d_2), do the following steps.

a. For $D_1 = (6000, 2000)$

3.1a Determine the amount of input materials for each production line by
$I_1 = 6000/0.76082 = 7886.23$ and
$I_2 = 2000/0.75877 = 2635.84$.

3.2a For demand pair $D_1 = (6000, 2000)$, the input flow of each station is shown in Table 4.

3.3a Transform input flows from both general production path and reworking path into the stations' loading vector W_1 = (7886.23, 7728.51, 7535.29, 7233.88, 7125.37, 6888.29, 6598.98, 6440.60, 6363.32, 6060.61, 2635.84, 2564.67, 2515.94, 2410.27, 2362.07, 2312.38, 2203.70, 2146.41, 2105.62, 2026.35). The calculation process is summarized in Table 4.

3.4a For demand pair D_1 = (6000, 2000), the minimal capacity vector Y_1 = (9000, 10000, 10000, 10500, 7500, 9000, 7500, 7000, 9000, 8000, 4500, 5000, 5000, 3500, 3000, 3000, 2500, 3500, 4500, 4000). The calculation process is summarized in Table 4.

b. For $D_2 = (5000, 3000)$

3.1b Determine the amount of input materials for each production line by
$I_1 = 5000/0.76082 = 6571.82$ and
$I_2 = 3000/0.75877 = 3953.75$.

$$\vdots$$

Step 4: Five minimal capacity vectors for d = 8000 are obtained from Step 3. The results are summarized in Table 5.

In this case, five minimal capacity vectors for d = 8000 are generated and thus the system reliability R_{8000} = 0.92668 is derived by the RSDP algorithm. It means

Table 3 The station data of PBC manufacturing system

Station (a_i)	Success rate (p_i)	Capacity^a	Probability	Station (a_i)	Success rate (p_i)	Capacity^a	Probability
a_1/a_{11}	0.980/0.973	0	0.001/0.002	a_6/a_{16}	0.958/0.953	0	0.002/0.002
		4500	0.002/0.012			3000	0.002/0.002
		9000	0.010/0.015			6000	0.003/0.008
		13500	0.987/0.971			9000	0.005/0.014
						12000	0.988/0.974
a_2/a_{12}	0.975/0.981	0	0.005/0.007	a_7/a_{17}	0.976/0.974	0	0.001/0.005
		5000	0.025/0.010			2500	0.002/0.005
		10000	0.970/0.983			5000	0.002/0.002
						7500	0.013/0.008
						10000	0.982/0.890
a_3/a_{13}	0.960/0.958	0	0.001/0.003	a_8/a_{18}	0.988/0.981	0	0.002/0.001
		2500	0.001/0.004			3500	0.007/0.008
		5000	0.003/0.009			7000	0.019/0.014
		7500	0.015/0.011			10500	0.972/0.977
		10000	0.980/0.973				
a_4/a_{14}	0.985/0.980	0	0.002/0.005	a_9/a_{19}	0.943/0.950	0	0.003/0.002
		3500	0.003/0.008			4500	0.004/0.005
		7000	0.011/0.008			9000	0.010/0.005
		10500	0.984/0.979			13500	0.983/0.988
a_5/a_{15}	0.956/0.962	0	0.004/0.003	a_{10}/a_{20}	0.990/0.987	0	0.001/0.001
		1500	0.006/0.003			4000	0.005/0.006
		3000	0.006/0.012			8000	0.015/0.015
		4500	0.010/0.015			12000	0.979/0.978
		6000	0.015/0.018				
		7500	0.020/0.021				
		9000	0.957/0.928				

^aThe unit of capacity is boards/day

Table 4 The results of Step 3 in Case 1

	A_1	a_2	a_3	a_4	a_5	a_6	a_7	a_8	a_9	a_{10}
$f_{1,i}^{(G)}$	7886.23	7728.51	7535.29	7233.88	7125.37	6811.86	6525.76	6369.14	6292.71	5934.03
$f_{1,i}^{(R|a_8 \to a_6)}$	0.00	0.00	0.00	0.00	0.00	76.43	73.22	71.46	70.60	66.58
$f_{1,i}^{(R|a_{10} \to a_{10})}$	0.00	0.00	0.00	0.00	0.00	0.00	0.00	0.00	0.00	59.34
$f_{1,i}^{(R|a_8 \to a_6, a_{10} \to a_{10})}$	0.00	0.00	0.00	0.00	0.00	0.00	0.00	0.00	0.00	0.67
w_i	7886.23	7728.51	7535.29	7233.88	7125.37	6888.29	6598.98	6440.60	6363.32	6060.61
y_i	9000	10000	10000	10500	7500	9000	7500	7000	9000	8000

	a_{11}	a_{12}	a_{13}	a_{14}	a_{15}	a_{16}	a_{17}	a_{18}	a_{19}	a_{20}
$f_{2,i}^{(G)}$	2635.84	2564.67	2515.94	2410.27	2362.07	2272.31	2165.51	2109.21	2069.13	1965.68
$f_{2,i}^{(R|a_{18} \to a_{16})}$	0.00	0.00	0.00	0.00	0.00	40.07	38.19	37.20	36.49	34.67
$f_{2,i}^{(R|a_{20} \to a_{20})}$	0.00	0.00	0.00	0.00	0.00	0.00	0.00	0.00	0.00	25.55
$f_{2,i}^{(R|a_{18} \to a_{16}, a_{20} \to a_{20})}$	0.00	0.00	0.00	0.00	0.00	0.00	0.00	0.00	0.00	0.45
w_i	2635.84	2564.67	2515.94	2410.27	2362.07	2312.38	2203.70	2146.41	2105.62	2026.35
y_i	4500	5000	5000	3500	3000	3000	2500	3500	4500	4000

Table 5 The minimal capacity vectors for d in Case 1

Demand pair	Minimal capacity vector
$D_1 = (6000, 2000)$	$Y_1 = (9000, 10000, 10000, 10500, 7500, 9000, 7500, 7000, 9000, 8000, 4500,$ $5000, 5000, 3500, 3000, 3000, 2500, 3500, 4500, 4000)$
$D_2 = (5000, 3000)$	$Y_2 = (9000, 10000, 7500, 7000, 6000, 6000, 7500, 7000, 9000, 8000, 4500,$ $5000, 5000, 7000, 4500, 6000, 5000, 3500, 4500, 4000)$
$D_3 = (4000, 4000)$	$Y_3 = (9000, 10000, 7500, 7000, 6000, 6000, 5000, 7000, 4500, 8000, 9000,$ $10000, 7500, 7000, 6000, 6000, 5000, 7000, 4500, 8000)$
$D_4 = (3000, 5000)$	$Y_4 = (4500, 5000, 5000, 7000, 4500, 6000, 5000, 3500, 4500, 4000, 9000,$ $10000, 7500, 7000, 6000, 6000, 7500, 7000, 9000, 8000)$
$D_5 = (2000, 6000)$	$Y_5 = (4500, 5000, 5000, 3500, 3000, 3000, 2500, 3500, 4500, 4000, 9000,$ $10000, 10000, 10500, 7500, 9000, 7500, 7000, 9000, 8000)$

that the PCB manufacturing network can produce 8000 boards per day with a probability 0.92668.

The utility of demand satisfaction under each demand assignment is addressed in this subsection. The possibility of demand satisfaction under each demand pair is defined as satisfaction probability herein. According to the satisfaction probability, the production manager can decide a better strategy (demand assignment) to produce products. The proposed algorithm can be easily executed only for a specified demand pair. That is, the production manager could evaluate the satisfaction probability for each demand pair in terms of its corresponding minimal capacity vector. This probability is denoted as R_{d_1,d_2} for each demand pair where $d_1 + d_2 = d$. In the PCB manufacturing network, five minimal capacity vectors are obtained for $d = 8000$. For demand pair $D_1 = (6000, 2000)$, the corresponding minimal capacity vector is $Y_1 = (9000, 10000, 10000, 10500, 7500, 9000, 7500, 7000, 9000, 8000, 4500, 5000, 5000, 3500, 3000, 3000, 2500, 3500, 4500, 4000)$ and the satisfaction probability $R_{6000,2000} = \Pr\{X|X \geq Y_1\} = 0.83204$. Table 6 provides R_{d_1,d_2} for different demand pairs and ranks the satisfaction probability of each demand assignment. The results indicate that $D_2 = (5000, 3000)$ with $R_{5000,3000} = 0.84679$ would be a reliable strategy to produce products since it has higher satisfaction probability. Thus, the production manger may decide to produce 5000 boards/day by L_1 and to produce 3000 boards/day by L_2.

Table 6 Demand pair and the corresponding R_{d_1,d_2} in Case 1

Demand pair	Probability	Rank
$D_1 = (6000, 2000)$	$R_{6000,2000} = 0.83204$	2
$D_2 = (5000, 3000)$	$R_{5000,3000} = 0.84679$	1
$D_3 = (4000, 4000)$	$R_{4000,4000} = 0.80168$	4
$D_4 = (3000, 5000)$	$R_{3000,5000} = 0.82386$	3
$D_5 = (2000, 6000)$	$R_{2000,6000} = 0.79701$	5

6.2 Case 2: Touch Panel Manufacturing System

With the rapid development of wireless communication applications, the information requirements from people are increasing greatly. Portable consumer electronic products, such as cell phone, GPS, digital camera, tablet PC, and notebook are rising in demand. In particular, touch panel is an intuitive input interface which is also a critical technology for the mentioned devices. Due to the trend of human-machine interface being friendlier, the application of touch panel is going to be more widely. Hence, evaluating system reliability of touch panel manufacturing system can provide managers with an understanding of the system capability and can indicate possible improvements.

The studied company, named C Company in this example, is a professional supplier of touch panel in Taiwan. Following is the touch panel production process of the company. The compositions of a touch panel are Indium Tin Oxide (ITO), ITO Glass, and ITO Film. These three parts are separated by spacer. A touch panel is driven by a plurality of first pixels and second pixels. Each first pixel has a light sensing component and generating a first sensing signal and a switch. The switch is composed of a first end coupled to the light sensing component, a control end for receiving a first gate driving signal, and a second end for transmitting the first sensing signal according to the first gate driving signal. Figure 5 shows a module process control plan of touch panel and steps are explained as follows.

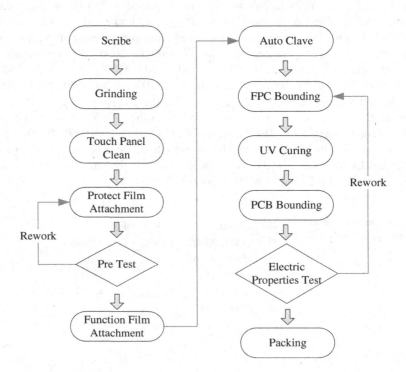

Fig. 5 The touch panel production process

Scribe Using the wheel type to cut the materials according to the specifications.

Grinding In order to reduce occurrence of edge crack, using the automatic grinding machine to grind the edge and corner.

Clean Avoiding glass cullet and particle scratching the sensor in the subsequent process.

Protect film attachment Affixing the protect film for scratching the sensor.

Pre-test This step is the most important. The aim of pre-test is to avoid wasting module material, for example, flexible printed circuit (FPC) and printed circuit board (PCB).

Function film attachment Function film is put into upper stage and sensor is put into lower stage. Then combine these two stages by functional film machine.

FPC bounding and PCB bounding This two steps are the keys to the module process of touch panel. Use ACF (Anisotropic Conductive Film) to conduct the conductive particles in the FPC or PCB bounding process.

Electric properties test Test the sensitivity after bounding process and confirm the function of the touch panel.

Because the pretest can avoid wasting materials in the follow-up process; while electric properties test can confirm the function of the touch panel. Hence, defective WIP output from these two stations are reworked. According to the process control plan diagram, WIPs are checked-out by these two tests, there are two situations— one is go to next station and the other is rework. Hence, these two stations would affect the production capability for the touch panel manufacturing system. The touch panel manufacturing system with two production lines and each production line with 12 stations is represented as Fig. 6.

In this case study, the supplier has set up two production lines to meet orders. The capacity and probability distribution of each station are provided in Table 7. The capacity of each station can be measured by batch size. For instance, the maximal capacity of the Scribe station a_1 is 400 pcs/day and the batch size is 100, then the capacity of the scribe station may be 0, 100, 200, 300, and 400. In the touch panel manufacturing system, defective WIP output from station $r_1 = 5$ (resp. $r_2 = 11$) is reworked starting from station $k_1 = 4$ (resp. $k_2 = 8$). The touch panel manufacturing system has to satisfy demand $d = 400$ (pcs/day) in which output touch panels are batched by 100 pcs.

The minimal capacity vectors for d are derived by the following steps.

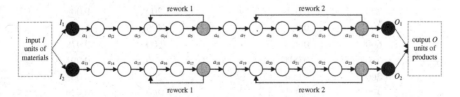

Fig. 6 A touch panel manufacturing system

Table 7 The capacity distribution of touch panel manufacturing system

Station	Capacity	Probability	Station	Capacity	Probability
a_1	0	0.001	a_{13}	0	0.002
	100	0.002		100	0.002
	200	0.015		200	0.002
	300	0.020		300	0.003
	400	0.962		400	0.991
a_2	0	0.050	a_{14}	0	0.005
	250	0.050		250	0.020
	500	0.900		500	0.975
a_3	0	0.001	a_{15}	0	0.001
	100	0.001		100	0.001
	200	0.001		200	0.001
	300	0.002		300	0.003
	400	0.002		400	0.003
	500	0.003		500	0.003
	600	0.003		600	0.003
	700	0.987		700	0.985
a_4	0	0.002	a_{16}	0	0.002
	300	0.010		300	0.003
	600	0.988		600	0.995
a_5	0	0.001	a_{17}	0	0.008
	200	0.001		200	0.009
	400	0.001		400	0.012
	600	0.997		600	0.971
a_6	0	0.030	a_{18}	0	0.010
	250	0.060		250	0.010
	500	0.910		500	0.980
a_7	0	0.002	a_{19}	0	0.001
	100	0.002		100	0.001
	200	0.010		200	0.003
	300	0.010		300	0.004
	400	0.040		400	0.010
	500	0.936		500	0.981
a_8	0	0.010	a_{20}	0	0.005
	200	0.015		200	0.008
	400	0.017		400	0.008
	600	0.958		600	0.979
a_9	0	0.004	a_{21}	0	0.001
	200	0.004		200	0.001
	400	0.992		400	0.998
a_{10}	0	0.001	a_{22}	0	0.001

(continued)

Table 7 (continued)

Station	Capacity	Probability	Station	Capacity	Probability
	250	0.003		250	0.001
	500	0.996		500	0.998
a_{11}	0	0.003	a_{23}	0	0.008
	200	0.005		200	0.014
	400	0.992		400	0.978
a_{12}	0	0.010	a_{24}	0	0.015
	100	0.010		100	0.018
	200	0.011		200	0.020
	300	0.013		300	0.028
	400	0.956		400	0.919

Step 1: Find the maximum output for each path.
$O_{1,max} = \min\{400, 500, 700, \ldots, 400, 400\} \times 0.76338$
$\qquad = 400 \times 0.76338 = 305.35$ and
$O_{2,max} = 400 \times 0.76338 = 305.35$.
Step 2: Find the demand assignment (d_1, d_2) satisfying $d_1 + d_2 = 400$ under constraints $d_1 \leq 305.35$ and $d_2 \leq 305.35$. Since the output touch panels are batched by 100 pcs, the feasible demand pairs are $D_1 = (300, 100)$, $D_2 = (200, 200)$, and $D_3 = (100, 300)$.
Step 3: For each demand pair (d_1, d_2), find the corresponding minimal capacity vector.

a. For $D_1 = (300, 100)$

3.1a The amount of input materials for each production line is
$\qquad I_1 = 300/0.76338 = 392.99$ and
$\qquad I_2 = 100/0.76338 = 130.99$.
3.2a For demand pair $D_1 = (300, 100)$, the input flow of each station is shown in Table 8.
3.3a The stations' loading vector transformed from input flows is
$\qquad W_1 = (392.9900, 385.1302, 377.8127, 378.4196, 372.7433, 356.3426, 341.3763, 353.1588, 348.9209, 329.0324, 325.7421, 304.5689, 130.9970, 128.3771, 125.9379, 126.1402, 124.2481, 118.7812, 113.7924, 117.7199, 116.3073, 109.6778, 108.5810, 101.5232)$. The calculation process is summarized in Table 8.
3.4a The minimal capacity vector is thus $Y_1 = (400, 500, 400, 600, 400, 500, 400, 400, 400, 500, 400, 400, 200, 250, 200, 300, 200, 250, 200, 200, 200, 250, 200, 200)$

Table 8 The results of Step 3 in Case 1

	A_1	a_2	a_3	a_4	a_5	a_6	a_7	a_8	a_9	a_{10}	a_{11}	a_{12}
$f_{1,i}^{(G)}$	130.9970	128.3771	125.9379	120.9004	119.0869	113.8471	109.0655	106.4479	105.1705	99.1758	98.1841	91.8021
$f_{1,i}^{(R|a_5 \to a_4)}$	0.0000	0.0000	0.0000	5.2398	5.1612	4.9341	4.7269	4.6135	4.5581	4.2983	4.2553	3.9787
$f_{1,i}^{(R|a_1 \to a_8)}$	0.0000	0.0000	0.0000	0.0000	0.0000	0.0000	0.0000	6.3820	6.3054	5.9460	5.8865	5.5039
$f_{1,i}^{(R|a_5 \to a_4, a_{11} \to a_8)}$	0.0000	0.0000	0.0000	0.0000	0.0000	0.0000	0.0000	0.2766	0.2733	0.2577	0.2551	0.2385
w_i	130.9970	128.3771	125.9379	126.1402	124.2481	118.7812	113.7924	117.7199	116.3073	109.6778	108.5810	101.5232
y_i	200.0000	250.0000	200.0000	300.0000	200.0000	250.0000	200.0000	200.0000	200.0000	250.0000	200.0000	200.0000

	a_{13}	a_{14}	a_{15}	a_{16}	a_{17}	a_{18}	a_{19}	a_{20}	a_{21}	a_{22}	a_{23}	a_{24}
$f_{2,i}^{(G)}$	392.9900	385.1302	377.8127	362.7002	357.2597	341.5403	327.1956	319.3429	315.5108	297.5267	294.5514	275.4056
$f_{2,i}^{(R|a_{17} \to a_{16})}$	0.0000	0.0000	0.0000	15.7194	15.4836	14.8024	14.1807	13.8403	13.6742	12.8948	12.7659	11.9361
$f_{2,i}^{(R|a_{23} \to a_{20})}$	0.0000	0.0000	0.0000	0.0000	0.0000	0.0000	0.0000	19.1458	18.9161	17.8379	17.6595	16.5116
$f_{2,i}^{(R|a_{17} \to a_{16}, a_{23} \to a_{20})}$	0.0000	0.0000	0.0000	0.0000	0.0000	0.0000	0.0000	0.8298	0.8198	0.7731	0.7654	0.7156
w_i	392.9900	385.1302	377.8127	378.4196	372.7433	356.3426	341.3763	353.1588	348.9209	329.0324	325.7421	304.5689
y_i	400.0000	500.0000	400.0000	600.0000	400.0000	500.0000	400.0000	400.0000	400.0000	500.0000	400.0000	400.0000

Table 9 Minimal capacity vectors for d in Case 2

Demand pair	Minimal capacity vector
$D_1 = (300, 100)$	$Y_1 = (400, 500, 400, 600, 400, 500, 400, 400, 400, 500, 400, 400, 200, 250,$ $200, 300, 200, 250, 200, 200, 200, 250, 200, 200)$
$D_2 = (200, 200)$	$Y_2 = (300, 500, 300, 300, 400, 250, 300, 400, 400, 250, 400, 300, 300, 500,$ $300, 300, 400, 250, 300, 400, 400, 250, 400, 300)$
$D_3 = (100, 300)$	$Y_3 = (200, 250, 200, 300, 200, 250, 200, 200, 200, 250, 200, 200, 400, 500,$ $400, 600, 400, 500, 400, 400, 400, 500, 400, 400)$

Table 10 Demand pair and the corresponding R_{d_1,d_2} in Case 2

Demand pair	Probability	Rank
$D_1 = (300, 100)$	$R_{300,100} = 0.63515$	3
$D_2 = (200, 200)$	$R_{200,200} = 0.66345$	2
$D_3 = (100, 300)$	$R_{100,300} = 0.70637$	1

b. For $D_2 = (200, 200)$

  3.1b Determine the amount of input materials for each production line by
  $I_1 = 200/0.76338 = 261.99$ and
  $I_2 = 200/0.76338 = 261.99$.

  Step 4: Five minimal capacity vectors for $d = 8000$ are obtained from Step 3. The results are summarized in Table 9.

  In this case study, we obtain three minimal capacity vectors for $d = 400$ and the system reliability $R_{400} = 0.79001$. That is, we have the possibility of 0.79001 to produce 400 pcs/day in the touch panel manufacturing system. We further focus on the satisfaction probability for each specific demand pair. The production manager could evaluate the satisfaction probability for each demand pair in terms of its corresponding minimal capacity vector. According to Table 9, we may find that the larger assigned demand in L_2, the higher satisfaction probability. This phenomenon is caused by the capacity probability distribution of each station. Overall speaking, the demand pair $D_3 = (100, 300)$ with satisfaction probability 0.70637 is the best strategy to produce 400 products. The production manager may easily make a decision to produce products according to the Table 10.

7 Conclusion

This chapter addresses the system reliability evaluation to measure the performance of a manufacturing system with multiple production lines in parallel. We model the manufacturing system as a multistate manufacturing network (MMN) by the revised graphical transformation and decomposition. Based on the transformed MMN and decomposed paths, the probability that the MMN can provide sufficient capacities to satisfy a given demand d is evaluated. Such a probability is referred to as the system reliability. First, the prior-set technique is developed to evaluate the input flow of each station. Thus, the input flow of each station can be represented in terms of prior-set. Based on the concept of prior-set, an algorithm is utilized to generate the minimal capacity vectors that stations should provide to satisfy demand d. In terms of such vectors, the system reliability is derived by applying the RSDP algorithm. In addition, a decision-making issue based on the minimal capacity vectors derived is further addressed. The production manager may determine a reliable strategy to assign the amount of output that each production line should produce to fulfill demand.

According to the system reliability, the production manager could conduct a sensitivity analysis to investigate the most important station in the manufacturing system to improve the reliability. When application, the sensitivity analysis could be conducted by increasing the capacity (machine) of a station at a time (the other stations are retained as the same conditions). Hence, the production manager can find the most sensitive station that increases the system reliability most, and such a station is the most important part of the MMN. Similarly, improving the capacity probability distribution of a station is also another measurement to determine the importance of an MMN.

References

Alexopoulos, C. (1995). A note on state-space decomposition methods for analyzing stochastic flow networks. *IEEE Transactions on Reliability, 44*, 354–357.

Aven, T. (1985). Reliability evaluation of multistate systems with multistate components. *IEEE Transactions on Reliability, 34*, 473–479.

Buscher, U., & Lindner, G. (2007). Optimizing a production system with rework and equal sized batch shipment. *Computers & Operations Research, 34*, 515–535.

Chen, M. S., & Lan, C. H. (2001). The maximal profit flow model in designing multiple-production-line system with obtainable resource capacity. *International Journal of Production Economics, 70*, 175–184.

Ford, L. R., & Fulkerson, D. R. (1962). *Flows in network*. New Jersey: Princeton University Press.

Hudson, J. C., & Kapur, K. C. (1985). Reliability bounds for multistate systems with multistate components. *Operations Research, 33*, 153–160.

Jane, C. C., & Laih, Y. W. (2008). A practical algorithm for computing multi-state two-terminal reliability. *IEEE Transactions on Reliability, 57*, 295–302.

Kemmoe, S., Pernot, P. A., & Tchernev, N. (2013). Model for flexibility evaluation in manufacturing network strategic planning. *International Journal of Production Research,*. doi:10.1080/00207543.2013.845703.

Lan, C. H. (2007). The design of multiple production lines under deadline constraint. *International Journal of Production Economics, 106*, 191–203.

Lee, H., & Garcia-Diaz, A. (1993). A network flow approach to solve clustering problems in group technology. *International Journal of Production Research, 31*, 603–612.

Lee, H., & Garcia-Diaz, A. (1996). Network flow procedures for the analysis of cellular manufacturing systems. *IIE Transactions, 28*, 333–345.

Li, J., & Li, J. (2013). Integration of manufacturing system design and quality management. *IIE Transactions, 45*, 555–556.

Lin, Y. K. (2009a). Two-commodity reliability evaluation of a stochastic-flow network with varying capacity weight in terms of minimal paths. *Computers & Operations Research, 36,* 1050–1063.

Lin, Y. K. (2009b). Optimal routing policy of a stochastic flow network. *Computers & Industrial Engineering, 56,* 1414–1418.

Lin, Y. K., & Chang, P. C. (2012a). Evaluate the system reliability for a manufacturing network with reworking actions. *Reliability Engineering & System Safety, 106,* 127–137.

Lin, Y. K., & Chang, P. C. (2012b). System reliability of a manufacturing network with reworking action and different failure rates. *International Journal of Production Research, 50,* 6930–6944.

Lin, Y. K., & Chang, P. C. (2012c). Reliability evaluation for a manufacturing network with multiple production lines. *Computers & Industrial Engineering, 63,* 1209–1219.

Lin, Y. K., & Chang, P. C. (2013). A novel reliability evaluation technique for stochastic flow manufacturing networks with multiple production lines. *IEEE Transactions on Reliability, 61,* 92–104.

Lin, Y. K., Chang, P. C., & Chen, J. C. (2012). Reliability evaluation for a waste-reduction parallel-line manufacturing system. *Journal of Cleaner Production, 35,* 93–101.

Lin, Y. K., Chang, P. C., & Chen, J. C. (2013). Performance evaluation for a footwear manufacturing system with multiple production lines and different station failure rates. *International Journal of Production Research, 51,* 1603–1617.

Listeş, O. (2007). A generic stochastic model for supply-and-return network design. *Computers & Operations Research, 34,* 417–442.

Liu, N., Kim, Y., & Hwang, H. (2009). An optimal operating policy for the production system with rework. *Computers & Industrial Engineering, 56,* 874–887.

Mourtzis, D., Doukas, M., & Psarommatis, F. (2013). Design and operation of manufacturing networks for mass customization. *CIRP Annals—Manufacturing Technology, 62,* 467–470.

Paquet, M., Martel, A., & Montreuil, B. (2008). A manufacturing network design model based on processor and worker capabilities. *International Journal of Production Research, 46,* 2009–2030.

Pillai, V. M., & Chandrasekharan, M. P. (2008). An absorbing Markov chain model for production systems with rework and scrapping. *Computers & Industrial Engineering, 55,* 695–706.

Teunter, R., Kaparis, K., & Tang, O. (2008). Multi-product economic lot scheduling problem with separate production lines for manufacturing and remanufacturing. *European Journal of Operational Research, 191,* 1241–1253.

Xue, J. (1985). On Multistate System Analysis. *IEEE Transactions on Reliability, 34,* 329–337.

Yarlagadda, R., & Hershey, J. (1991). Fast algorithm for computing the reliability of communication network. *International Journal of Electronics, 70,* 549–564.

Zuo, M. J., Tian, Z., & Huang, H. Z. (2007). An efficient method for reliability evaluation of multistate networks given all minimal path vectors. *IIE Transactions, 39,* 811–817.

Systemability: A New Reliability Function for Different Environments

Alessandro Persona, Fabio Sgarbossa and Hoang Pham

1 Introduction: How Environmental Aspects Influence Lifetime of Systems

Industrial applications often observe the difference between laboratory reliability test in standard conditions and component or system reliability when it is set in motion through different environments and real-world conditions. As a matter of fact reliability variable is considerably influenced by environmental factors. Environmental factors may change failure rate, reliability, and availability of systems. When a component or a system works in an operative plant, it reflects a reliability function that is usually different from the theory reliability but also from all its similar applications in other industrial plants. This concept also concerns a service or a logistic system and all the other systems in which the reliability calculation is extremely necessary: it is confirmed in several remote maintenance applications (Persona et al. 2007). Usually, parameters of survival function of a component or system are calculated during the testing phases, but environmental factors (i.e., operating temperature, vibrations, possible shocks, moisture, etc.) can change the hazard rate of the components/systems and consequently their reliability functions during their production time. For this reason, it is difficult to estimate the real-lifetime distributions of the system products. The operating environment is often unknown and it is different from the laboratory or standard environment. Therefore, in reliability engineering, conversion problems and synthesis of test results are often faced from different environments.

A. Persona · F. Sgarbossa (✉)
Department of Management and Engineering, University of Padova,
Stradella San Nicola, 3, Vicenza 36100, Italy
e-mail: fabio.sgarbossa@unipd.it

H. Pham
Department of Industrial and Systems Engineering, Rutgers, The State University of New
Jersey, 96 Frelinghuysen Road, Piscataway, NJ 08855-8018, USA

© Springer-Verlag London 2016
H. Pham (ed.), *Quality and Reliability Management and Its Applications*,
Springer Series in Reliability Engineering, DOI 10.1007/978-1-4471-6778-5_6

Incorrect estimation of reliability function could lead to the wrong functional design of the system and an incorrect definition of the appropriate maintenance policies to improve the efficiency of industrial systems.

The aim of this chapter is to overlap this problem by operational study of the new parameter introduced by Pham (2005a, b, c) called systemability. This approach is very suitable for theoretical modelling of manufacturing systems and its application is convenient and simple particularly in presence of components with Weibull distribution lifetime (Battini et al. 2007, 2008).

In the next section, some models are just briefly mentioned in order to define the state of the art and mathematical models available: a literature analysis is summarized in order to show this innovative concept. Then, we introduce the mathematical function called systemability. Moreover, theoretical application of the systemability to several system configurations is illustrated later. Moreover, it is reported the systemability approach applied to a real case study regarding motorcycle components, demonstrating the goodness-of-fit of this approach. At the end of the chapter, some careful consideration about the capability of the application and how to pursue with the research in that field are shown.

2 Scientific Contributions on Environmental Effects and Reliability Estimation in Random Environments

Reliability is well known as the probability that a component (system) meets a determined mission, for a determined time and in determined environment (Pham 2005d). The operating environment is often unknown and it is different from the laboratory or standard environments. Most used reliability approaches suppose the lifetime distributions to be depending on time only, so test reliability functions are used to describe also the operating lifetimes. The environmental factors may change failure rate, reliability, and availability of components, so traditional approaches could cause the incorrect estimation of reliability. For this reason some researches create several models to estimate the reliability in operating environments, using the data collected during the test phases. Cox (1972) first studied the relation between the environmental conditions and the hazard rate, introducing the proportional and additive hazard rate. These models (Badia et al. 2002; Carroll 2003; Finkelstein 2003, 2006) have been widely used in several experiments where the time to failure depends on a group of covariates. These covariates are usually used to define qualitative and quantitative variables, representing different operating conditions, different environments, different treatments, and so on. Badia et al. (2002) applied the proportional and additive hazard rate for modeling the change of lifetime distribution of components. Environmental Factors (EF) have been defined to assess covariates. In fact, an environmental factor converts reliability data in one environmental condition into equivalent information in other ones, so EF are defined as the quotient of the mean lives X1 and X2 in two different environments (Elsayed and Wang 1996; Wang et al. 1992a, b, c, d). In scientific literature, the definition of

EF for different distributions has been developed and accepted. Wang et al. (1992a, b, c, d) have defined EF for gamma, normal, log-normal, inverse Gaussian distribution and Elsayed and Wang (1996) have studied environmental factors for the binomial distribution.

In the software reliability field, many studies have developed several models to estimate the reliability of the software in different working conditions (Pham 2003; Tamura et al. 2006; Teng and Pham 2004, 2006; Zhang et al. 2001; Zhang and Pham 2006; Zhao et al. 2006). In this field, non-homogeneous Poisson process (NHPP) has been successfully applied to model the software failure, and it is widely used to determine when stop testing and release the software. Generally, these models, however, assume that the field environments are the same as a testing environment. So, models with environmental factor consideration are developed: Zhang and Pham (2006) propose a general NHPP model, based on proportional hazard rate, considering constant η the environmental factor. Pham (2003) introduce a new model, called random field environment model (RFE), where they describe the η environmental factor by gamma and beta distribution.

In mechanical engineering, several contributions have been given and widely accepted. Oh and Bai (2001) have proposed some models to study the lifetime distributions based on the test data, the warranty data and additional field data after the warranty expires. They have illustrated the methods to estimate the maximum likelihood and so defined specific formulas for Weibull distribution. Attardi et al. (2005) have studied the survival characteristics of a component installed in two different cars with different working conditions. They have introduced a mixed Weibull distribution which depends on the covariates through the Weibull scale parameters.

Abbassi et al. (2006) have introduced an approach based to simulated annealing algorithm to estimate the parameters of Weibull distribution. Sohn et al. (2007) proposed a random effects Weibull regression model for forecasting the occupational lifetime of the employees who join another company, based on their characteristics. Advantage of using such a random effects model is the ability of accommodating not only the individual. Ram and Tiwari (1989) have estimated the reliability of a component through a Monte-Carlo simulation, with the introduction of the factor η that is greater than 1 if the operating condition is more stressful than the testing ones, otherwise, it is less than 1.

Sun et al. have introduced a new model enables maintenance personnel to predict the reliability of pipelines with different preventive maintenance (PM) strategies, and hence effectively assists them in making optimal PM decisions.

Pham (2005a, b, c) recently introduced an innovative approach, called systemability. It is very innovative and interesting because it is quite different from the literature studies described before; in fact, it calculates the reliability in random environment using, as starting data, the reliability obtained during the test, and processing it using a gamma distribution in particular, or a distribution that represents operating environments in general, which takes into consideration the EF. This is a fundamental condition for applications in real contexts.

3   A New Approach: Systemability Function

The traditional reliability definitions and its calculations have commonly been carried out through the failure rate function within a controlled laboratory-test environment. In other words, such a reliability function is applied to the failure testing data and, with the help of parameter estimation approaches, it then can be used to make predictions on the reliability of the system used in the field. The underlying assumption for such calculation is that the field (or operating) environments and the testing environments are the same.

By definition, a mathematical reliability function is the probability that a system will be successful in the interval from time 0 to time t, given by:

$$R(t) = \int_t^\infty f(s)\mathrm{d}s = e^{-\int_o^t h(s)\mathrm{d}s} \tag{1}$$

where $f(s)$ and $h(s)$ are the failure time density and failure rate function, respectively.

The operating environments are often unknown and yet different due to the uncertainties of environments in the field. A new look at how reliability researchers can take account of the randomness of the field environments into mathematical reliability modeling covering system failure in the field is great interest.

Pham (2005a, b, c) recently developed a new mathematical function, called *systemability*, considering the uncertainty of the operating environments in the function for predicting the reliability of systems.

Notation

$h(t_i)$ ith component hazard rate function
$R(t_i)$ ith component reliability function
λ_i Intensity parameter of Weibull distribution for ith component
$\underline{\lambda}$ $\underline{\lambda} = (\lambda_1, \lambda_2, \lambda_3 \ldots, \lambda_n)$
γ_i Shape parameter of Weibull distribution for ith component
$\underline{\gamma}$ $\underline{\gamma} = (\gamma_1, \gamma_2, \gamma_3 \ldots, \gamma_n)$
η A common environment factor
$G(\eta)$ Cumulative distribution function of η
α Shape parameter of Gamma distribution
β Scale parameter of Gamma distribution

Definition (Pham 2005a) *Systemability* is defined as the probability that the system will perform its intended function for a specified mission time under the random operating environments.

In a mathematical form, the *systemabililty* function is given by:

$$R_s(t) = \int_\eta e^{-\eta \int_o^t h(s)ds} dG(\eta)$$ (2)

where η is a random variable that represents the system operating environments with a distribution function G.

This new function captures the uncertainty of complex operating environments of systems in terms of the system failure rate. It also would reflect the reliability estimation of the system in the field.

If it assumes that η has a gamma distribution with parameters α and β, i.e., $\eta \sim \text{gamma}(\alpha, \beta)$ where the pdf of η is given by:

$$f_\eta(x) = \frac{\beta^\alpha x^{\alpha-1} e^{-\beta x}}{\Gamma(\alpha)} \quad \text{for} \quad \alpha, \beta > 0; \quad x \geq 0$$ (3)

then the systemability function of the system in Eq. 2 using the Laplace transform is given by:

$$R_s(t) = \left[\frac{\beta}{\beta + \int_0^t h(s)ds} \right]^\alpha$$ (4)

4 Using Systemability in Different System Configurations

This section presents several systemability results and its variances of some system configurations, such as series, parallel, and k-out-of-n systems. Considering the following assumptions:

1. A system consists of n independent components where the system is subject to a random operational environment η.
2. ith component lifetime is assumed to follow the Weibull density function, i.e.,

   - Component hazard rate

$$h_i(t) = \lambda_i \gamma_i t^{\gamma_i - 1}$$ (5)

- Component reliability

$$R_i(t) = e^{-\lambda_i t^{\gamma_i}} \text{ for } t > 0 \tag{6}$$

Given common environment factor $\eta \sim \text{gamma}(\alpha, \beta)$, the series systemability function can be calculated as follows.

4.1 Systemability Calculations: Series System Configuration

Now a specific systemability calculation for a series system configuration is presented. In a series system, all components must operate successfully if the system is to function. The conditional reliability function of series systems subject to an actual operating environment η is given by:

$$R_{\text{Series}} = \left(t|\eta, \underline{\lambda}, \underline{\gamma}\right) = \exp\left(-\eta \sum_{i=1}^{n} \lambda_i t^{\gamma_i}\right) \tag{7}$$

Therefore, from Eq. 2, the series systemability is given as follows:

$$R_{\text{Series}} = \left(t|\eta, \underline{\lambda}, \underline{\gamma}\right) = \int_{\eta} \exp\left(-\eta \sum_{i=1}^{n} \lambda_i t^{\gamma_i}\right) dG(\eta) = \left[\frac{\beta}{\beta + \sum_{i=1}^{n} \lambda_i t^{\gamma_i}}\right]^{\alpha} \tag{8}$$

The variance of a function $R(t)$ is given by:

$$\text{Var}[R(t)] = E[R^2(t)] - (E[R(t)])^2 \tag{9}$$

Given $\eta \sim \text{gamma}(\alpha, \beta)$, the variance of systemability for any system structure can be easily obtained. Therefore, the variance of series systemability can be obtained:

$$\text{Var}\left[R_{\text{Series}} = \left(t|\eta, \underline{\lambda}, \underline{\gamma}\right)\right] = \left[\frac{\beta}{\beta + 2\sum_{i=1}^{n} \lambda_i t^{\gamma_i}}\right]^{\alpha} - \left[\frac{\beta}{\beta + \sum_{i=1}^{n} \lambda_i t^{\gamma_i}}\right]^{2\alpha} \tag{10}$$

Figures 1 and 2 show the reliability and systemability functions of a series system (here $k = 5$) for $\alpha = 2$, $\beta = 3$, and for $\alpha = 2$, $\beta = 1$, respectively.

Fig. 1 Comparisons of series system reliability versus systemability functions for $\alpha = 2$ and $\beta = 3$ (Pham 2005)

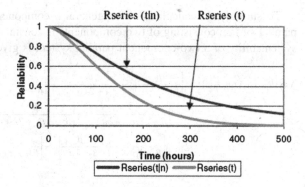

Fig. 2 Comparisons of series system reliability versus systemability functions for $\alpha = 2$ and $\beta = 1$ (Pham 2005)

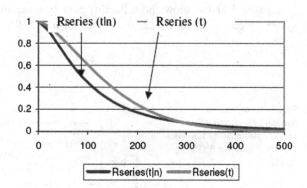

4.2 Systemability Calculations: Parallel System Configuration

Similarly, the systemability of parallel systems (conditional reliability function subject to a randomly operating environment) is given by:

$$R_{\text{Parallel}} = \left(t|\eta, \underline{\lambda}, \underline{\gamma}\right) = \exp(-\eta\lambda_i t^{\gamma_i}) - \sum_{i_1, i_2 = 1; i_1 \neq i_2}^{n} \exp(-\eta(\lambda_{i1} t^{\gamma_{i1}} + \lambda_{i2} t^{\gamma_{i2}}))$$

$$+ \sum_{i_1, i_2, i_3 = 1; i_1 \neq i_2 \neq i_3}^{n} \exp(-\eta(\lambda_{i1} t^{\gamma_{i1}} + \lambda_{i2} t^{\gamma_{i2}} + \lambda_{i3} t^{\gamma_{i3}})) - \cdots + (-1)^{n-1} \exp\left(-\eta \sum_{i=1}^{n} + \lambda_i t^{\gamma_i}\right)$$

(11)

Hence, the parallel systemability is given by:

$$R_{\text{Parallel}} = \left(t|\eta, \underline{\lambda}, \underline{\gamma}\right) = \sum_{k=1}^{n} (-1)^{k-1} \sum_{i_1, i_2, \ldots i_k = 1; i_1 \neq i_2 \ldots \neq i_k}^{n} \left[\frac{\beta}{\beta + \sum_{j=i_1, \ldots i_k} \lambda_j t^{\gamma_j}}\right]^{\alpha}$$ (12)

To simplify the calculation of a general n-component parallel system, here a parallel system consisting of two components is considered. The variance of series systemability of a two-component parallel system is given by:

$$
\begin{aligned}
\mathrm{Var}\left[R_{\text{parallelo}}\left(t|\eta,\underline{\lambda},\underline{\gamma}\right)\right] = & \left[\frac{\beta}{\beta+2\lambda_1 t^{\gamma_1}}\right]^{\alpha} + \left[\frac{\beta}{\beta+2\lambda_2 t^{\gamma_2}}\right]^{\alpha} + \left[\frac{\beta}{\beta+2\lambda_1 t^{\gamma_1}+2\lambda_2 t^{\gamma_2}}\right]^{\alpha} + \left[\frac{\beta}{\beta+\lambda_1 t^{\gamma_1}+\lambda_2 t^{\gamma_2}}\right]^{\alpha} \\
& - \left[\frac{\beta}{\beta+2\lambda_1 t^{\gamma_1}+\lambda_2 t^{\gamma_2}}\right]^{\alpha} - \left[\frac{\beta}{\beta+\lambda_1 t^{\gamma_1}+2\lambda_2 t^{\gamma_2}}\right]^{\alpha} \\
& - \left[\left[\frac{\beta}{\beta+\lambda_1 t^{\gamma_1}}\right]^{\alpha} + \left[\frac{\beta}{\beta+\lambda_2 t^{\gamma_2}}\right]^{\alpha} - \left[\frac{\beta}{\beta+\lambda_1 t^{\gamma_1}+\lambda_2 t^{\gamma_2}}\right]^{\alpha}\right]^{2}
\end{aligned}
$$

(13)

Figures 3 and 4 show the reliability and systemability functions of a parallel system (here $k = 1$) for $\alpha = 2$, $\beta = 3$ and for $\alpha = 2$, $\beta = 1$, respectively.

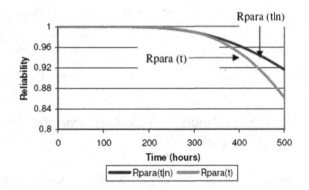

Fig. 3 Comparisons of parallel system reliability versus systemability functions for $\alpha = 2$ and $\beta = 3$ (Pham 2005)

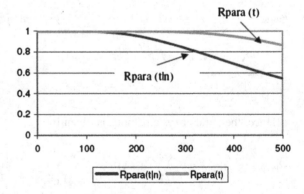

Fig. 4 Comparisons of parallel system reliability versus systemability functions for $\alpha = 2$ and $\beta = 1$ (Pham 2005)

4.3   Systemability Calculations: k-Out-of-n System Configuration

For k-out-of-n system configuration, in order to simplify the complexity of the systemability function (Pham 2005a), it assumes that all the components in the k-out-of-n systems are identical. Therefore, the conditional reliability function of a component subject to a randomly operating environment can be written as:

$$R(t|\eta, \lambda, \gamma) = e^{-\eta\lambda t^{\gamma}} \tag{14}$$

The systemability of k-out-of-n systems is given by

$$R_{k-\text{out-of}-n}(t|\mu, \lambda, \gamma) = \sum_{j=k}^{n} \binom{n}{j} \sum_{l=0}^{n-j} \binom{n-j}{l} (-1)^{l} e^{-\eta(j+l)\lambda t^{\gamma}} \tag{15}$$

Note that

$$\left(1 - e^{-\eta\lambda t^{\gamma}}\right)^{(n-j)} = \sum_{l=0}^{n-j} \binom{n-j}{l} \left(-e^{-\eta\lambda t^{\gamma}}\right)^{l} \tag{16}$$

The conditional reliability function of k-out-of-n systems, from Eq. 15, can be rewritten as:

$$R_{k/n}(t|\eta, \lambda, \gamma) = \sum_{j=k}^{n} \binom{n}{j} \sum_{l=0}^{n-j} \binom{n-j}{l} (-1)^{l} e^{-\eta(j+l)\lambda t^{\gamma}} \tag{17}$$

Then if $\eta \sim \text{gamma}(\alpha, \beta)$ then the k-out-of-n systemability is given by:

$$R_{(T_1,\ldots,T_n)}(t|\eta, \lambda, \gamma) = \sum_{j=k}^{n} \binom{n}{j} \sum_{l=0}^{n-j} \binom{n-j}{l} (-1)^{l} \left[\frac{\beta}{\beta + \lambda(j+l)t^{\gamma}}\right]^{\alpha} \tag{18}$$

It can be easily shown that:

$$R_{k/n}^{2}(t|\eta, \lambda, \gamma) = \sum_{i=k}^{n} \binom{n}{j} \sum_{j=k}^{n} \binom{n}{j} e^{-\eta(i+j)\lambda t^{\gamma}} \left(1 - e^{-\eta\lambda t^{\gamma}}\right)^{(2n-i-j)} \tag{19}$$

Since

$$\left(1 - e^{-\eta\lambda t^{\gamma}}\right)^{(2n-i-j)} = \sum_{l=0}^{2n-i-j} \binom{2n-i-j}{l} \left(-e^{-\eta\lambda t^{\gamma}}\right)^{l} \tag{20}$$

Equation 19 can be rewritten, after several simplifications, as follows:

$$R^2_{k/n}(t|\eta, \lambda, \gamma) = \sum_{i=k}^{n} \binom{n}{i} \sum_{j=k}^{n} \binom{n}{j} (-1)^l \sum_{l=0}^{2n-i-j} \binom{2n-i-j}{l} e^{-\eta(i+j+l)\lambda t^\gamma} \quad (21)$$

Therefore, the variance of k-out-of-n systemability function is given by

$$\mathrm{Var}(R_{k/n}(t|\lambda, \gamma) = \int_{\eta} R^2_{k/n}(t|\eta, \lambda, \gamma)dG(\eta) - \left[\int_{\eta} R^2_{k/n}(t|\eta, \lambda, \gamma)dG(\eta) \right]^2$$

$$= \sum_{i=k}^{n} \binom{n}{i} \sum_{j=k}^{n} \binom{n}{j} \sum_{l=0}^{2n-i-j} \binom{2n-i-j}{l} (-1)^l \left(\frac{\beta}{\beta+(i+j+l)\lambda t^\gamma} \right)^2$$

$$- \left(\sum_{j=k}^{n} \binom{n}{j} \sum_{l=0}^{n-j} \binom{n-j}{l} (-1)^l \left(\frac{\beta}{\beta+(j+l)\lambda t^\gamma} \right)^2 \right)^2$$

$$(22)$$

    Figures 5 and 6 show the reliability and systemability functions of a 3-out-of-5 system for $\alpha = 2$, $\beta = 3$ and for $\alpha = 2$, $\beta = 1$, respectively.
    Assume $\lambda = 0.00001$, $\gamma = 1.5$, $n = 3$, $k = 2$, and $\eta \sim$ gamma(α, β), the systemability and its confidence intervals of a 2-out-of-3 system for $\alpha = 2$, $\beta = 1$ and $\alpha = 2$, $\beta = 2$, are shown in Figs. 7 and 8, respectively. Figure 9 is the same calculations for $\alpha = 3$ and $\beta = 2$.

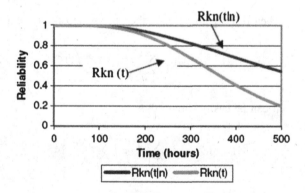

Fig. 5 Comparisons of k-out-of-n system reliability versus systemability functions for $\alpha = 2$ and $\beta = 3$ (Pham 2005)

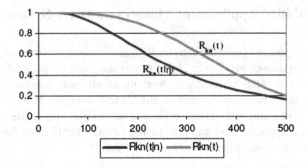

Fig. 6 Comparisons of k-out-of-n system reliability versus systemability functions for $\alpha = 2$ and $\beta = 1$ (Pham 2005)

Fig. 7 A 2-out-of-3 systemability and its 95 % confidence interval ($\alpha = 2$, $\beta = 1$) (Pham 2005)

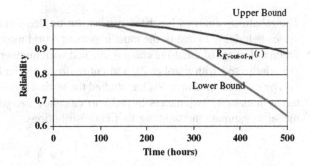

Fig. 8 A 2-out-of-3 systemability and its 95 % confidence interval ($\alpha = 2$, $\beta = 2$) (Pham 2005)

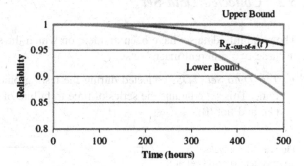

Fig. 9 A 2-out-of-3 systemability and its 95 % confidence interval ($\alpha = 3$, $\beta = 2$) (Pham 2005)

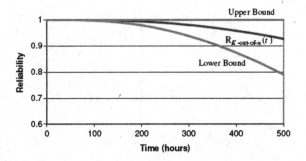

5 Systemability Application: Motorcycle Drive-System

Motorcycle components present different lifetime distributions between the testing environment and the operating one. This aspect makes difficult the prediction of the reliability of these components in the working conditions, therefore motorcycle manufacturers meet troubles to estimate the correct warranty policies. After the validation of systemability function, explained in the previous section, this innovative concept is applied in order to predict and estimate the lifetime distribution of several components and systems (Persona et al. 2009).

5.1 Examined Components and System

The reliability analysis has been carried out on drive-systems (Fig. 10) of motorcycles belonging to one of the most important world motorcycle manufacturer. The lifetime data set of the drive chain (indicated with number 1 in the Fig. 10), relative gear (indicated with number 2), and entire drive-system have been examined.

The experimental analysis has studied the testing and operating data set to define the systemability parameters in order to calculate the goodness-of-fit of this new model to estimate the working lifetime distributions.

5.2 Collected Data Set

The reliability data set have been divided, on which the study has been developed, in these two different macro-sets:

- *Test Data Set (TDS)* collected during the testing phases, in well-known conditions. This set contains the series of time to failure of the components, censured (1), and not (0).

Fig. 10 Motorcycle drive-system (Persona et al. 2009)

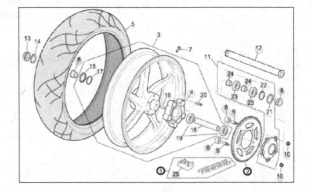

- *Work Data Set (WDS)* related to the lifetime data during the operating conditions. This set contains the reliability of the components and system calculated using the sold spare parts in specific year and the number of motorcycles on field at the beginning of each year. It is important to notice that the operating conditions are the same for the two studied components, because they belong to the same system.

In the next page, the collected data set for each component is illustrated (Table 1). The data related to the time have been normalized in order to guarantee the privacy policy of the manufacturer.

5.3 Application of Systemability Concept

With the collected data set, the application of systemability have been analysed through the following steps:

- modeling of all testing data set by Weibull distribution;
- estimation of the systemability parameters using the testing and operating data set of one component (drive-chain);
- estimation of the operating lifetime distribution of the other component (gear) and calculation of the goodness-of-fit of these predicted data, using the systemability parameters, previously calculated;
- estimation of the operating reliability of the entire drive-system using the systemability series calculations and evaluation of its goodness-of-fit.

5.4 Modeling of the Testing Data

Using the reliability and survival tool of Minitab® software, the parameters of the Weibull distribution have been calculated to model the testing data of each component where the Weibull distribution is given.

Figure 11a, b shows the results of the Minitab® elaborations while Table 2 reports the values of the Weibull parameters for the different components.

5.5 Systemability Parameters Estimation

Using the data set of the drive-chain components, the systemability parameters are calculated in order to model the data set related to the operating conditions (WDS) of this component.

Table 1 Collected data set
(Persona et al. 2009)

Test data set (TDS)			
Drive-chain		Gear	
ttf	Censured	ttf	Censured
0.363	1	0.151	1
0.418	1	0.363	1
0.351	1	0.418	1
0.289	1	0.351	1
0.355	1	0.289	1
0.320	1	0.355	1
0.352	1	0.352	1
0.283	1	0.305	1
0.334	1	0.283	1
0.212	1	0.295	1
0.190	1	0.334	1
0.226	1	0.212	1
0.349	1	0.190	1
0.400	1	0.226	1
0.270	0	0.349	1
0.398	0	0.400	1
0.295	0	0.398	0
0.391	0	0.664	0
0.257	0	0.263	0
0.350	0	0.823	0
0.472	0	0.221	0
0.221	0	0.321	0
0.051	0	0.418	0
0.204	0	0.218	0
0.214	0	0.320	0
0.218	0	0.435	0
0.242	0	0.370	0
0.151	0	–	–
0.236	0	–	–
0.435	0	–	–
0.370	0	–	–

Work data set (WDS)					
Drive-chain		Gear		Drive-system	
t	$R(t)$	t	$R(t)$	t	$R(t)$
0.173	0.992	0.173	0.981	0.173	0.973
0.212	0.980	0.212	0.977	0.212	0.958
0.280	0.971	0.280	0.969	0.280	0.940
0.345	0.960	0.345	0.955	0.345	0.915
0.407	0.941	0.407	0.934	0.407	0.874
0.471	0.910	0.471	0.899	0.471	0.810

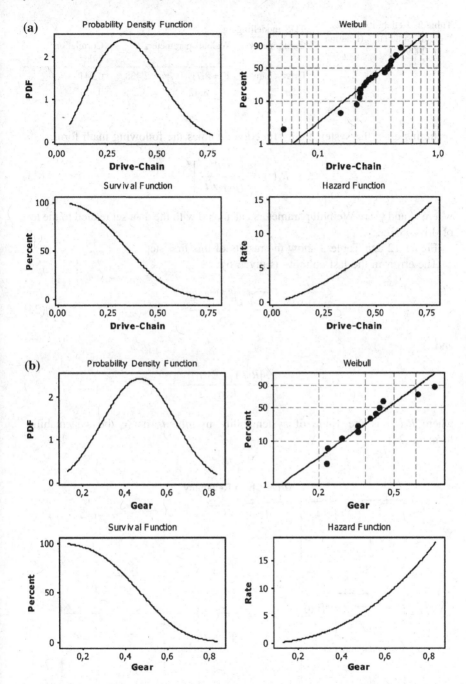

Fig. 11 Reliability analysis and functions for drive-chain (**a**) and gear (**b**) (Persona et al. 2009)

Table 2 Reliability results: Weibull parameters (Persona et al. 2009)

TDS modelling			
Components	Weibull parameters		Correlation index
Drive-cham	$\lambda' = 8.71$	$\gamma' = 2.428$	0.941
Gear	$\lambda'' = 8.42$	$\gamma'' = 3.288$	0.951

In this case, the systemability function assumes the following math form:

$$R_s(t) = \left[\frac{\beta}{\beta + \lambda' t^{\gamma'}}\right]^\alpha \tag{23}$$

where λ' and γ' are Weibull parameters calculated with the data set related to the test of drive-chain.

Figure 12 and Table 3 show the results of this first step.

The error, in the last column, is given by:

$$e = \frac{\sum_{i=1}^{n} \frac{|R(t_i) - R_s(t_i)|}{R(t_i)}}{n} \tag{24}$$

and

$$e_i = \frac{|R(t_i) - R_s(t_i)|}{R(t_i)} \tag{25}$$

where $R_s(t_i)$ is the value of systemability in time t_i, using the systemability formula (23).

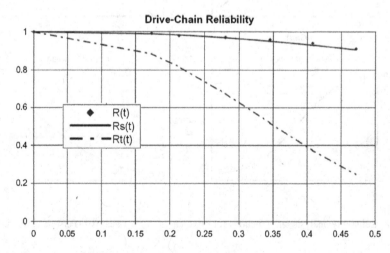

Fig. 12 Drive-chain reliability: systemability versus work data set (Persona et al. 2009)

Table 3 Reliability and systemability values of drive-chain component (Persona et al. 2009)

Drive-chain: reliability values			
t_i	$R(t)$	$Rs(t)$	e_i %
0.173	0.992	0.990	0.23
0.213	0.980	0.984	0.36
0.280	0.971	0.970	0.16
0.345	0.960	0.951	0.88
0.408	0.941	0.930	1.11
0.472	0.90	0.905	0.55

	Weibull		Systemability		
	λ'	γ'	α	β	e %
Drive-chain	8.71	2.428	0.4	5	0.55

Moreover, in formula (25), $R(t_i)$ is the value of the work data set (WDS) collected in the operating conditions. Finally, n is the number of values.

5.6 Application of Systemability to Gear Component

Using the data related to the gear component and the parameters of systemability calculated here, the goodness-of-fit has been estimated using the absolute mean error defined before (25). The same values of α, β have been used, because drive-chain and gear belong to the same systems, so they work in the same operating conditions.

Table 4 illustrates the results of this step, also shown in Fig. 13.

In this case, the systemability values have been calculated with:

$$R_s(t) = \left[\frac{\beta}{\beta + \lambda'' t^{\gamma''}} \right]^{\alpha} \qquad (26)$$

Table 4 Reliability and systemability values of gear component (Persona et al. 2009)

Gear: reliability values			
t_i	$R(t)$	$Rs(t)$	e_i %
0.173	0.981	0.998	1.73
0.213	0.977	0.996	1.90
0.280	0.969	0.990	2.17
0.345	0.955	0.980	2.69
0.408	0.934	0.967	3.55
0.472	0.899	0.948	5.43

	Weibull		Systemability		
	λ''	γ''	α	β	e %
Gear	8.42	3.288	0.4	5	2.91

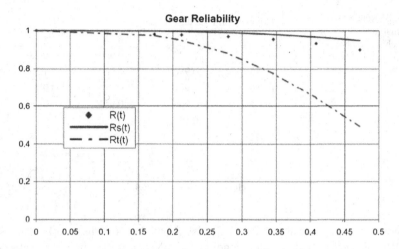

Fig. 13 Gear reliability: systemability versus work data set (Persona et al. 2009)

where λ'' and γ'' are Weibull parameters calculated with the data set related to the test of gear.

The errors have been given using formulas (24) and (25).

5.7 Application of Systemability to Drive-System

Using the data collected about drive-system, shown in Table 1, and using the Weibull parameters calculated before, the systemability function is applied in order to predict the system lifetime.

The drive-system can be studied as a series-system configuration, composed by the gear and drive-chain components (Fig. 14).

Therefore, the systemability of drive-system, consisting of the gear and drive-chain components, is given as follows:

$$R_S\left(t\big|\underline{\lambda},\underline{\gamma}\right) = \left[\frac{\beta}{\beta + (\lambda' t^{\gamma'} + \lambda'' t^{\gamma''})}\right]^{\alpha} \tag{27}$$

where λ', λ'' and γ', γ'' are the intensity and the shape parameters of Weibull distribution of the components.

As in previous application, the mean absolute error is calculated in order to estimate the goodness-of-fit of the systemability function using formulas (24) and (25).

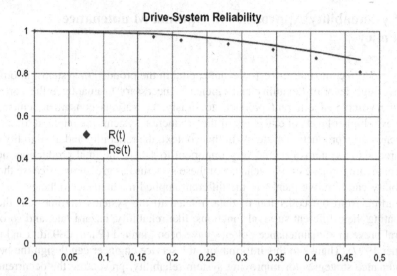

Fig. 14 Drive-system reliability: systemability versus work data set (Persona et al. 2009)

6 Several Considerations About Systemability Approach

In this Chapter, the reliability modeling of practical application-systems with respect to random field environments is discussed, addressing the fact that the difference between the lifetime distributions collected during the test and the distributions related to the operating environment is well known. For this reason, it is difficult to estimate the reliability of a component when it works in different environments.

Starting from Pham's paper (2005a) the concept of systemability, then, in this chapter, this approach is utilized to examine real-world application case studies in automatic manufacturer packaging machines as well as motorcycle systems in determining the reliability of customer products. The case study has been carried out in order to estimate the reliability and systemability of several motorcycle components and drive system in operative conditions, starting from the data collected during the tests.

This research addresses three important aspects:

(a) the study of the performance of the systemability when its parameters change, especially in the field environments;
(b) the definition of quantitative relationships between systemability parameters and different environment conditions; and
(c) the definition of a general model for the estimation of reliability function in order to design and produce components or system products and reflect ways to define the best maintenance policies.

7 Systemability Approach for Optimal Maintenance Policy

Many modern companies invest relevant capital in the production systems, in order to have high level of flexibility and efficiency, necessary to guarantee the variety and the volume of different products to satisfy the various consumer demands. Moreover higher levels of efficiency of the production systems are required to make convenient the production systems. In this context, the reliability and availability of the production systems become very important to satisfy the final production rate. Hence, the importance of maintenance of these systems has grown up, because their reliability can improve thanks to the different applied maintenance policies.

Starting from the collection of data, related to the systems lifetime and then estimating their different survival functions, like reliability, hazard rate, and so on, several preventive maintenance policies have been planned (Pham 2005d; Tan and Kramer 1997). Thanks to this information, the service engineer can design the best maintenance strategies for improving system reliability, preventing the occurrence of system failures, and reducing maintenance costs of systems (Faccio et al. 2014).

When a component or a system works in an operative plant, it reflects a reliability function that is usually different from the theory reliability but also from all its similar applications in other industrial plants. Environmental factors may change failure rate, reliability, and availability of systems. Incorrect estimation of reliability function could lead to the wrong functional design of the system and an incorrect definition of the appropriate maintenance policies.

In the next section, a literature review is illustrated in order to discuss the importance of this work. Then, in Sect. 9, the age replacement policy (ARP), introduced by Barlow and Hunter (1960) is discussed considering also the environmental factors and their effects on ARP are investigated. In this part of chapter, the new concept systemability, introduced by Pham (2005a, b) is used to model the age replacement policy. It allows considering the environmental effects in the reliability estimation, separating them from the intrinsic survival performances of the component.

A sensitivity analysis is conducted in order to study the cost curves in function of change on systemability parameters. Several graphics are introduced in order to analyze the changing of the optimal time of preventive maintenance and to estimate the difference between the UEC considering the environmental factors or not.

The proposed methodology is applied to an interesting case study in the automatic packaging machines for beer production and its performance is discussed.

The same section-structure is used for periodic replacement policy (PRP) (Sect. 12), starting from a literature review, illustrating the Classical approach versus the systemability one, making a sensitive analysis of cost curves in function of systemability parameters and applying the new approach to an interesting real case study, finally discussing the results.

8 Maintenance Policies Types: State of Art and Future Researches

In the last decades, most scientific contributions on maintenance policies have been developed and widely used. The first study of preventive maintenance was carried out by Barlow and Hunter (1960). Their research introduced two different maintenance policies to maximize 'limiting efficiency,' i.e., fractional amount of up-time over long intervals. The first policy is called policy *type I*, or also age replacement policy, while the second one is called policy *type II* or periodic replacement. The optimum policies are determined, in each case, as unique solutions of certain integral equations depending on the failure distribution.

In the past several decades, maintenance, and replacement problems of deteriorating systems have been extensively studied in the literature. A survey of all maintenance policies has been developed by Wang (2002). The author has summarized, classified, and compared various existing maintenance policies in several group, like: age replacement policy, random age replacement policy, block replacement policy, periodic preventive maintenance policy, failure limit policy, sequential preventive maintenance policy, repair cost limit policy, repair time limit policy, repair number counting policy, reference time policy, mixed age policy, preparedness maintenance policy, group maintenance policy, opportunistic maintenance policy, etc. (Murthy and Whang 1996; Pham 1996; Pham and Wang 1997; Wang and Pham 1996).

In some early works, the age replacement policy was extensively studied. Under this policy, a unit is always replaced at its age T or failure, whichever occurs first, where T is a constant (Barlow and Proshan 1965). Later, as the concepts of minimal repair and especially imperfect maintenance became more and more established various extensions and modifications of the age replacement policy were proposed. For this class of policy, various maintenance models can be constructed according to different types of preventive maintenances (minimal, imperfect, perfect), corrective maintenances (minimal, imperfect, perfect), cost structures, etc. (Carroll 2003; Finkelstein 2003, 2006; Elsayed and Wang 1996; Wang et al. 1992a, b, c, d; Pham 2003; Tamura et al. 2006; Teng and Pham 2004, 2006; Zhang et al. 2001; Zhang and Pham 2006; Zhao et al. 2006; Oh and Bai 2001; Attardi et al. 2005; Abbasi et al. 2006; Sohn et al. 2007; Ram and Tiwari 1989; Persona et al. 2009; Tan and Kramer 1997; Faccio et al. 2014; Barlow and Hunter 1960; Wang 2002; Murthy and Whang 1996; Pham and Wang 1997; Wang and Pham 1996). Recently other studies have been developed about the replacement policies and spare parts management under different conditions (Chien 2008; Chien et al. 2009; Kenzin and Frostig 2009; Wang et al. 2009).

In the periodic preventive maintenance policy, a unit is preventively maintained at fixed time intervals kT ($T = 1, 2, \ldots$) independent of the failure history of the unit, and repaired at intervening failures where T is a constant. The basic periodic preventive maintenance policy is "periodic replacement with minimal repair at failures" policy under which a unit is replaced at predetermined times kT ($T = 1, 2,$ ...) and failures are removed by minimal repair (Barlow and Hunter 1960, *Type II*). As the concepts of minimal repair and especially imperfect maintenance (Pham 1996) became more and more established, various extensions and variations of these two policies were proposed.

The effects of the environmental factors in the development of preventive maintenance models are not considered in the previous scientific contributions, as shown in the analysis of this survey.

Notation

$f(t)$	Probability density function of failure
$h(t)$	Hazard rate function
$R(t)$	Reliability function
$F(t)$	Failure function
λ	Intensity parameter of Weibull distribution
γ	Shape parameter of Weibull distribution
η	A common environment factor
$G(\eta)$	Cumulative distribution function of η
α	Shape parameter of Gamma distribution
β	Scale parameter of Gamma distribution
c_p	Cost of a preventive maintenance operation
c_f	Cost of a breakdown maintenance operation
UEC	Unit expected cost of the replacement policy
t_p	Planned time of replacement

9   New Approach: Systemability Age Replacement Policy (S-ARP)

As described before, the mathematical form of systemability function is given by:

$$R_s(t) = \left[\frac{\beta}{\beta + \int_0^t h(s)ds} \right]^{\alpha} \qquad (28)$$

Introducing systemability approach into the traditional formulas developed by Barlow and Hunter (1960), the UEC(t_p) assumes this mathematical expression (Sgarbossa et al. 2010):

$$\text{UEC}_{\alpha\beta}(t_p) = \frac{c_p * \left[\frac{\beta}{\beta + \lambda t^\gamma}\right]^\alpha + c_f * \left(1 - \left[\frac{\beta}{\beta + \lambda t^\gamma}\right]^\alpha\right)}{\int_0^{t_p} \left[\frac{\beta}{\beta + \lambda t^\gamma}\right]^\alpha dt} \tag{29}$$

Also in this case the t_{sp}^* is a finite and unique solution if the hazard rate function $h_s(t)$ of systemability is strictly increasing.

If the systemability formula (28) is considered with Weibull distribution of testing data, with hazard rate $h(t)$, the systemabilty function is as follows:

$$R_s(t) = \left[\frac{\beta}{\beta + \lambda t^\gamma}\right]^\alpha \tag{30}$$

and known that $h_s(t) = \frac{1}{R_s(t)} \times \left(-\frac{\partial R_s(t)}{\partial t}\right)$,

where $\frac{\partial R_s(t)}{\partial t} = -\frac{\alpha\lambda\gamma \times \left(\frac{\beta}{\beta + \lambda t^\gamma}\right)^\alpha}{\beta + \lambda t^\gamma} \times t^{\gamma-1}$ so the related hazard rate function $h_s(t)$ becomes:

$$h_s(t) = \frac{\alpha\lambda\gamma \times t^{\gamma-1}}{\beta + \lambda t^\gamma} \tag{31}$$

To find a unique optimum solution, the function (31) has to be increasing. If this function is studied, it can be found that the $\lim_{t\to 0} h_s(t) = 0$ and $\lim_{t\to\infty} h_s(t) = 0$. Moreover, the function $h_s(t)$ is major than 0 for all values of variable t.

Differentiating the $h_s(t)$ with respect of time t and setting it equal to zero:

$$h_s'(t) = \frac{\partial h_s(t)}{\partial t} = \frac{\partial}{\partial t}\left(\frac{\alpha\lambda\gamma \times t^{\gamma-1}}{\beta + \lambda t^\gamma}\right) = 0 \tag{32}$$

The following solution is given:

$$t^* = \sqrt[\gamma]{\frac{\beta(\gamma - 1)}{\lambda}} \tag{33}$$

Considering the $h_s'(t)$, it can be noticed that $h_s'(t = 0) = 0$ and $h_s'(t = \infty) = 0$. Hence, if $0 \leq t \leq t^*$ the hazard rate function $h_s(t)$ is strictly increasing, while on the other hand, if $t \geq t^*$ the hazard rate function $h_s(t)$ is strictly decreasing, due also to $h_s'(t = 0) \geq 0$ and $h_s\prime(t = \infty) \geq 0$.

It is interesting to notice that t^* depends only on a parameter of systemability, that is β one. Moreover, increasing β value, the value of t^* increases, as shown in the following Fig. 15.

To find a unique and finite solution t_{sp}^*, which minimizes $\text{UEC}_{\alpha\beta}(t_p)$, the value of the unique optimal solution calculated in $t_p \leq t^*$, based on the hazard rate function

Fig. 15 $t^*(\beta)$ versus β (Sgarbossa et al. 2010)

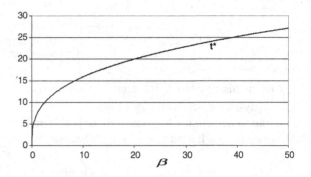

$h_s(t)$ of systemability is strictly increasing, has to be compared with each values of $UEC_{\alpha\beta}(t_p)$ for $t_p \geq t^*$ as follows:

Algorithm A1 Given c_p, c_f, λ, γ, α, β the optimal value of t_p, say t^*_{sp}, which minimizes the unit expected cost $UEC_{\alpha\beta}(t_p)$ for age replacement policy using systemabilty function is as follows:

- Step 1: Set-up Phase
  Set c_p, c_f, λ, γ, α, β real positive constants
  Set t_p, t real positive variables.
- Step 2: Calculation Phase
  Calculate $h_s(t) = \frac{\alpha\lambda\gamma \times t^{\gamma-1}}{\beta + \lambda t^\gamma}$

  Calculate $R_s(t) = \left[\frac{\beta}{\beta + \lambda t^\gamma}\right]^\alpha$

  Calculate $UEC_{\alpha\beta}(t_p) = \frac{c_p \times \left[\frac{\beta}{\beta + \lambda t^\gamma}\right]^\alpha + c_f \times \left(1 - \left[\frac{\beta}{\beta + \lambda t^\gamma}\right]^\alpha\right)}{\int_0^{t_p} \left[\frac{\beta}{\beta + \lambda t^\gamma}\right]^\alpha dt}$

  Calculate $t^* = \sqrt[\gamma]{\frac{\beta(\gamma-1)}{\lambda}}$
- Step 3: Decision-Making Phase
  For $t_p \leq t^*$, for any $t_p \in \left(0, t^*_p\right]$ and $t_p \in \left(t^*_p, t^*\right)$, then $UEC_{\alpha\beta}(t_p) > UEC_{\alpha\beta}(t^*_p)$, then t^*_p minimizes $UEC_{\alpha\beta}(t^*_p)$.
  Condition A_C1: If $UEC_{\alpha\beta}(t^*_p) < UEC_{\alpha\beta}(t_p)$ for any $t_p \geq t^*$, $t^*_{sp} = t^*_p$ minimizes $UEC_{\alpha\beta}(t_p)$, then t^*_{sp} is the unique optimal solution;

  Condition A_C2: If $UEC_{\alpha\beta}(t^*_p) \geq UEC_{\alpha\beta}(t^i_p)$, where for any $t_p \in \left(t^*, t^i_p\right]$ and $t_p \in \left(t^i_p, \infty\right)$, then $UEC(t_p) > UEC(t^i_p)$, then $t^*_{sp} = t^i_p$ minimizes $UEC_{\alpha\beta}(t_p)$, then t^*_{sp} is the unique optimal solution.

9.1 Age Replacement Policy Using Systemability Function: UEC Curves in Function of Systemability Parameters

In this section, the function of $UEC(t_p)$ are tested with the introduction of systemability formula in order to study the effect of environmental factors, using Algorithm A.

Here below, several graphics show the function $UEC_{\alpha\beta}(t_p)$ for different values of the systemability parameters α and β, fixing the values of the costs parameters, c_p and c_f.

This series of graphics illustrates systemability, $R_s(t_p)$, and $UEC_{\alpha\beta}(t_p)$ functions, fixing one of the systemability parameters to a defined value (1, 10, 50) and changing the other one in a series of values between 1, 5, 10, 20, 50.

Let us consider the following values of the parameters in the cost function to carry out this study: $c_p = 5$; $c_f = 50$; $\lambda = 0.005$, and $\gamma = 3$.

Using Algorithm A1, proposed in the previous section, let us see how the following steps as the example with $\alpha = 10$ and $\beta = 50$:

- Step 1: Set-up Phase
  Set $c_p = 5$; $c_f = 50$; $\lambda = 0.005, \gamma = 3$, $\alpha = 10$, $\beta = 50$
  Set t_p, t real positive variables.
- Step 2: Calculation Phase
  Calculate $h_s(t) = \frac{\alpha\lambda\gamma \times t^{\gamma-1}}{\beta+\lambda t^{\gamma}} = \frac{10 \times 0.005 \times 3 \times t^{3-1}}{50+0.005t^3} = \frac{0.15 \times t^2}{50+0.005t^3}$
  Calculate $R_s(t) = \left[\frac{\beta}{\beta+\lambda t^{\gamma}}\right]^{\alpha} = \left[\frac{50}{50+0.005t^3}\right]^{10}$
  Calculate

$$UEC_{\alpha\beta}(t_p) = \frac{c_p \times \left[\frac{\beta}{\beta+\lambda t^{\gamma}}\right]^{\alpha} + c_f \times \left(1-\left[\frac{\beta}{\beta+\lambda t^{\gamma}}\right]^{\alpha}\right)}{\int_0^{t_p}\left[\frac{\beta}{\beta+\lambda t^{\gamma}}\right]^{\alpha} dt} = \frac{5 \times \left[\frac{50}{50+0.005t^3}\right]^{10} + 50 \times \left(1-\left[\frac{50}{50+0.005t^3}\right]^{10}\right)}{\int_0^{t_p}\left[\frac{50}{50+0.005t^3}\right]^{10} dt}$$

  Calculate $t^* = \sqrt[\gamma]{\frac{\beta(\gamma-1)}{\lambda}} = \sqrt[3]{\frac{50(3-1)}{0.005}} = 27.144$ months
- Step 3: Decision-Making Phase
  For $t_p \le t^*$, for any $t_p \in \left(0, t_p^*\right]$ and $t_p \in \left(t_p^*, t^*\right)$, then $UEC_{\alpha\beta}(t_p) > UEC_{\alpha\beta}(t_p^*)$, then t_p^* minimizes $UEC_{\alpha\beta}(t_p^*)$.

  Condition A1_C1: If $UEC_{\alpha\beta}(t_p^*) < UEC_{\alpha\beta}(t_p)$ for any $t_p \ge t^*$, $t_{sp}^* = t_p^*$ minimizes $UEC_{\alpha\beta}(t_p)$, then t_{sp}^* is the unique optimal solution;

  Condition A1_C2: If $UEC_{\alpha\beta}(t_p^*) \ge UEC_{\alpha\beta}(t_p^i)$, where for any $t_p \in \left(t^*, t_p^i\right]$ and $t_p \in \left(t_p^i, \infty\right)$, then $UEC(t_p) > UEC(t_p^i)$, then $t_{sp}^* = t_p^i$ minimizes $UEC_{\alpha\beta}(t_p)$, then t_{sp}^* is the unique optimal solution.

Results:

$$t_p^* = 3.83 \, \text{months}$$

$\text{UEC}_{\alpha\beta}(t_p^*) = 1.973 < \text{UEC}_{\alpha\beta}(t_p)$: Condition A1_C1 = TRUE $\rightarrow t_{\text{sp}}^* = t_p^*$ minimizes $\text{UEC}_{\alpha\beta}(t_p)$, then t_{sp}^* is the unique optimal solution.

Figures 16, 17 and 18 show the systemability function $R_s(t_p)$ and the cost function $\text{UEC}_{\alpha\beta}(t_p)$ for different values of α, varying the β parameter, in comparison with $R(t_p)$ and $\text{UEC}(t_p)$.

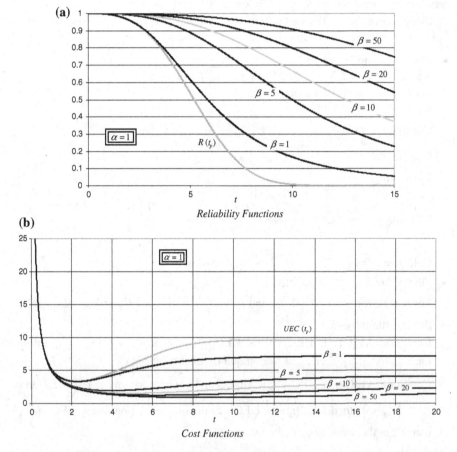

Fig. 16 **a** Comparison of reliability versus systemability function for $\alpha = 1$ and $\beta = 1, 5, 10, 20$ and 50. **b** Comparison of cost function $\text{UEC}(t_p)$ versus $\text{UEC}_{\alpha\beta}(t_p)$ for $\alpha = 1$ and $\beta = 1, 5, 10, 20$ and 50 (Sgarbossa et al. 2010)

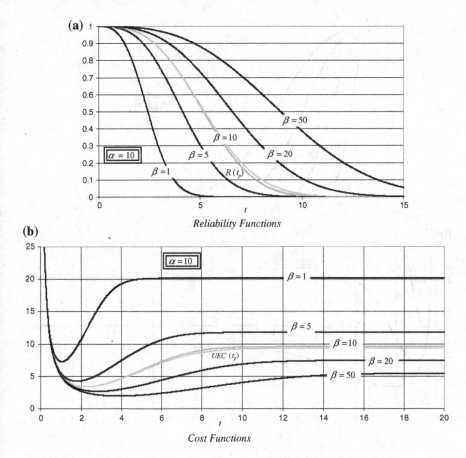

Fig. 17 **a** Comparison of reliability versus systemability function for $\alpha = 10$ and $\beta = 1, 5, 10, 20$ and 50. **b** Comparison of cost function $UEC(t_p)$ versus $UEC_{\alpha\beta}(t_p)$ for $\alpha = 10$ and $\beta = 1, 5, 10, 20$ and 50 (Sgarbossa et al. 2010)

Furthermore, Figs. 19, 20 and 21 show the systemability function $R_s(t_p)$ and the cost function $UEC_{\alpha\beta}(t_p)$ for different values of β, varying the α parameter, in comparison with $R(t_p)$ and $UEC(t_p)$.

As well shown in these graphics, two scenarios can be defined:

- Scenario 1: *hard environment effects*: when $\alpha > \beta$, the $UEC_{\alpha\beta}(t_p)$ is higher than $UEC(t_p)$ because $R_s(t)$ is worse than $R(t)$.
- Scenario 2: *soft environment effects*: when $\alpha < \beta$, the $UEC_{\alpha\beta}(t_p)$ is lower than $UEC(t_p)$ because $R_s(t)$ is better than $R(t)$.

The previous figures show the effects of the environmental factors, represented with the systemability parameters α and β, on reliability and on Expected Unit Cost functions. In fact, as explained before, if the operating environment is different from the testing one, the reliability will be different from the one calculated during the

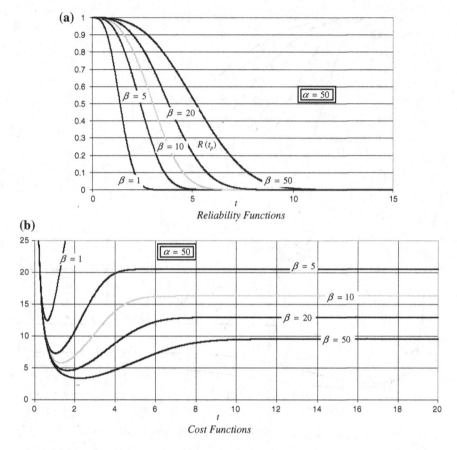

Fig. 18 **a** Comparison of reliability versus systemability function for $\alpha = 50$ and $\beta = 1, 5, 10, 20$ and 50. **b** Comparison of cost function $UEC(t_p)$ versus $UEC_{\alpha\beta}(t_p)$ for $\alpha = 50$ and $\beta = 1, 5, 10, 20$ and 50 (Sgarbossa et al. 2010)

test phases. As a consequence, also the Expected Unit Cost will be different and it could be higher or lower, depending on the effects of environmental factors, described by α and β parameters, as shown before.

9.2 Further Analysis About S-ARP: t^*_{sp} and $\%UEC(t^*_p)$ Curves in Function of Systemability Parameters

The second group shows several graphics in order to analyse the changing of the optimal time of preventive maintenance and to estimate the difference between the UEC considering the environmental factors or not.

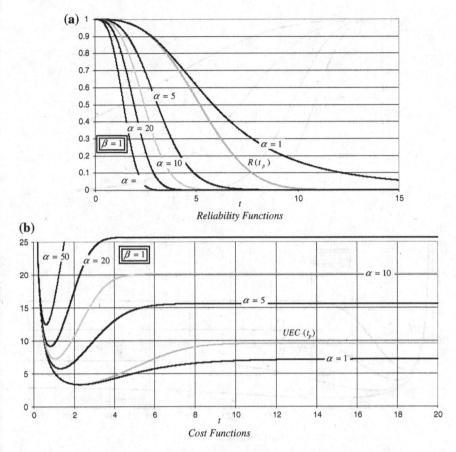

Fig. 19 **a** Comparison of reliability versus systemability function for $\beta = 1$ and $\alpha = 1, 5, 10, 20$ and 50. **b** Comparison of cost function $\text{UEC}(t_p)$ versus $\text{UEC}_{\alpha\beta}(t_p)$ for $\beta = 1$ and $\alpha = 1, 5, 10, 20$ and 50 (Sgarbossa et al. 2010)

Let us consider, the following values of the parameters: $\lambda = 0.005$ and $\gamma = 3$, for different values of costs parameters, c_p and c_f, and it has been illustrated:

(a) t^*_{sp} curve for $0 < \alpha < \infty$ and $\beta = 1, 5, 10, 20, 50$;
(b) t^*_{sp} curve for $0 < \beta < \infty$ and $\alpha = 0.5, 1, 5, 10, 20, 50$;
(c) $\%\text{UEC}(t^*_p)$ curve for $0 < \alpha < \infty$ and $\beta = 1, 5, 10, 20, 50$;
(d) $\%\text{UEC}(t^*_p)$ curve for $0 < \beta < \infty$ and $\alpha = 0.5, 1, 5, 10, 20, 50$;
(e) t^*_{sp} surface for $0 < \alpha < \infty$ and $0 < \beta < \infty$;
(f) $\%\text{UEC}(t^*_p)$ surface for $0 < \alpha < \infty$ and $0 < \beta < \infty$.

where t^*_{sp} is the value which minimized the $\text{UEC}_{\alpha\beta}(t_p)$, calculated with Eq. 29.

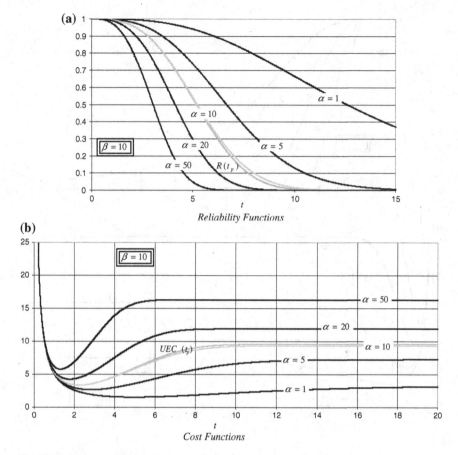

Fig. 20 **a** Comparison of reliability versus systemability function for $\beta = 10$ and $\alpha = 1, 5, 10, 20$ and 50. **b** Comparison of cost function UEC(t_p) versus UEC$_{\alpha\beta}(t_p)$ for $\beta = 10$ and $\alpha = 1, 5, 10, 20$ and 50 (Sgarbossa et al. 2010)

Then $\%\mathrm{UEC}(t_p^*)$ is calculated as:

$$\%\mathrm{UEC}(t_p^*) = 100 \times \left(\frac{\mathrm{UEC}_{\alpha\beta}(t_p^*)}{\mathrm{UEC}_{\alpha\beta}(t_{\mathrm{sp}}^*)} - 1 \right) \tag{34}$$

It indicates the percent difference between the real unit expected cost $\mathrm{UEC}_{\alpha\beta}(t_p^*)$, using a planned time t_p^* calculated with only Weibull distribution, without consider the environmental effects, and the optimum value of unit expected cost $\mathrm{UEC}_{\alpha\beta}(t_{\mathrm{sp}}^*)$ estimated considering also the environmental effects with systemability parameters, calculated in the real optimum time value t_{sp}^*.

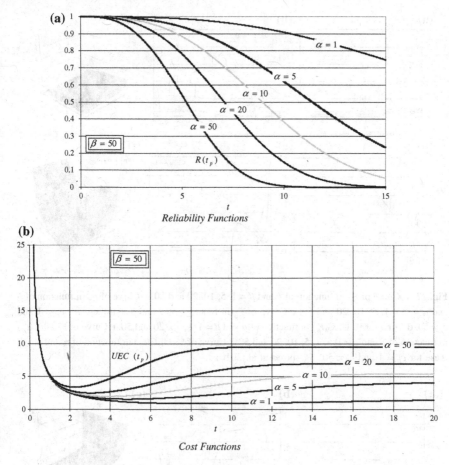

Fig. 21 **a** Comparison of reliability versus systemability function for $\beta = 50$ and $\alpha = 1, 5, 10, 20$ and 50. **b** Comparison of cost function $\text{UEC}(t_p)$ versus $\text{UEC}_{\alpha\beta}(t_p)$ for $\beta = 50$ and $\alpha = 1, 5, 10, 20$ and 50 (Sgarbossa et al. 2010)

All calculated t_{sp}^* are minor than t^*, given by Eq. 33, and satisfy condition (a) of Algorithm A hence t_{sp}^* can be considered the unique and finite solution of $\text{UEC}_{\alpha\beta}(t_{\text{sp}}^*)$.

Figures 22 and 23 show the curve of t_{sp}^* and $\%\text{UEC}(t_P^*)$ for different values of c_p, fixed $c_f = 50$, varying α and β values. For each figure, the optimal times t_p^* has been shown which values have been calculated with the traditional Weibull-based formula.

From the analysis of the previous graphics some considerations could be made:

- fixed parameter β, t_{sp}^* is a decreasing function of parameter α;
- fixed parameter α, t_{sp}^* is an increasing function of parameter β;

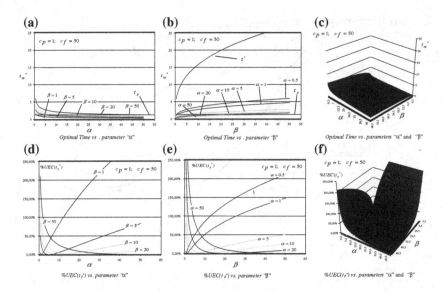

Fig. 22 **a** Curve of t_{sp}^* in function of α and $\beta = 1, 5, 10, 20$ and 50. **b** Curve of t_{sp}^* in function of β and $\alpha = 0.5, 1, 5, 10, 20$ and 50. **c** Curve of t_{sp}^* in function of α and β. All these for $c_p = 1$ and $c_f = 50$. **d** Curve of % UEC(t_p^*) in function of α and $\beta = 1, 5, 10, 20$ and 50. **e** Curve of % UEC(t_p^*) in function of β and $\alpha = 0.5, 1, 5, 10, 20$ and 50. **f** Curve of % UEC(t_p^*) in function of α and β. All these for $c_p = 1$ and $c_f = 50$ (Sgarbossa et al. 2010)

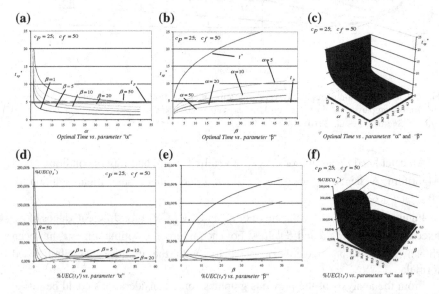

Fig. 23 **a** Curve of t_{sp}^* in function of α and $\beta = 1, 5, 10, 20$ and 50. **b** Curve of t_{sp}^* in function of β and $\alpha = 0.5, 1, 5, 10, 20$ and 50. **c** Curve of t_{sp}^* in function of α and β. All these for $c_p = 25$ and $c_f = 50$. **d** Curve of % UEC(t_p^*) in function of α and $\beta = 1, 5, 10, 20$ and 50. **e** Curve of % UEC(t_p^*) in function of β and $\alpha = 0.5, 1, 5, 10, 20$ and 50. **f** Curve of % UEC(t_p^*) in function of α and β. All these for $C_p = 25$ and $C_f = 50$ (Sgarbossa et al. 2010)

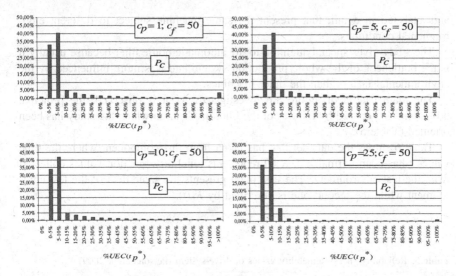

Fig. 24 Probability distribution of % UEC(t_p^*) for different values of c_p (Sgarbossa et al. 2010)

- overall, a general increasing of t_{sp}^* is well shown for increasing values of couple $(\alpha; \beta)$, from $(0,0)$ to $(50,50)$.
- increasing the cost parameter c_p, the %UEC(t_p^*) is less influenced by parameter α and more influenced by the other systemability parameter β;

Furthermore, the %UEC(t_p^*) values have been collected in different intervals, every five points percent, and for each interval, the probability distribution Pc of %UEC(t_p^*) has been calculated in order to define the mean value of %UEC(t_p^*) and its distribution.

Here below, several graphics (Fig. 24) show the probability distribution Pc of %UEC(t_p^*) for different values of the cost parameter c_p, respectively 1, 5, 10, 25, while the cost parameter c_f is equal to 50.

The mean value for each case is about 10 %, so if a preventive maintenance time is design using the test data set, the mean major cost, due to the environmental effects, is about 10 %.

10 Real Application: S-ARP on Automatic Packaging Machines

In this section, the proposed methodology applied to an interesting case study in the automatic packaging machines for beer production is illustrated. Some data related to different applications of the same model of machine, in several customer plants located in different countries, have been collected. The two plants have been visited and analysed in detail, that present the same machine subject to different environmental conditions.

The study gives special attention to the automatic bottle filler because usually it is the bottleneck of the whole production system and it is more influenced by the environmental factors than the other machines.

The main causes of failure are investigated. As it is observed, the principal downtimes are caused by failures of the filler heads upon which the study has been oriented (Table 5).

Table 6 shows the data set that has taken from this component where the time values t has been normalized. $R_a(t)$ represents reliability function values collected from the test environment. The first plant presents the same data of the test environment while the second one is described by $R_b(t)$.

Table 5 Reliability and systemability values of drive-system (Persona et al. 2009)

Drive-system: reliability values			
t_i	$R(t)$	$Rs(t)$	e_i %
0.173	0.973	0.988	1.51
0.213	0.958	0.980	2.33
0.280	0.940	0.961	2.16
0.345	0.915	0.935	2.21
0.408	0.874	0.905	3.39
0.472	0.810	0.869	7.17

	Weibull				Systemability		
	Drive-chain		Gear		α	β	$è$ %
	λ	γ	λ''	γ''			
Drive-system	8.71	2.428	8.24	3.288	0.4	5	3.13

Table 6 Testing and operating data set (Sgarbossa et al. 2010)

T	0.00	0.03	0.07	0.10	0.13	0.17	0.20	0.23	0.27	0.30	0.33
$R_a(T)$	100.00	99.98	99.95	99.92	99.86	99.76	99.59	99.33	98.94	98.36	97.52
$R_b(T)$	100.00	99.88	99.79	99.64	99.40	99.03	98.50	97.73	96.66	95.20	93.25
T		0.37	0.40	0.43	0.47	0.50	0.53	0.57	0.60	0.63	0.67
$R_a(T)$		96.34	94.71	93.21	89.59	85.81	82.30	75.17	66.79	60.10	52.21
$R_b(T)$		90.70	87.44	83.36	78.37	72.44	65.59	57,91	49.61	41.00	32.47
T		0.70	0.73	0.77	0.80	0.83	0.87	0.90	0.93	0.97	1.00
$R_a(T)$		42.21	33.20	23.45	14.88	9.68	4.50	2.56	1.06	0.37	0.10
$R_b(T)$		24.46	17.37	11.52	7.05	3.94	1.98	0.88	0.34	0.11	0.03

Table 7 Parameters of Weibull and systemability function and cost parameters (Sgarbossa et al. 2010)

Weibull parameters		Systemability parameters		Cost parameters	
λ	γ	α	β	c_p	c_f
0.0000915	8.10	3.10	1.42	3	30

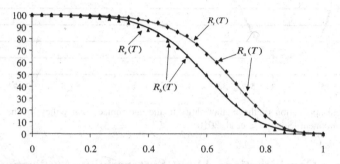

Fig. 25 Reliability during testing ($R_a(t)$), during operation ($R_b(t)$), Weibull function ($R_t(t)$) and systemability function ($R_s(t)$) (Sgarbossa et al. 2010)

Fitting the data using Weibull distribution and systemability function, the parameters illustrated in Table 7 and Fig. 25 have been obtained. Also the cost parameters given from the maintenance service of the company are included.

Considering $c_p = 3$, $c_f = 30$, $\lambda = 0.0000915$, $\gamma = 8.1$, $\alpha = 3.10$, $\beta = 1.42$, where the parameters values of systemability have been estimated using the least squares estimate method (Battini et al. 2007), the age replacement policy of the second plant has been optimized thanks to the consideration of the environmental factors with systemability function.

Considering only the Weibull parameters, the application of traditional approach has given us the replacement time related to the testing data, $t_p^* = 2.24$ months, and the unit expected cost is UEC(t_p^*) $= 1.184$ k€/months.

Using Algorithm A, the optimal time to replacement has been calculated considering also the environmental effects to the survival function, applying the systemability function. In this case, the optimal solution is $t_{sp}^* = 1.71$ months and the Unit Expected Cost is UEC$_{\alpha\beta}(t_{sp}^*) = 1.998$ k€/months. If it is calculated the Unit Expected Cost in the optimal solution of testing data, but taking into consideration the environmental factors, its value will be UEC$_{\alpha\beta}(t_p^*) = 2.392$ k€/month (Fig. 26).

As well demonstrated, the Unit Expected Costs are different between these cases and the optimal solution, using Algorithm A and considering the environmental effects, allows to saving about the 20 %.

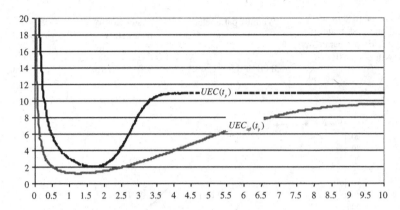

Fig. 26 Real application of the methodology to the age replacement policy on the automatic machine component (Sgarbossa et al. 2010)

11 Some Considerations About Systemability Application on ARP

Many maintenance models have been developed and widely used in scientific literature. The age replacement policy is one of the most important.

Incorrect estimation of reliability function could lead to the wrong functional design of the system and an incorrect definition of the appropriate maintenance policies.

In the previous sections, the effects of the different operating environments in the age replacement policy have been demonstrated using a new concept, called systemability (Pham 2005a, b, c).

The aim of this work is to introduce the systemability concept in the maintenance policies design in order to show its benefits, because it considers also the environmental conditions in which the systems operate.

With its parameters, systemability calculates the reliability in random environments, using as starting data, the reliability obtained during the test, and processing it using a gamma distribution in particular, or a distribution that represents operating environments in general, which takes into consideration the environmental factors. In other words, each operating environment is characterized by related systemability parameters calculated before with several failure data. These parameters allow to estimate the reliability behavior of each system operating in such environment conditions, using systemability function.

Hence, the purpose of this section is to evaluate the application of systemability function to the age replacement policy and highlight the benefits arisen with the use of this concept in comparison with the classical methodology.

The analytical behavior of the total cost function and a useful algorithm have been introduced in order to help the practitioners to apply this innovative model.

Furthermore, a sensitivity analysis of these benefits is conducted in order to show how the outcomes vary in function of the different environmental conditions, represented by systemability parameters α and β. In conclusions the application of systemability permits the correct estimation of time-replacement and consequently the minimization of global maintenance cost UEC. The results of this application have been illustrated with a series of graphics and summarized at the end. It's important to highlight that the mean value of $\%\text{UEC}(t_p^*)$ is about 10 %, that is the percent difference between the real Unit Expected Cost $\text{UEC}_{\alpha\beta}(t_p^*)$, using a planned time t_p^* calculated with only Weibull distribution, without consider the environmental effects, and the optimum value of Unit Expected Cost $\text{UEC}_{\alpha\beta}(t_{sp}^*)$ estimated considering also the environmental effects with systemability parameters, calculated in the real optimum time value t_{sp}^*. A real industrial application has demonstrated the importance of this study, and shown about a 20 % saving (about 5,000 €/years) using this methodology in comparison with the application of traditional age replacement policy.

In few words, if a preventive maintenance time is design using the test data set, the mean major cost, due to the environmental effects, will be about 10 % in comparison with the use of systemability on estimating of the time to replacement.

12 New Approach: Systemability Periodic Replacement Policy (S-PRP)

As above mentioned, this policy is one of the most used time-based preventive maintenance policies. It is also called policy *type II*. It involves a preventive maintenance action on the system after a operative time t_p, independent from the number of replacements at failure, occur in time t_p.

After each maintenance replacement action, the component becomes as good as new, hence the number of preventive maintenance actions do not affect to the failure rate of the entire system.

In other words the system failure rate remains undisturbed by the minimal repair.

Under minimal repair, if we assume that a complex system might have many failure modes, than its reliability, after a repair is the same as it was just before the failure. In this case, the sequence of failure at a system level follows a non-homogenous Poisson process (NHPP) (Sgarbossa et al. 2010; Crow 1974).

For periodic replacement policy characterized by the assumption of NHPP, the first failure is governed by a distribution $F(t)$ with failure rate $r(t)$ (Crow 1974).

Under the effect of the operational environment, using systemability formulation, we can introduce the failure distribution function $F_s(t)$ for the first failure

$$F_s(t) = 1 - \left[\frac{\beta}{\beta + \int_0^t r(s)ds} \right]^\alpha \tag{35}$$

where $r(s)$ is $r(s) = \lambda \cdot \gamma \cdot s^{\gamma-1}$

Then under the assumption of Power Law process for the baseline failure process $r(s)$ (Crow 1974), the systemability formula $R_s(t)$ is:

$$R_s(t) = \left[\frac{\beta}{\beta + \lambda \cdot t^\gamma} \right]^\alpha \tag{36}$$

and the first system failure is given by: $r_s(t) = \frac{dF_s(t)}{dt} = \frac{1}{R_s(t)} \cdot \left(-\frac{\partial R_s(t)}{\partial t} \right)$, where:

$$\frac{\partial R_s(t)}{\partial t} = -\frac{\alpha \cdot \lambda \cdot \gamma \cdot \left(\frac{\beta}{\beta + \lambda \cdot t^\gamma} \right)^\alpha}{\beta + \lambda \cdot t^\gamma} \cdot t^{\gamma-1} \tag{37}$$

Under this assumption, the systemability failure rate function $r_s(t)$ is modelled by:

$$r_s(t) = \frac{\alpha\lambda\gamma \times t^{\gamma-1}}{\beta + \lambda t^\gamma} \tag{38}$$

As explained in (Crow 1974) each succeeding failure follows the intensity function $u_s(t)$ of the process. Then we can assume that $u_s(t) = r_s(t)$, where $r_s(t)$ is the failure rate for the distribution function of the first system failure, because the NHPP has the same functional form as the failure rate governing the first system failure (Crow 1974).

Then, under minimal repair, we first define the systemability intensity function as follows:

$$u_s(t) = \frac{\alpha \cdot \lambda \cdot \gamma \cdot t^{\gamma-1}}{\beta + \lambda \cdot t^\gamma} \tag{39}$$

Using the traditional formulation of $UEC(t_p)$, under minimal repair, it assumes now this mathematical expression:

$$UEC_{\alpha\beta}(t_p) = \frac{c_p + c_f \cdot \int_0^{t_p} \frac{\alpha \cdot \lambda \cdot \gamma \cdot t^{\gamma-1}}{\beta + \lambda \cdot t^\gamma} dt}{t_p} = \frac{c_p + c_f \cdot \left[\alpha \cdot \ln\left(1 + \frac{\lambda}{\beta} t_p^\gamma \right) \right]}{t_p} \tag{40}$$

Also in this case t_{sp}^* is a finite and unique solution, if the systemability intensity function $u_s(t)$ is strictly increasing. Note that the $\lim_{t \to 0} u_s(t) = 0$, the $\lim_{t \to \infty} u_s(t) = 0$ and the function $u_s(t)$ is greater than 0 for all values of variable t (see Sect. 9).

To find a unique and finite solution t_{sp}^* which minimizes the $\text{UEC}_{\alpha\beta}(t_p)$, it is necessary to compare the value of the unique optimal solution calculated in $t_p \leq t^*$, based on the systemability intensity function $u_s(t)$, for each value of the $\text{UEC}_{\alpha\beta}(t_p)$ for $t_p \geq t^*$ as follows:

Algorithm A2 Given c_p, c_f, λ, γ, α, β, the optimal value of t_p, say t_{sp}^*, which minimizes the unit expected cost $\text{UEC}_{\alpha\beta}(t_p)$ for periodic replacement policy using systemability function is as follows:

Step 1: Set-up Phase
   Set c_p, c_f, λ, γ, α, β, real positive constants.
   Set t_p, t real positive variables.
Step 2: Calculation Phase
   Calculate $u_s(t) = \frac{\alpha \cdot \lambda \cdot \gamma \cdot t^{\gamma-1}}{\beta + \lambda \cdot t^\gamma}$

   Calculate $\text{UEC}_{\alpha\beta}(t_p) = \frac{c_p + c_f \cdot \int_0^{t_p} \frac{\alpha \cdot \lambda \cdot \gamma \cdot t^{\gamma-1}}{\beta + \lambda \cdot t^\gamma} dt}{t_p} = \frac{c_p + c_f \cdot \left[\alpha \cdot \ln\left(1 + \frac{\lambda t_p^\gamma}{\beta}\right)\right]}{t_p}$

   Calculate $t^* = \sqrt[\gamma]{\frac{\beta \cdot (\gamma - 1)}{\lambda}}$
Step 3: Decision-Making Phase
   For $t_p \leq t^*$, for any $t_p \in \left(0, t_p^*\right]$ and $t_p \in \left(t_p^*, t^*\right)$, then $\text{UEC}_{\alpha\beta}(t_p) > \text{UEC}_{\alpha\beta}(t_p^*)$, then t_p^* minimizes $\text{UEC}_{\alpha\beta}(t_p^*)$.

Condition A2_C1: If $\text{UEC}_{\alpha\beta}(t_p^*) < \text{UEC}_{\alpha\beta}(t_p)$ for any $t_p \geq t^*$, $t_{sp}^* = t_p^*$ minimizes $\text{UEC}_{\alpha\beta}(t_p)$, then t_{sp}^* is the unique optimal solution;

Condition A2_C2: If $\text{UEC}_{\alpha\beta}(t_p^*) \geq \text{UEC}_{\alpha\beta}(t_p^i)$, where for any $t_p \in \left(t^*, t_p^i\right]$ and $t_p \in \left(t_p^i, \infty\right)$, then $\text{UEC}(t_p) > \text{UEC}(t_p^i)$, then $t_{sp}^* = t_p^i$ minimizes $\text{UEC}_{\alpha\beta}(t_p)$, then t_{sp}^* is the unique optimal solution.

12.1 Periodic Replacement Policy Using Systemability Function: UEC Curves in Function of Systemability Parameters

Following the same steps carried out in Sect. 9, here below, several graphics show the function $\text{UEC}_{\alpha\beta}(t_p)$ for different values of the systemability parameters α and β, fixing the values of the costs parameters, c_p and c_f.

This series of graphics illustrates the systemability intensity function $u_s(t)$ and $\text{UEC}_{\alpha\beta}(t_p)$ functions, fixing one of the systemability parameters to a defined value (5, 20, 50) and changing the other one in a series of values between 1, 5, 10, 20, 50.

Let us now consider the following values to the parameters in the cost function:
$\lambda = 0.005$ and $\gamma = 3$,
$c_p = 20$; $c_f = 5$;
$\alpha = 5$ and $\beta = 25$:

Step 1: Set-up Phase
   Set $c_p = 20$ k€/action; $c_f = 5$ k€/action, $\lambda = 0.005$, $\gamma = 3$, $\alpha = 5$ and $\beta = 25$
   Set t_p, t real positive variables.
Step 2: Calculation Phase
   Calculate $u_s(t) = \frac{\alpha \cdot \lambda \cdot \gamma \cdot t^{\gamma-1}}{\beta + \lambda \cdot t^{\gamma}} = \frac{5 \cdot 0.005 \cdot 3 \cdot t^{3-1}}{25 + 0.005 \cdot t^3} = \frac{0.075 \cdot t^2}{25 + 0.005 \cdot t^3}$

   Calculate $UEC_{\alpha\beta}(t_p) = \frac{c_p + c_f \cdot \left[\alpha \cdot \ln\left(1 + \frac{\lambda}{\beta} t_p^{\gamma}\right)\right]}{t_p} = \frac{20 + 5 \cdot \left[5 \cdot \ln\left(1 + \frac{0.005}{25} t_p^3\right)\right]}{t_p}$

   Calculate $t^* = \sqrt[\gamma]{\frac{\beta \cdot (\gamma-1)}{\lambda}} = \sqrt[3]{\frac{25 \cdot (3-1)}{0.005}} = 21.5$
Step 3: Decision-Making Phase
   For $t_p \leq t^*$, for any $t_p \in \left(0, t_p^*\right]$ and $t_p \in \left(t_p^*, t^*\right)$, then $UEC_{\alpha\beta}(t_p) > UEC_{\alpha\beta}(t_p^*)$,
   then t_p^* minimizes $UEC_{\alpha\beta}(t_p^*)$.

Condition A2_C1: If $UEC_{\alpha\beta}(t_p^*) < UEC_{\alpha\beta}(t_p)$ for any $t_p \geq t^*$, $t_{sp}^* = t_p^*$ minimizes $UEC_{\alpha\beta}(t_p)$, then t_{sp}^* is the unique optimal solution;

Condition A2_C2: If $UEC_{\alpha\beta}(t_p^*) \geq UEC_{\alpha\beta}(t_p^i)$, where for any $t_p \in \left(t^*, t_p^i\right]$ and $t_p \in \left(t_p^i, \infty\right)$, then $UEC(t_p) > UEC(t_p^i)$, then $t_{sp}^* = t_p^i$ minimizes $UEC_{\alpha\beta}(t_p)$, then t_{sp}^* is the unique optimal solution.
Results:

$$t_{sp}^* = 16.95 \text{ weeks}$$

$UEC_{\alpha\beta}(t_p^*) = 2.183 < UEC_{\alpha\beta}(t_p)$: Condition A2_C1 = TRUE $\rightarrow t_{sp}^* = t_p^*$ minimizes $UEC_{\alpha\beta}(t_p)$, then t_{sp}^* is the unique optimal solution.

   Figures 27, 28, and 29 show the systemability intensity function $u_s(t)$ and the cost function $UEC_{\alpha\beta}(t_p)$ for different values of α, varying the β parameter. Furthermore, Figs. 30, 31 and 32 show the same functions for different values of β, varying the α parameter.
   As carried out in Sect. 9, we can define two scenarios, in function of the impact of environmental conditions, modelled by the systemability parameters:

- Scenario 1: *hard environment effects:* when $\alpha > \beta$, the $UEC_{\alpha\beta}(t_p)$ is higher than $UEC(t_p)$ because the $u_s(t)$ is higher than $u(t)$.
- Scenario 2: *soft environment effects:* when $\alpha < \beta$, the $UEC_{\alpha\beta}(t_p)$ is lower than $UEC(t_p)$ because the $u_s(t)$ is lower than $u(t)$.

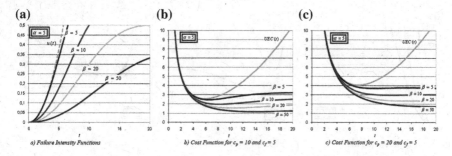

Fig. 27 a Comparison of failure intensity functions (PLP vs. systemability) for $\alpha = 5$ and $\beta = 5$, 10, 20 and 50. **b** Comparison of cost function $\mathrm{UEC}(t_p)$ versus $\mathrm{UEC}_{\alpha\beta}(t_p)$ for $\alpha = 5$ and $\beta = 5$, 10, 20 and 50, with $c_p = 10$ and $c_f = 5$. **c** Comparison of cost function $\mathrm{UEC}(t_p)$ versus $\mathrm{UEC}_{\alpha\beta}(t_p)$ for $\alpha = 5$ and $\beta = 5$, 10, 20 and 50, with $c_p = 20$ and $c_f = 5$ (Sgarbossa et al. 2015)

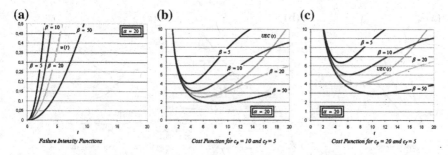

Fig. 28 a Comparison of failure intensity functions (PLP vs. systemability) for $\alpha = 20$ and $\beta = 5$, 10, 20 and 50. **b** Comparison of cost function $\mathrm{UEC}(t_p)$ versus $\mathrm{UEC}_{\alpha\beta}(t_p)$ for $\alpha = 20$ and $\beta = 5$, 10, 20 and 50, with $c_p = 10$ and $c_f = 5$. **c** Comparison of cost function $\mathrm{UEC}(t_p)$ versus $\mathrm{UEC}_{\alpha\beta}(t_p)$ for $\alpha = 20$ and $\beta = 5$, 10, 20 and 50, with $c_p = 20$ and $c_f = 5$ (Sgarbossa et al. 2015)

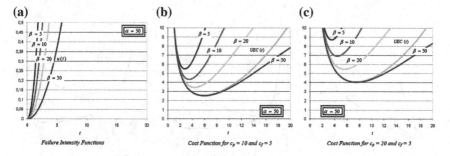

Fig. 29 a Comparison of failure intensity functions (PLP vs. systemability) for $\alpha = 50$ and $\beta = 5$, 10, 20 and 50. **b** Comparison of cost function $\mathrm{UEC}(t_p)$ versus $\mathrm{UEC}_{\alpha\beta}(t_p)$ for $\alpha = 50$ and $\beta = 5$, 10, 20 and 50, with $c_p = 10$ and $c_f = 5$. **c** Comparison of cost function $\mathrm{UEC}(t_p)$ versus $\mathrm{UEC}_{\alpha\beta}(t_p)$ for $\alpha = 50$ and $\beta = 5$, 10, 20 and 50, with $c_p = 20$ and $c_f = 5$ (Sgarbossa et al. 2015)

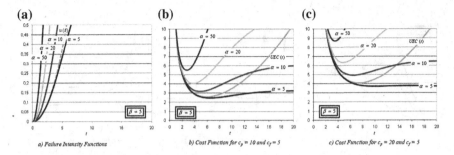

Fig. 30 **a** Comparison of failure intensity functions (PLP vs. systemability) for $\beta = 5$ and $\alpha = 5$, 10, 20 and 50. **b** Comparison of cost function $UEC(t_p)$ versus $UEC_{\alpha\beta}(t_p)$ for $\beta = 5$ and $\alpha = 5$, 10, 20 and 50, with $c_p = 10$ and $c_f = 5$. **c** Comparison of cost function $UEC(t_p)$ versus $UEC_{\alpha\beta}(t_p)$ for $\beta = 5$ and $\alpha = 5$, 10, 20 and 50, with $c_p = 20$ and $c_f = 5$ (Sgarbossa et al. 2015)

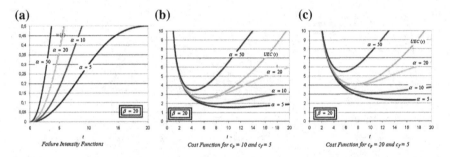

Fig. 31 **a** Comparison of failure intensity functions (PLP vs. systemability) for $\beta = 20$ and $\alpha = 5$, 10, 20 and 50. **b** Comparison of cost function $UEC(t_p)$ versus $UEC_{\alpha\beta}(t_p)$ for $\beta = 20$ and $\alpha = 5$, 10, 20 and 50, with $c_p = 10$ and $c_f = 5$. **c** Comparison of cost function $UEC(t_p)$ versus $UEC_{\alpha\beta}(t_p)$ for $\beta = 20$ and $\alpha = 5$, 10, 20 and 50, with $c_p = 20$ and $c_f = 5$ (Sgarbossa et al. 2015)

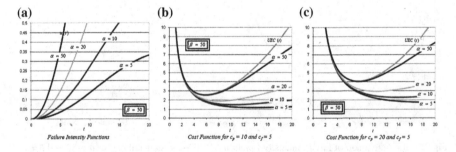

Fig. 32 **a** Comparison of failure intensity functions (PLP vs. systemability) for $\beta = 50$ and $\alpha = 5$, 10, 20 and 50. **b** Comparison of cost function $UEC(t_p)$ versus $UEC_{\alpha\beta}(t_p)$ for $\beta = 50$ and $\alpha = 5$, 10, 20 and 50, with $c_p = 10$ and $c_f = 5$. **c** Comparison of cost function $UEC(t_p)$ versus $UEC_{\alpha\beta}(t_p)$ for $\beta = 50$ and $\alpha = 5$, 10, 20 and 50, with $c_p = 20$ and $c_f = 5$ (Sgarbossa et al. 2015)

Generally speaking, increasing c_p the optimal value t_{sp}^* increases and the expected unit cost becomes higher.

Further Analysis About S-PRP: t_{sp}^* and $\%UEC(t_p^*)$ curves in function of systemability parameters

As carried out in Sect. 9, we have estimated the difference between the UEC, considering or not environmental factors, using the same scenarios defined before for S-ARP and the same Weibull shape and scale values $\lambda = 0.005$ and $\gamma = 3$, as baseline reliability parameters.

In detail, Figs. 33 and 34 show the curve of t_{sp}^* for different values of c_p, respectively 20, 50 fixed $c_f = 5$, and for each case:

(a) for different values of β (5, 10, 20, 50) and with α varying continuously from 0 to 50;
(b) for different values of α (5, 10, 20, 50) and β varying continuously from 0 to 50;
(c) varying continuously α and β from 0 to 50.

Moreover, Figs. 33 and 34 also illustrate the curve of $\%UEC(t_p^*)$ for different values of c_p, respectively 20, 50 fixed $c_f = 5$, and for each case:

(d) different values of β(5, 10, 20, 50) varying continuously α from 0 to 50;

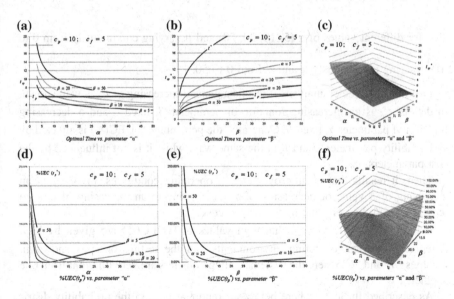

Fig. 33 **a** Curve of t_{sp}^* in function of α and $\beta = 5$, 10, 20 and 50. **b** Curve of t_{sp}^* in function of β and $\alpha = 5$, 10, 20 and 50. **c** Curve of t_{sp}^* in function of α and β. All these for $c_p = 10$ and $c_f = 5$. **d** Curve of % UEC(t_p^*) in function of α and $\beta = 5$, 10, 20 and 50. **e** Curve of % UEC(t_p^*) in function of β and $\alpha = 5$, 10, 20 and 50. **f** Curve of % UEC(t_p^*) in function of α and β. All these for $c_p = 10$ and $c_f = 5$ (Sgarbossa et al. 2015)

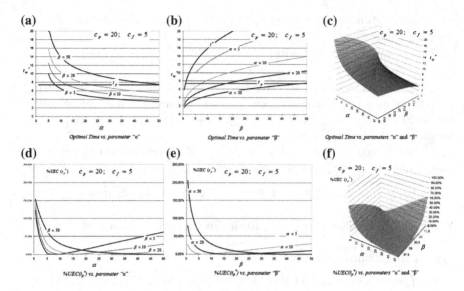

Fig. 34 **a** Curve of t_{sp}^* in function of α and $\beta = 5$, 10, 20 and 50. **b** Curve of t_{sp}^* in function of β and $\alpha = 5$, 10, 20 and 50. **c** Curve of t_{sp}^* in function of α and β. All these for $c_p = 20$ and $c_f = 5$. **d** Curve of % UEC(t_p^*) in function of α and $\beta = 5$, 10, 20 and 50. **e** Curve of % UEC(t_p^*) in function of β and $\alpha = 5$, 10, 20 and 50. **f** Curve of % UEC(t_p^*) in function of α and β. All these for $c_p = 20$ and $c_f = 5$ (Sgarbossa et al. 2015)

(e)   for different values of α (5, 10, 20, 50) and β varying continuously from 0 to 50;

(f)   varying continuously α and β from 0 to 50.

From this parameter analysis, it results a general increasing of t_{sp}^* is well shown. In details, t_{sp}^* is an increasing function of parameter β, fixed α and a decreasing function of parameter α, fixed β. Generally, the percentage %UEC(t_p^*) is affected by systemability parameter α and β in the same way, while it is not influenced by the cost parameters, c_p.

Then, it results that an improvement on operating conditions, involved in an increasing of β value or decreasing of α value, bring to an increasing of time to replace t_{sp}^* and then on total unit expected cost.

It is quite obvious to note that high values of %UEC(t_p^*) are given by high differences between operating environments and the testing ones as shown in graphics "a" and "b" where the systemability parameters are very different from one another.

As described in Sect. 9, here below we report in Fig. 35 the probability distribution Pc of the %UEC(t_p^*) for different interval of value, calculated before. Different distributions are plotted for different values of the cost parameter c_p, respectively 10, 20, and 50 while the cost parameter c_f is equal to 5. We can note

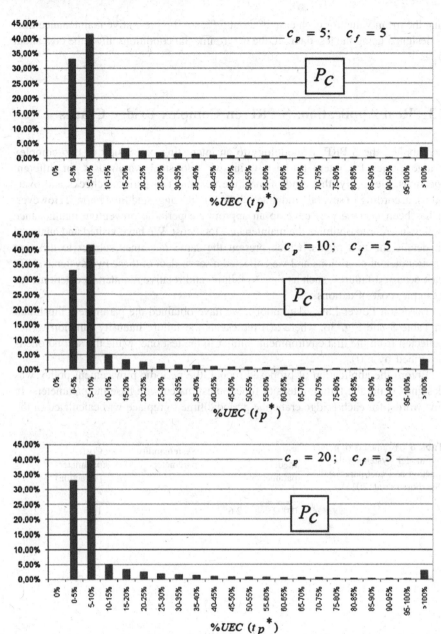

Fig. 35 Probability distribution (P_c) of % UEC(t_p^*) for different values of c_p (Sgarbossa et al. 2015)

that the mean value for each case is about 5 %, so, if a preventive maintenance time is designed using the test data set, the environmental conditions affect the cost about 5 %.

13 Real Application: S-PRP on Complex Bridge Cranes

We applied the S-PRP methodology to an interesting case study of two complex bridge cranes for internal logistics, in two two steel warehouses, in different locations. Generally, this kind of machine is subject to periodic checks and over-hauls, according to several guidelines about on the suggested time period. However, it has been necessary to develop an appropriate periodic preventive maintenance policy in order to optimize the maintenance total cost. We have visited and analysed in details the two plants, which present the same machines subject to different environmental conditions and we collected a set of data of these two systems about past occurred failures, such as time to failure, and relative maintenance actions, for example, cost of actions.

Based on Power Law assumptions, we have obtained the parameters illustrated in Table 8 and Fig. 36, where $u(t)$ represents the failure intensity function values collected from the first environment, similar to the test one, while the second one is described by $u_s(t)$.

Before the application of proposed model, the periodic replacement policy was designed using the failure intensity function estimated with baseline parameters. In few words, for each bridge crane the periodic time to replace was calculated in the

Table 8 Parameters of power law model, systemability function and cost parameters (Sgarbossa et al. 2015)

Power law model parameters		Systemability parameters		Cost parameters [k€/action]	
λ	γ	α	β	c_p	c_f
0.002	2.6	6.7	10.3	1.45	0.42

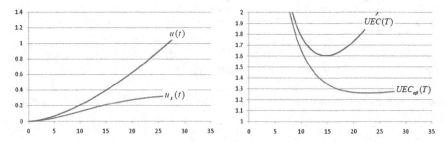

Fig. 36 Real application of the methodology to the periodic replacement policy on the complex bridge cranes (Sgarbossa et al. 2015)

same way and using the same failure intensity function, assessed during testing phase.

Considering $c_p = 1.45$ k€/action; $c_f = 0.42$ k€/action, $\lambda = 0.002$, $\gamma = 2.6$, the Power Law parameters of the baseline intensity function, the application of tradition approach (Sgarbossa et al. 2015) has given the replacement time related to the testing data, $t_p^* = 14.7$ weeks, and the unit expected cost is $\text{UEC}(t_p^*) = 0.161$ k€/ week.

Then, the systemability parameters $\alpha = 6.7$ and $\beta = 10.3$ have been obtained considering the based failure intensity function $u(t)$, the parameters' value of systemability have been estimated using the least squares estimate method, comparing the estimated value of systemability function and the observed value of reliability in the operational environment, where the failure intensity function is $u_s(t)$, different from the testing one.

Using Algorithm A2, we have calculated the optimal time for replacement, considering the environmental effects to the survival function, applying the systemability intensity function, with an optimal solution of $t_{\text{sp}}^* = 22.4$ weeks and an unit expected cost is $\text{UEC}_{\alpha\beta}(t_{\text{sp}}^*) = 0.126$ k€/week.

If we calculate the unit expected cost in the optimal solution of testing data, while taking into consideration the environmental factors, the value we obtain $\text{UEC}_{\alpha\beta}(t_{\text{sp}}^*) = 0.135$ k€/week (Fig. 36). As well demonstrated, the unit expected costs are different between these cases and the optimal solution, using Algorithm A2, and considering the environmental effects, allowing a 7 % saving, comparing the optimal solution with $\text{UEC}_{\alpha\beta}(t_{\text{sp}}^*) = 0.126$ k€/week to the cost $\text{UEC}(t_p^*) = 0.161$ k €/week, which is calculated without considering environmental conditions.

14 Some Considerations About Systemability Application on ARP

As carried out for S-ARP, in the previous sections we have investigated the impacts of different operational environments in periodic replacement policy, using a new concept called systemability (Pham2005). Systemability, with its parameters, allows considering the environmental effects in the survival functions estimation, separating them from the intrinsic survival performances of the component.

Several numerical examples have been carried out in order to validate the initial assumption. The results of this application have been illustrated with a series of graphics and summarized at the end.

It's important to highlight that the mean value of $\%\text{UEC}(t_p^*)$ is about 5 %. An real industrial application have demonstrated the importance of this study, shown a saving about 10 % using this methodology.

In few word, if a preventive maintenance time is design using the test data set, the mean major cost, due to the environmental effects, will be about 5 %.

References

Abbasi, B., Jahromi, A., H., E., Arkat, J., & Hosseinkouchack, M., (2006). Estimating the parameters of Weibull distribution using simulated annealing algorithm. *Applied Mathematics and Computation, 183*, 85–93.

Attardi, L., Guida, M., & Pulcini, G. (2005). A mixed-Weibull regression model for the analysis of automotive warranty data. *Reliability Engineering & System Safety, 87*, 265–273.

Badia, F. G., Berrade, M. D., & Campos, C. A. (2002). Aging properties of the additive and proportional hazard mixing models. *Reliability Engineering & System Safety, 78*, 165–172.

Barlow, R. E., & Hunter, L. C. (1960). Optimum preventive maintenance policies. *Operations Research, 8*(1), 90–100.

Barlow, R. E., & Proshan, F. (1965). *Mathematical theory of reliability*. New York: Wiley.

Battini, D., Faccio, M., Persona, A., & Sgarbossa, F., (2007). Reliability in random environment: Systemability and its application. In *Proceeding of 13th International Society of Science and Applied Technologies ISSAT (ISSAT 07)*, August 2–4, Seattle, Washington USA

Battini, D., Faccio, M., Persona, A., & Sgarbossa, F., (2008). Reliability of motorcycle components using systemability approach. In *Proceeding of 14th International Society of Science and Applied Technologies ISSAT (ISSAT 08)*, August 6–9, Orlando, Florida, USA

Carroll, K. J. (2003). On the use and utility of the Weibull model in the analysis of survival data. *Controlled Clinical Trials, 24*, 682–701.

Chien, Y. H. (2008). Optimal age-replacement policy under an imperfect renewing free-replacement warranty. *IEEE Transactions on Reliability, 57*(1), 125–133.

Chien, Y. H., Sheu, S. H., & Chang, C. C. (2009). Optimal age-replacement time with minimal repair based on cumulative repair cost limit and random lead time. *International Journal of System Science, 40*(7), 703–715.

Cox, D. R. (1972). Regression models and life-tables. *Journal of the Royal Statistical Society: Series B (Methodological), 34*(2), 187–220.

Crow, L. H. (1974). Reliability analysis for complex, repairable systems. In F. Proschan & R. J. Serfing (Eds.), *Reliability and biometry* (pp. 379–410). Philadelphia: SIA.

Elsayed, E. A., & Wang, H. Z. (1996). Bayes and classical estimation of environmental factors for the binomial distribution. *IEEE Transaction on Reliability, 45*(4), 661–665.

Faccio, M., Persona, A., Sgarbossa, F., & Zanin, G. (2014). Industrial maintenance policy development: A quantitative framework. *International Journal of Production Economics, 147* (Part A), 85–93.

Finkelstein, M. S. (2003). The expected time lost due to an extra risk. *Reliability Engineering & System Safety, 82*, 225–228.

Finkelstein, M. S. (2006). On relative ordering of mean residual lifetime functions. *Statistics & Probability Letters, 26*, 939–944.

Kenzin, M., & Frostig, E. (2009). M out of n inspected systems subject to shocks in random environment. *Reliability Engineering and System Safety, 94*(8), 1322–1330.

Murthy, D. N. P., & Whang, M. C. (1996). Optimal discrete and continuous maintenance policy for a complex unreliable machine. *International Journal of Systems Science, 27*(5), 483–495.

Oh, Y. S., & Bai, D. S. (2001). Field data analysis with additional after-warranty failure data. *Reliability Engineering and System Safety, 72*.

Persona, A., Pham, H., Regattieri, A., & Battini, D. (2007). Remote control and maintenance outsourcing networks and its applications in supply chain management. *Journal of Operations Management, 25*, 1275–1291.

Persona, A., Pham, H., & Sgarbossa, F. (2009). Systemability function to optimisation reliability in random environment. *International Journal of Mathematics in Operational Research, 1*(3), 397–417. ISSN:1757-5850.

Pham, H. (2003). Software reliability and cost models. Perspectives, comparison and practice. *European Journal of Operational Research, 149*, 475–489.

Pham, H. (2005a). A new generalized systemability model. *International Journal of Performability Engineering, 1*(2), 145–155.

Pham, H. (2005b). A novel systemability function and its application. *International Topical Meeting on Environmental Reliability Risk Studies*, February 24–25, Seoul, Korea.

Pham, H. (2005c). A new generalized systemability model. In *Proceedings of the 11th International Society of Science and Applied Technologies ISSAT (ISSAT 2005)*, August 4–6, St. Louis, Missouri, USA.

Pham, H. (2005d). *Springer handbook of engineering statistic.* Springer.

Pham, H., & Wang, H. (1996). A quasi-renewal process for software reliability and testing costs. *IEEE Transactions on Systems, Man and Cyernetics– Part A, 31*(6).

Pham, H., & Wang, H. (1997). Changes to: A quasi renewal process and its applications in imperfect maintenance. *International Journal of Systems Science, 28*(12), 1329–13210.

Ram, S. K. C., & Tiwari (1989). Bayes estimation of reliability under a random environment. *IEEE Translaction of Reliability, 38.*

Sgarbossa, F., Persona, A., & Pham, H. (2010). Age replacement policy in random environment using systemability. *International Journal of System Science, 41*(11), 1383–1397. ISSN: 0020-7721.

Sgarbossa, F., & Pham, H. (2010). A cost analysis of systems subject to random field environments and reliability. *IEEE Transactions on Systems, Man and Cyernetics– Part C: Applications and Reviews, 40*(4), 429–437.

Sgarbossa, F., Persona, A., & Pham, H. (2015). Using Systemability Function for Periodic Replacement Policy in Real Environments. *Quality and Reliability Engineering International, 31*(4), 617–633.

Sohn, S. Y., Chang, I. S., & Moon, T. H. (2007). Random effects Weibull regression model for occupational lifetime. *European Journal of Operational Research, 179,* 124–131.

Tamura, Y., Yamada, S., & Kimura, M. (2006). Software reliability modelling in distributed development environment. *Journal of Quality in Maintenance Engineering, 12*(4), 425–432.

Tan, J. S., & Kramer, M. A. (1997). A general framework for preventive maintenance optimization in chemical process operations. *Computer & Chemical Engineering, 21*(12), 1451–1469.

Teng, X., & Pham, H. (2004). A software cost model for quantifying the gain with considerations of random field environments. *IEEE Transaction on Computer, 53*(3), 380–384.

Teng, X., & Pham, H. (2006). A new methodology for predicting software reliability in the random field environments. *IEEE Transaction on Reliability, 45*(4), 458–468.

Wang, H. (2002). A survey of maintenance policies of deteriorating systems. *European Journal of Operational Research, 139*(3), 469–489.

Wang, H., & Pham, H. (1996). Optimal maintenance policies for several imperfect repair models. *International Journal of Systems Science, 27*(6), 543–550.

Wang, H. Z., Ma, B. H., & Shi, J. S. (1992a). A research study of environmental factors for the gamma distribution. *Microelectronic Reliability, 32*(3), 331–335.

Wang, H. Z., Ma, B. H., & Shi, J. S. (1992b). Estimation of environmental factors for the normal distribution. *Microelectronic Reliability, 32*(4), 457–463.

Wang, H. Z., Ma, B. H., & Shi, J. S. (1992c). Estimation of environmental factors for the log normal distribution. *Microelectronic Reliability, 32*(5), 679–685.

Wang, H. Z., Ma, B. H., & Shi, J. S. (1992d). Estimation of environmental factors for the inverse gaussian distribution. *Microelectronic Reliability, 32*(7), 931–934.

Wang, L., Chu, J., & Mao, W. L. (2009). A condition-based replacement and spare provisioning policy for deteriorating systems with uncertain deterioration to failure. *European Journal of Operational Research, 149*(1), 184–205.

Zhang, X., & Pham, H. (2006). Software field failure rate prediction before software deployment. *The Journal of Systems and Software, 79,* 291–300.

Zhang, X., Shin, M. Y., & Pham, H. (2001). Exploratory analysis of environmental factors for enhancing the software reliability assessment. *The Journal of Systems and Software, 57,* 73–78.

Zhao, J., Liu, H. W., Cui, G., & Yang, X. Z. (2006). Software reliability growth model with change-point and environmental function. *The Journal of Systems and Software, 7,* 1578–1587.

Part III
Maintenance Management

Innovative Maintenance Management Methods in Oil Refineries

M. Bevilacqua, F.E. Ciarapica, G. Giacchetta, C. Paciarotti
and B. Marchetti

This chapter describes the relevant steps for the design of a preventive maintenance program in an oil refinery plant and its application. The method was developed during a period of 3 years in one of the main Italian refinery.

The preventive maintenance program for the most critical equipments of the reference plant is developed analyzing the used oil sector standards (ASA, API, UNI, ISO, ASME). These standards have been used for several oil refinery equipments such as pumps, tanks and processing furnaces.

The research has been focused on the turnaround process that has been analyzed by two different approaches: a risk-based method and an innovative criticality index. Both have been used as decision tools for clearly assessing the maintenance tasks and equipment to include in the process. The relevant technical specifications (operating conditions, process fluids, and safety system configuration) have been gathered for each component. The data have then been used to assess the failure mode effects and severity so that to calculate the criticality index. The outcome of the analysis has highlighted important relations among variable, confirming, and validating the results obtained through other reliability assessment techniques (see Bevilacqua et al. 2000, 2012).

The results have highlighted a clear improvement in terms of resources usage optimization, outage duration, production loss and total costs reduction, and increase of the interval between two shutdowns.

M. Bevilacqua · F.E. Ciarapica · G. Giacchetta · C. Paciarotti
Dipartimento di Energetica, Università Politecnica della Marche,
via Brecce Bianche, Ancona, Italy

B. Marchetti (✉)
Università degli Studi ECampus, via Isimbardi 10, Novedrate, Como, Italy
e-mail: barbara.marchetti@uniecampus.it

© Springer-Verlag London 2016   197
H. Pham (ed.), *Quality and Reliability Management and Its Applications*,
Springer Series in Reliability Engineering, DOI 10.1007/978-1-4471-6778-5_7

1 Introduction

In the contemporary highly challenging environment, a reliable production system is a crucial factor for competitiveness, almost all the processing and manufacturing sectors are required to maximize availability and efficiency of equipment, controlling failure and deterioration, guarantee a safe and correct operation and minimizing the costs (Zhaoyang et al. 2011). In case of oil and gas industry, the challenge it is even harder due to the highly critical nature of its activities that cannot afford unexpected failures (Telford et al. 2011).

For refineries maintenance, inspection functions are the backbone for safe and reliable plant operations and play a pivotal role in efficiently achieving the desired production target and profitability for the company. Maintenance functions in the refinery include mechanical, electrical, instrumentation, and civil functions, which are responsible for monitoring, repair, and maintenance of equipment in the respective defined areas. Inspection is responsible for certification of quality adherence on the repair and maintenance activities through monitoring of static equipment functioning during plant operations. Fulfillment of statutory requirements and liaison with regulatory bodies, failure analysis, and remaining life assessment of plant equipment to establish a repair replacement plan in advance for reliable plant operations, metallurgy upgrading, etc., are also performed by inspection. On the basis of the process condition, on-stream monitoring of all critical equipment is decided by an individual group and monitored religiously. Preventive maintenance, predictive maintenance, a structured repair system, and full-fledged plant shutdown management must be reliable (Kosta and Keshav 2013).

Among others the maintenance of refineries raises the issue that, in the current international context, average downtime in such industries can reach 10 % of the production time, and refineries are often used only at 60 % of their capacity (Dossier special Algérie/Special Issue for the Algerian Petroleum 2002). Thus, important financial gains and safety improvements can be affected by optimizing maintenance tasks. In this paper, we focus on the dynamic scheduling of these maintenance tasks (Aissani et al. 2009).

Planned refinery turnarounds are major maintenance or overhaul activities and represent the most critical and expensive maintenance task for industries. A turnaround is a global maintenance operation that involves the partial or total shutdown of the plant and involves either the critical equipment, that is not possible to isolate during the normal functioning of the plant, and those that have shown problems or need a periodical inspection. The frequency of this operations vary by type of unit, and often requires 1–2 years of planning and preparation, and sometimes longer when major capital equipment changes are needed. A major unit turnaround may last about 20–60 days and involves as many as 1500–2000 skilled contractor workers brought on-site to perform a myriad of interrelated jobs that require significant coordination and safety measures.

In principle it could be possible to shut down only the portion of the plant needing maintenance, but the work is normally too disruptive to continue operating. Moreover, labor assets to perform the daily maintenance work in the operating portion of the plant will be in short supply. Thus, the maintenance outage most often involves a total plant shutdown. A turnaround in any major refinery unit can affect production of finished products, such as gasoline or distillate. Safety is a major concern when implementing refinery turnarounds. Refineries run with materials at high temperatures and high pressures, and some of the materials themselves are caustic or toxic and must be handled appropriately. Maintenance is required to assure safe operations, and turnarounds themselves require extra safety precautions.

The turnaround is a common process in the oil and gas industries so that many best practices and case studies are available on literature as well as the related regulations and standards; however, there are few research works dealing with the development of innovative tools for managing and optimizing the correlated operations.

The research wok presented in this chapter had the aim of filling this gap proposing an innovative maintenance program applied to the turnaround management and based on two different methodologies: the risk analysis and the application of an innovative criticality index that has been developed by the authors with the purpose of being an economic, flexible, simple, and complete decision tool to apply for evaluating the criticalities of equipments and plants in order to optimize the use of economical, human, and instrumental resources needed for the refinery maintenance activities.

2 Relevant Literature

Several papers related to maintenance practices, methodologies, and tools have been examined, as well as regulations and standards.

In the last decade, the Total Productive Maintenance (TPM) and reliability centered maintenance (RCM) methodologies have been accepted as the most promising strategies for improving maintenance performances (Ahuja and Khamba 2008, 2009); advantages and issues related to its application have been addressed in many studies (Khanna 2008; Bhangu et al. 2011).

Innovative maintenance strategies and methods applied to a wide variety of fields have been proposed by several authors in the past few years.

Savino et al. (2011) developed a modified FMECA methodology in which the criticality evaluation is made considering both production performances and users/workers safety. Tsakatikas et al. (2008), demonstrated the use of FMECA together with a Decision Support System (DSS) for the establishment of spare parts criticality with a focus on industrial maintenance needs. Sachdeva et al. (2008), proposed a new maintenance decision strategy alternative to traditional FMEA approach and based on Analytical Hierarchical Process (AHP) technique which provides an aid to the maintenance managers/analysts to formulate an efficient and effective priority ranking of the various components/failure modes based on a number of maintenance issues.

Ghosh and Roy (2010) presented a Multiple-Criteria Decision-Making (MCDM) methodology for selecting the optimal mix of maintenance approaches—Corrective Maintenance (CM), Time-Based Preventive Maintenance (TBPM) and Condition-Based Predictive Maintenance (CBPM)—for different equipment in a typical process plant. According to Rosmaini et al. (2011), preventive maintenance (PM) strategies have been as well addressed in many studies. One of the popular strategies that is widely used is the preventive replacement (PR), which aims to determine the optimum replacement time. The critical issue, however, is that most studies assume the aging of a component to be time dependent. In reality, the failure of a component is influenced by an external factor. They present the process of revising or updating the PR time by considering external factors by using the proportional hazard model (PHM). Zhou et al. (2010) proposed an opportunistic preventive maintenance policy for multiunit series systems based on dynamic programming, with integrating the online information of the intermediate buffers. An optimal preventive maintenance practice is determined by maximizing the short-term cumulative opportunistic maintenance cost savings for the whole system which is a combination of the maintenance cost saving, the downtime cost saving, the penalty cost for advancing the preventive maintenance action, and the penalty cost for work in process.

Conn et al. (2010) analyzed the maintenance issues when dealing with multiple heterogeneous systems, a mix of maintenance strategies (replacement and inspection), economic dependence among maintenance activity, etc. With this initial determining factor, the design or reorganization of a maintenance plan must begin with an exhaustive study of all the equipments and facilities, the purpose of which is to obtain all the information necessary to justify and analyze the viability of each maintenance task, thus to decide the best maintenance strategy for the plant. Such a study must begin with a detailed inventory of equipments and facilities, including their characteristics and functional interrelationships; records of past failures, if they exist; the cost of acquisition and supply; the direct and indirect cost of maintenance, if this information is available; needs and operational factors; the type of maintenance to be carried out and any legal or contractual obligations concerning maintenance (such as periodic inspections subject to regulation, guarantee periods, etc.); the legal or contractual obligations of the company, such as those related with legal sanctions or penalties regarding the quantity or quality of the production; the means (tools, auxiliary equipment, etc.) available for maintenance; human resources, and the qualifications of the personnel available; the maintenance tasks that can or must be contracted out and any other aspects that are relevant for the case in question, Gomez et al. (2006).

Peres and Noyes (2003) present a methodology for the evaluation of maintenance strategies by taking into consideration the effect of certain variables on the dynamic of maintenance, on its structure and its context of evolution. This factual approach considers the return to an operational state as a point of entry into the evaluation of a strategy. It is based on the treatment of data collected from the history of the behavior of equipment on which the strategy to be evaluated can be applied.

You-Tern et al. (2004) focused their work of the idea that the development of maintenance strategies must take into account that resources are limited and, therefore, maintenance will be imperfect.

Several papers on the evaluation of components reliability have also been taken into account. Reliability is a fundamental aspect for proper maintenance execution, current reliability evaluation methods are based on the availability of knowledge about component states. However, component states are often uncertain or unknown, especially during the early stages of the development of new systems. In such cases it is important to understand how uncertainties will affect system reliability assessment. Pham (2013) developed a new software reliability model incorporating the uncertainty of system fault detection rate per unit of time subject to the operating environment. Brissaud et al. (2011) presented a model of failure rates as a function of time and influencing factors that allows to represent the system life phases considering a large variety of qualitative or quantitative, precise or approximate influencing factors. Zhang and Mostashari (2011) proposed a method to assess the reliability of systems with continuous distribution of component based on Monte Carlo simulation. They demonstrated that component uncertainty has significant influence on the assessment of system reliability.

Garg et al. (2010) developed a reliability model for systems that undergoes partial as well as direct total failure, for calculating both time dependent and steady state availability under idealized as well as faulty Preventive Maintenance (PM).

Some of the relevant literature analyzed focused in maintenance strategies applied to oil refineries.

Prabhakar and Jagathy Raj (2013) proposed a maintenance approach based on the implementation of RCFA programs and reliability-centered maintenance, alternative to traditional methods of reliability assurance, like preventive maintenance, predictive maintenance, and condition-based maintenance that they consider inadequate to face the extreme demands on reliability of the plant. The authors developed an accelerated approach to RCM implementation, that, while ensuring close conformance to the standard, does not require large amount of analysis, long implementation, and a high number of skilled people as the traditional ones.

Aissani et al. (2009) presented a multiagent approach for the dynamic maintenance task scheduling for a petroleum industry production system. Agents simultaneously ensure effective maintenance scheduling and the continuous improvement of the solution quality by means of reinforcement learning, using the SARSA algorithm. Reinforcement learning allows the agents to adapt, learning the best behaviors for their various roles without reducing the performance or reactivity. The results obtained in a petroleum refinery demonstrated the innovation of their approach.

Christen et al. (2011), a competing risk model, namely a Random Sign model, is considered to relate failure and maintenance times. They proposed a novel Bayesian analysis of the model and applied it to actual data from a water pump in an oil refinery, developed an optimal maintenance policy under a formal decision theoretic approach, using a competing risk model, namely a Random Sign model, to relate failure and maintenance times. They proposed a novel Bayesian analysis of the model and validated it with data from a water pump in an oil refinery.

Laggoune et al. (2009) proposed a maintenance plan based on opportunistic multigrouping replacement optimization for multicomponent systems applied to continuous operating units such as oil refinery. Their approach is based on the analysis of individual components, according to the well-known age-based model. The optimization algorithm allows rearranging the optimal individual replacement times in such a way that all component times become multiple of the smallest one to allow for joint replacements. In this way, the times obtained by the multigrouping approach do not give individual optimality conditions, but satisfies the optimal cost regarding the whole system.

In this scenario, the optimization of major maintenance activities represents a key factor for improving safety and economic aspects. Turnaround in particular, is the most challenging within the maintenance tasks: industry surveys report that between 35 and 52 % of maintenance budgets are expended in individual area or whole plant shutdowns.

Kister and Hawkins (2006) stated that the majority of preventive and planned maintenance work is performed while the manufacturing plant is in operation but major maintenance works that cannot be performed while the plant is operating are also periodically required.

An optimal maintenance approach is a key support to industrial production in the contemporary process industry and many tools have been developed for improving and optimizing this task.

As for the standard to consider, in the oil sector the current regulations to elaborate a preventive maintenance program, (ASA, API, UNI, ISO, ASME) represents the reference for the key equipments of the plant: pumps, tanks, reservoirs, and processing furnaces.

3 The Test Plant

The study was carried out in a pilot refinery located in central Italy that operates directly in the supply of crude oil and semi-refined products destined for processing holding one of the most modern and technologically advanced refineries in the country. It is certified to ISO 14001 for environmental protection, OHSAS 18001 for safety and ISO 9002 for quality. Built in 1950 the refinery covers an area of over 700,000 m^2, it has a refining capacity of 3,900,000 tons/annum, equivalent to approximately 85,000 barrels per day, and a storage capacity of over 1,500,000 m^3. The refinery is also equipped with a dispatch facility by land which has a potential capacity of over 12,000 tons/day. It is connected to the sea through a combination of marine terminals which can accommodate tankers from 1,000 to 400,000 tons.

The company owns also one of the first IGCC (integrated gasification combined cycle) plants constructed in Europe.

3.1 The Production Cycle

The refinery offers a production system that ensures flexibility in operations on the basis of the crude oil used and the best possible yield in distillates. The system employs "Hydroskimming", together with a conversion system "Thermal Cracking/Visbreaking": thus high-quality products can be made that meet environmental specifications that are required by the regulations.

The processes of refining as the following sequence: Topping—Catalytic Reforming—Isomerization—Vacuum—Visbreaking—Thermal Cracking. A synthetic scheme of the process is presented in the following Fig. 1.

3.2 The Turnaround Process

The turnaround is a global maintenance operation that involves the partial or total shutdown of the plant and consists in modifying and/or restoring its working conditions through the intervention on its components with the aim of increasing

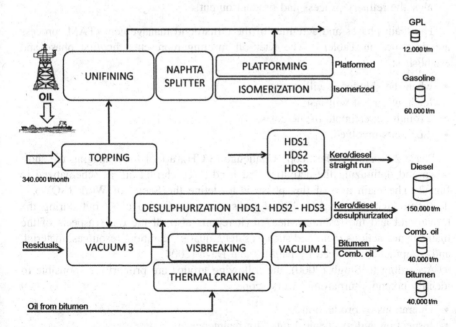

Fig. 1 Scheme of the refining process in the analyzed refinery

the energetic efficiency, guarantee a homogeneous functioning and the integrity of the security systems, limit the wear to increase the service life. It is a maintenance process that involves either the critical equipment, that is not possible to isolate during the normal functioning of the plant, and those that have shown problems or need a periodical inspection. The turnaround management is very challenging since all the interventions have to be executed in a very strict time span (4 weeks at maximum), in which all the resources (human, materials, technical, economical) have to be effectively organized and planned.

Maintenance activities during a planned turnaround might include:

- Routine inspections for corrosion, equipment integrity or wear, deposit formation, integrity of electrical and piping systems;
- Special inspections (often arising from anomalies in the prior operating period) of major vessels or rotating equipment or pumps to investigate for abnormal situations;
- Installation of replacement equipment for parts or entire pumps or instruments that wear out;
- Replacement of catalysts or process materials that have been depleted during operations.

Improvement activities could include:

- Installation of new, upgraded equipment or technology to improve the refinery processing;
- Installation of new, major capital equipment or systems that may significantly alter the refinery process and product output.

The main phases and activities of the turnaround management (TAM) process are described in Table 1. The Kick-off meeting represents the first phase and establishes:

- when the shutdown will start;
- for how long it will last;
- a rough cut estimate of the costs;
- the plants involved.

Phase 2 represents the Scope Challenge (S/CH), a tool for managing the shutdown and optimizing the activities that need to be carried out as schematized in Fig. 2. The main aim of this phase is to define the Scope of Work (SOW), a document that contains the list of tasks and activities to carry out during the turnaround and the resources needed; it reports also all the other aspects of the maintenance actions identified such as safety, quality, duration, resources, material, and equipment as defined by Duffuaa and Daya (2004).

According to Singh (2000), the following groups are primarily responsible to identify potential turnaround work scope:

- Operations or production
- Inspection and mechanical integrity maintenance

Table 1 Main phases of a turnaround process

Phase 1 definition and approach	Kick-off meeting ✓ Management strategy definition; ✓ Scope of Work definition; ✓ Team, organization, responsibilities, and roles approval; ✓ Determination of communication matrix, of contracts strategies, of the Health, Safety, Environmental, and Quality (HSEQ) general plan
Phase 2 scope definition	Scope challenge ✓ Preliminary scope identification; ✓ Scope improvement; ✓ Definition of the operative system and needs; ✓ Development of initial plans and cost estimation; ✓ Definition of materials need and of items with long waiting list; ✓ Strategies compilation; ✓ Risk analysis
Phase 3 Scope improvement	✓ Scope tune up and evaluation of improvement opportunities; ✓ Development of works assignment; ✓ Programs and scheduling evaluation; ✓ Identification of the pre-shutdown works; ✓ Development of subcontracting plans; ✓ Tune up of HSEQ plans; ✓ Training needs development; ✓ Start of the logistic and materials planning; ✓ Compilation of the resources plan and integration with contractors
Phase 4 budget & scheduling	✓ Fine regulation of timetable and record of communication sections; ✓ Re-improvement of scope and methods; ✓ Critical path/Density analysis/Workload; ✓ Contingencies programming; ✓ Definition of procedures coordination; ✓ System variations approval; ✓ Exercitations HSE/quality; ✓ Details
Phase 5 Mobilization and execution	✓ Execution of all preparatory works; ✓ Subcontractors integration; ✓ Collocation of material equipments; ✓ Assembling and scaffoldings access; ✓ Isolation removal; ✓ Production and preassembling; ✓ Cleaning and controls; ✓ Temporary installations
Phase 6 demobilization and closedown	✓ Critics and lessons learned after TAM; ✓ Scheduling and budget control; ✓ Updating of documents with modifications; ✓ Shutdown reports; ✓ Verification of employed teams; ✓ Preparation of working list for next TAM

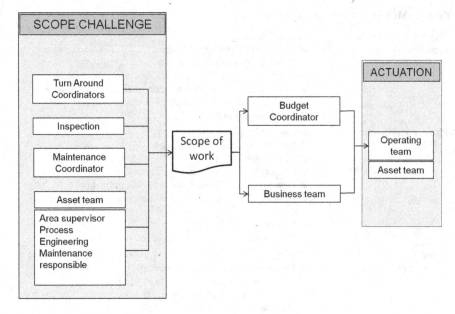

Fig. 2 Scope challenge scheme

- Rotating and machinery
- Instrument, electrical and analyzer
- Process and technical
- Capital projects and engineering
- Turnaround planning
- Safety, health, and environment
- Outside parties, e.g., vendors, regulatory agencies, insurance, etc.

In order to achieve world-class results on plant shutdowns and turnarounds, companies must strive to finalize the work scope 8–12 months before the turnaround.

The poor exercise of work scope identification and validation done during the preparation phase results in a lot of additional and emergent work creeping up during the execution period. This can cause a lot of constraints on the available resources resulting in schedule slippages and escalation of costs.

Phase 3 is used to tune up and optimize the schedule of the activities defined in the SOW, to define the labor (internal and from subcontracts) and the training needed, and the logistic and resources plan.

In phase 4 the detailed budget plan and activities scheduling is prepared as well as a contingency program; moreover the coordination of the different procedures is defined.

In phase 5, first the preparatory works and then the planned maintenance activities are carried out.

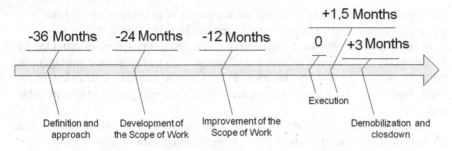

Fig. 3 Turnaround timeline

Phase 6 closes the process, summarizes the results obtained and starts the preparation of the working list for the next turnaround.

The shutdowns are scheduled on a 10-year basis and this program is shared within the refinery personnel through the Multiyear Shutdown Plan.

Although TAM can cause a production plant availability reduction of about 2–3 % on a yearly basis and its economical benefits due to reduction in equipment failures can increase the firm profits due to the minor loss production up to 15 %.

As an example the last oil refinery TAM lasted 18 days, with a medium use of more than 500 workers/day, a total of 100,000 worked hours and about 10 million euro of investment. To reach all the planned objectives of the TAM it was necessary the following timeline has been drawn as shown in Fig. 3.

4 Materials and Methods

Bahrami and Price (2000) demonstrated that different methods have been formulated for identifying the critical equipments of a process, some centerd exclusively on the effect of failure on the service and others are based on the involved risk, such as the HAZOPs (Hazard and Operability Studies) method, Casal et al. (1999), or safety equipment (Cepin 2002; Hokstad et al. 1995). In other cases, the aim is to classify the maintenance activities to be carried out rather than to classify the equipment itself, e.g., the FMEA (Failure Mode and Effects Analysis) method (Bevilacqua et al. 2000). Risk-Based Inspection (RBI) is proposed as a sound methodology for identifying and assessing the risks of operating plant equipment and is widely used by many upstream and downstream oil and gas, petrochemical companies worldwide to various degrees (Arunraj and Maiti 2007). The RBI implemented in the analyzed refinery is the Shell RBI method (S-RCM Training Guide 2000). It uses equipment history and the likely consequences of equipment failure to determine Inspection regimes focused on actual risks, so as to prevent unsafe incidents from occurring, and is based on API 581 base resource document which involved representatives from a number of major oil and petrochemical companies, and a comprehensive statistical analysis of petrochemical facilities over a number of years (API Recommended Practice 580–581 2002a, b).

In this paper we propose a methodology that, using the synergies provided by the simultaneous adoption of risk-based analysis and maintenance management methods, enables a preventive maintenance program to be made with a view to the production of servicing plans that ensure greater reliability at the lowest possible cost.

The method applied in the refinery for managing the TAM (Shell method) has allowed to reorganize the whole process highlighting measures to achieve better results in terms of performance and flexibility. The validation of the method focused on two main aspects: the achievement of the defined objectives and its economic convenience.

The process plants of the analyzed refinery include more than 10,000 equipments. After the collection period of 18 months, the technical specifications (operating conditions, process fluids, and safety system configuration) have been collected for each component and, in a following step, the failure mode effects and severity have been analyzed so that to calculate the criticality and reliability indexes.

The development and the implementation of both approach (risk-based method and Criticality Index), and the application to a specific stage in the maintenance activities of a medium-sized refinery was carried out by a panel of experts.

The panel was made up of 10 participants, and included 2 academics, whose research studies are mainly focused on risk analysis and maintenance management, 3 technical operators and 3 managerial operators involved in the maintenance processes, 2 operators from an external service company (ESC) called into manage the maintenance activities on the basis of a global service contract.

The following steps were involved in the criticality analysis of the analyzed plant:

- identify and list all the equipment and components liable to maintenance;
- define the relevant data, the preventive maintenance plans, and the recording of any faults;
- attribute the factors value to the risk items considered;
- extract only the objects that can be effectively monitored using Computer Maintenance Management System (CMMS);
- determine the criticality of the single item and select the candidates for monitoring.

The most critical items have been classified in four severity classes according to their possible effects:

- *First class* catastrophic; complete failure of the equipment or component, with very high risk for workers and environment.
- *Second class* critical; the failure mode limits significantly equipment performances.
- *Third class* marginal; the failure mode causes a degradation of the performances.

- *Fourth class* small; the failure mode causes small problems to the users, but it is not possible to notice an important deterioration.

The equipment selected for predictive maintenance was then given ad input in a machine-monitoring matrix in order to attribute the related monitoring activities (Ai) and their frequency for each machine Mj.

5 Risk-Based Methodology

5.1 Scope of Work (SOW) Development

The elaboration of the SOW list has to be based on a rational evaluation of the costs–risks–benefits that, starting from a structured analysis of each alternative, then defining the TAM duration, the costs and the availability of plants relative to the inclusion of the different items in the turnaround.

A strategic issue to develop an effective SOW is the project team: in the presented case, it was composed by the shutdown manager and leader, the operation lines manager, the technology, inspections, planned services, costs and programming responsible, as well as the engineers team and the members of the senior management team. The methodology for assessing the best alternative is composed by three main processes:

- Simple process;
- Extended process;
- Shell Global process.

5.1.1 The Simple Process: Application of the RAM Matrix for the Risks Management

In this type of process the risk is defined as:

$$\text{RISK} = (P) \text{ x } (E) \text{ x } (C)$$

Being P = probability of the thread occurrence; E = level of exposure to the thread; C = thread consequences. The quantification of the probability is quite complex and if reliable data are not available it is necessary to rely on keywords; an example for the probability assessment is presented in Table 2.

Keywords are used also to quantify the exposition, the consequences of the health and safety, the loss of image for the company, the environmental impact, the economical consequences (the relative tables are not reported here).

Table 2 Keywords to quantify the probability parameter (P)

Probability			
Keyword	Probability	Description	Example
Very rare—probability almost null	0.001	A	Fire from pump with severe damages
Unlikely but possible	0.01	B	Explosion of gas cloud
Rare but possible	0.1	C	Security valve blocked
Quite probable—isolated possibilities	0.5	D	Fire with moderated economical damages
Possible occurrence— possibility of repetition	1	E	Unexpected pump stop

In order to manage the risk, the R&M (reliability and maintenance) management team of the refinery applied the decision matrix presented in Table 3. This matrix is used to choose the best alternative in a set characterized by a calculated level of risk.

If the risk level assigned with the RAM fells within the critical range, the R&M team performs a Root Cause Analysis (RCA), of the failure for assessing causes and avoid repetitions of the event. For each improvement choice, it is necessary to recalculate probability and consequences to ascertain the acceptability of the new level of risk.

5.2 Extended Process: J-Factor Application

In addition to the conditions above defined, the team has to evaluate the most profitable alternative in terms of maximum risk reduction efficacy.

The Alternative 0 is the reference alternative and it is used to compare all the others alternatives. The index for ranking the different solution is called Justification Factor (J- Factor) and is given by:

$$J - Factor = \frac{OR - NR}{CAA}$$

where the OR = original risk, represents the initial risk calculated for the thread (Alternative 0); NR = new risk for implementing a different solution (Alternative 1); CAA = Cost of the alternative solution. The J-Factor gives an indication about the amount of risk reduction for each euro invested, the higher the J-Factor, the better is the alternative.

Table 3 Decision matrix (RAM) applied in the analyzed refinery

Decision matrix							Probability				
Index			Consequences								
Severity	Keyword		Health and safety	Economical	Environment	Image	A	B	C	D	E
							Very rare: probability almost null	Unlikely but possible	Rare but possible	Quite probable Isolated possibilities	Possible occurrence Possibility of repetition
1	Minor		Limited consequences discomfort: medication/accident 1–3 days	Negligible damage <10,000 €	Negligible effects	Negligible impact within company confines	1	2	3	4	10
2	Moderate		Poor health 3–10 days	Minor damage >10,000 € <100,000 €	Short-term effects	Limited impact surrounding areas	2	4	6	16	30
3	Severe		Occupational disease Reversible in 10–30 days	Localized damage >100,000 € <1 M €	Short-term effects noted outside	Local territory	3	6	18	24	60
4	Very severe		Permanent damage to health: >30 days Accident involving several people	Important damage >1 M € <10 M €	Transient reversible damage	Regional level impact	4	16	36	64	80
5	Catastrophic		Lethal exposure: fatal accident	Extended damage >10 M €	Permanent environmental damage	National level impact	10	30	60	80	100

5.3   Shell Global Process

This method consists of five main steps:

1. Risk definition by using criticality matrix and confidence rating;
2. Generation of a list of alternatives for each item falling in the high criticality' area;
   Alternative 0 = no action
   Alternative 1 = selection of action 1
   Alternative 2 = selection of action 2
   Alternative n = selection of action n
3. Calculation of risk reduction (J-Factor) for each alternative selected;
4. Definition of the MUST (mandatory actions) and WANTS (suggested actions; desirable conditions associated with convenience indexes);
5. Classifying and choosing the best alternative.

5.4   Decision-Making Method Applied for an Optimal Definition of TAM Equipment List

A key factor for the execution of a TAM capable to ensure proper maintenance actions, to respect the defined timeline, the estimated costs and resources usage, is the definition of the equipments and components that have to be included in the process. A common mistake in the management of this kind of actions is the inclusion of items that could be treated during plant working conditions with ordinary maintenance. To limit their number to the minimum possible means to increase the possibility of being compliant with the TAM schedule and avoiding unnecessary loss of production.

Following the Shell global process it has been possible to classify items that have necessarily to be included in TAM (i.e., equipment that cannot be isolated while the plant is in normal operation), and items that can be subjected to other maintenance alternatives.

This method allowed also to evaluate the failure risk and the benefits deriving from any preventive measures (the product of the cost of the measures multiplied by the new probability of the failure's occurrence) and thus compare the failure risks. In Fig. 4 a scheme of the decision-making method applied is presented.

Other than identifying the phenomena of critical failure to analyze, the risk matrix determines the priorities, i.e., the deadline for starting analysis of the event and maintenance plans.

The decision-making process carried out using the risk matrix require the involvement of an experts team able to evaluate every aspects of the different operations examined. In this study, they defined the likelihood of failure (LOF) and the consequence of failure (COF) of each item.

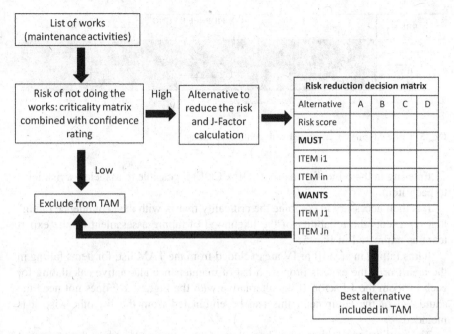

Fig. 4 Decision-making method

Table 4 Criticality matrix

Risk of not doing the work (likelihood of failure)	Criticality classes				
H	L	H	VH	U	U
M	L	M	H	VH	U
L	N	L	M	H	VH
N	N	N	L	M	H
Consequences	N	L	M	H	VH
Economical	<10 k $	10–100 k $	0.1–1 M$	1–10 M$	>10 M$
Safety	No cons.	Slight cons.	Major cons.	Single fatality	Multiple fatality
Environment	No effect	Slight effect	Localized effect	Extensive effect	Very severe effect

LOF is divided in 4 classes: N = negligible, L = low, M = medium, H = high; COF are evaluated taking into account economical, safety, and environmental aspects and divided in 6 classes: N = negligible, L = low, M = medium, H = high, VH = very high, U = unacceptable as expressed in Table 4.

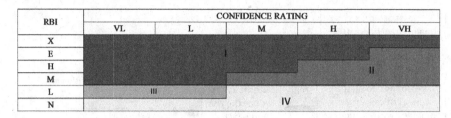

Fig. 5 Criticality matrix combined with confidence rating

Entering in the criticality matrix (LOF x COF) it possible to associate a risk level to each item.

It is then necessary to combine the criticality matrix with the "confidence rating" that represents the reliability of the likelihood of failure assessment by the expert team as expressed in Fig. 5.

Items falling in area III or IV are excluded from the TAM list; for items falling in the area I or II the experts propose a list of maintenance alternative calculating for each of them the J-Factor. If the alternative with the highest J-F does not need the shutdown of the equipment, this can be eliminated from the list otherwise it is included.

The application of this method has allowed the company to reduce the amount of turnaround equipment by 24 % with respect to the previous TAM, with a consequent reduction of costs of about 18 % and of resources usage of about 20 %. The item excluded were subjected to current maintenance due to their low associated risk.

6 Criticality Index-Based Method

This main aim in developing this innovative index was to create an economic, flexible, simple, and complete decision tool to apply for evaluating the criticalities of equipments and plants in order to optimize the use of economical, human, and instrumental resources needed for the maintenance of the refinery as schematized in Fig. 6. The objective was not only to use it for TAM optimization but as an instrument to also manage current maintenance actions. It represents an alternative to the risk-based method; nevertheless it could be also possible to apply them in combination to better exploit their strongest characteristics and verify the results of the analysis.

The questions to which this tool has to respond are: where, when, and how to intervene.

To reach this goal it is necessary to know the initial condition of each equipment and update them yearly. This allows to assess the critical equipment for the maintenance of safety and environmental standard and of production performances analyzed in the refinery, to define the causes of such criticality (failures, costs,

Fig. 6 CI characteristics and output management

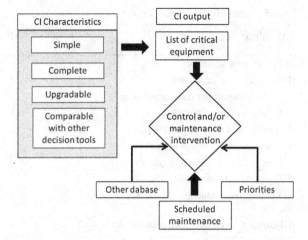

accidents, etc.) and the preventive and mitigating actions necessary to restore an acceptable level of criticality.

The CI differs from the common risk analysis index and is obtained by adding, with a defined weight, data intrinsic to each equipment related to:

- temperature;
- pressure;
- fluid;
- complexity;
- presence of spare equipment;
- Effects of failure on the plant;
- Failures (from historical data available on the plant);
- Environmental inconvenient;
- Working inconvenient;
- Accidents;
- Results from inspections;
- Improvement actions;

In order to calculate the index, 15 operating factors have been collected for each item.

$$\mathrm{CI}_m = \sum_{i=1}^{15} f_{im} \cdot p_i$$

where:
CI_m criticality index for the m item
f_{im} value of factor i for the component m
p_i weight of the factor i

Table 5 Factors f_i and relative weight

FACTOR f_i	P_i (weight of each factor i)	Positive or negative effects
(A1) corrosion sensitivity	4	+
(B1) economical sustainability	3	+
(C1) sensitivity to get dirty or scraping	1	+
(D1) process criticality (RBI)	1	+
(E1) safety and environmental impact	6	+
(F1) warning of technical committee	1	+
(G1) sensitivity to transitory phases	3	+
(H1) equipment complexity	2	+
(I1) impact on production efficiency (RCM)	2	+
(L1) impact on maintenance plan (work orders)	2	+
(M1) sensitivity to serious failure[a]	4	+
(N1) sensitivity to slight failure[b]	1	+
(O1) possibility of spare	−3	−
(P1) item under preventive–predictive maintenance	−4	−
(Q1) item under improving maintenance	−6	−

[a]*Shutdown* total shutdown of operations in a plant due to any type of anomaly
[b]*Slowdown* reduced working capacity of the plant, less than 75 % of the expected value

According to the panel of expert the most important factors to consider are listed in Table 5. The panel also indentified the weight of each factor, on a 6-point scale from low importance to high importance, and the possible contribution to CI (positive or negative).

Figure 7 represents the scheme used for the CI development.

Each factor f_i was then divided into subclasses and the relative factor value was defined as described in Table 6.

Four classes of criticality were defined for each item: Iper critical (CI ≥ 50); Critical (40 ≤ CI < 50); Subcritical (35 ≤ CI < 40); Noncritical (CI < 35).

The experimental phase was carried out for a period of 2 years for evaluating the results of the maintenance actions implemented. In the following Tables 7 and 8 the CI calculated for the first and the second years is reported.

In the following Table 9 the comparison of the index obtained in the first and second years shows that one equipment (White) maintained the same level of criticality; 6 items increased their CI passing from the subcritical class to the iper critical (Gray I); 6 items decreased their CI going from iper critical to subcritical class due to the maintenance program implemented (Gray III); finally two items remained in the same class even if with a lower CI due to the maintenance actions carried out (Gray II).

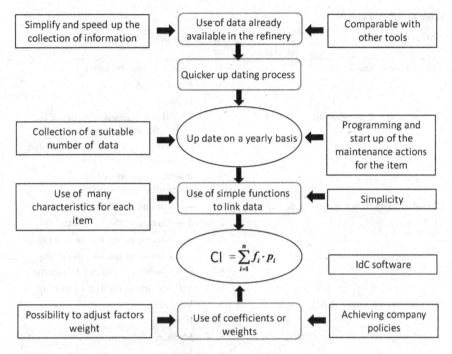

Fig. 7 Scheme representing the CI development steps

The maintenance activities implemented on the items that in the first year fell in the iper critical or critical class produced a decrease of their criticality level. In the second year of the study were identified:

- 12 iper critical equipments;
- 29 critical equipments;
- 27 subcriticals.

In order to define which item include in the turnaround other simulations have been carried out; in particular the weight of factors f_i linked to a particular specification established by the panel of experts have been modified. In the first simulation it was decided to modify the weight of the cost related parameters (B1, G1, H1, M1, N1, defined in Table 6). The weight value has been increased by a factor of two. In the second simulation a different weight has been given to parameters related to the refinery production process (D1, I1, E1 of Table 6). In this case it was incremented by a factor of 4. In the last simulation, only the weight of the factor E1, safety and environment, has been halved.

The simulation demonstrated that the most critical items remained the same and only the CI value changed but without modifying the criticality level. The example of the first simulation is shown in Table 10.

Table 6 f_i factor classes

FACTOR f_i	Number of classes	Factor value	Item description
(A1) corrosion sensitivity	3	1	Item with low thickness
		3	Slight corrosion
		6	Drilling
(B1) economical sustainability	3	1	10,000 € < maintenance cost of last 18 months < 100,000 €
		3	100,000 € < maintenance cost of last 18 months < 1,000,000 €
		6	maintenance cost of last 18 months > 1,000,000 €
(C1) sensitivity to get dirty or scraping	6	1	1–2 warning in the last 18 months
		2	3–4 warning in the last 18 months
		3	5–6 warning in the last 18 months
		4	7–8 warning in the last 18 months
		5	9–10 warning in the last 18 months
		6	>10 warning in the last 18 months
(D1) process criticality (RBI)	4	1,5	Very high
		3	High
		4,5	Medium
		6	Low
(E1) safety and environmental impact	6	1	Minor
		2	Moderate
		3	Medium
		4	Severe
		5	Very severe
		6	Catastrophic
(F1) warning of technical Committee	6	1	1–20 warning in the last 18 months
		2	21–40 warning in the last 18 months
		3	41–60 warning in the last 18 months
		4	61–80 warning in the last 18 months
		5	81–100 warning in the last 18 months
		6	>100 warning in the last 18 months
(G1) sensitivity to transitory phases	6	1	1–4 stopping events in the last 18 months
		2	5–8 stopping events in the last 18 months
		3	9–12 stopping events in the last 18 months
		4	13–16 stopping events in the last 18 months
		5	17–20 stopping events in the last 18 months

(continued)

Table 6 (continued)

FACTOR f_i	Number of classes	Factor value	Item description
		6	>20 stopping events in the last 18 months
(H1) equipment complexity	3	1	Equipment value <= 10,000 €
		3	10,000 < Equipment value <= 100,000 €
		6	Equipment value > 100,000 €
(I1) impact on production efficiency (RCM)	4	1, 5	Very high
		3	High
		4, 5	Medium
		6	Low
(L1) impact on maintenance plan (work orders)	6	1	1–6 work orders in the last 18 months
		2	7–12 work orders in the last 18 months
		3	13–18 work orders in the last 18 months
		4	19–24 work orders in the last 18 months
		5	25–30 work orders in the last 18 months
		6	>30 work orders in the last 18 months
(M1) sensitivity to serious failure	6	1	1–25 events in the last 18 months
		2	25–50 events in the last 18 months
(N1) sensitivity to slight failure		3	51–100 events in the last 18 months
		4	101–150 events in the last 18 months
		5	151–200 events in the last 18 months
		6	>200 events in the last 18 months
(O1) possibility of spare	3	1	No automatic standby
		3	1 spare
		6	2 spare
(P1) item under preventive–predictive maintenance	2	0	No preventive–predictive maintenance in the last 18 months
		−6	Yes preventive–predictive maintenance in the last 18 months
(Q1) item under improving maintenance	3	0	No improving maintenance in the last 18 months
		−3	1 improving maintenance in the last 18 months
		−6	2 or more improving maintenance in the last 18 months

Table 7 Example of CI calculated for some selected equipment in the first year

Item	C8901	C8903	R80011	E8105	R80012	E8001A	FV81004	E8002B	R8602	D8102
CI year I	83	71	62	59	51	49	49	48	52	47
Sensitivity to corrosion			3	6	3	3	3	3	6	3
Economical sustainability										
Sensitivity to dirt and abrasion	1									
Process criticality			2							
Safety and environmental impact	6	5								
Operating work load	3	1	2	1	1	1	1	1	1	1
Sensitivity to transient state	5	5	5	5	5	5	5	5	5	5
Equipment complexity	6	6	6	6	6	6	3	6	6	6
Impact on plant productivity	4, 5	4, 5				1, 5	4, 5	1, 5		
Impact on maintenance programming	1		2	1	3	1	1	1		1
Severe sensitivity to failure	1	1	1	1	1	1	1	1		1
Low sensitivity to failure	1		1	1	1			1		1
Availability of spare parts										
Subject to preventive maintenance										
Subject to improvement maintenance										
Ranking	1	2	3	4	5	6	7	8	9	10

Table 8 Example of CI calculated for some selected equipment in the second year

Item	C8901	P8004A	R8602	E8920	TK8101	F3751	C8903	D8902B	F1401	R80011
CI year II	97	77	70	61	61	58	58	56	55,5	52
Sensitivity to corrosion				3						1
Economical sustainability	3	1	1		1	1		3	3	3
Sensitivity to dirt and abrasion	3	1								
Process criticality									4,5	
Safety and environmental impact	6	5	6		3	3	3			
Operating work load	3	1	1	1	1		1	1		2
Sensitivity to transient state	6	6	6	6	6	4	6	6	2	6
Equipment complexity	6	3	6	6	6	6	6	6	6	6
Impact on plant productivity	4,5	6		6			4,5	4,5	6	
Impact on maintenance programming	1	2		1	2	4		1	3	1
Severe sensitivity to failure	1	1		1	1	1		1	1	1
Low sensitivity to failure	1	1			1	1		1	2	1
Availability of spare parts		1								
Subject to preventive maintenance										
Subject to improvement maintenance										
Ranking	1	2	3	4	5	6	7	8	9	10

Table 9 Comparison between CI of the first and second years of the study

Item	IdC year I	Ranking year I	IdC year II	Ranking year II	Color
C8901	83	1°	97	1°	White
P8004A	36	Sub critical	77	2°	Gray I
R8602	52	9°	70	3°	
E8920	39	Sub critical	61	4°	
TK8101	28	Sub critical	61	5°	
F3751	38	Sub critical	58	6°	
C8903	71	2°	58	7°	Gray II
D8902B	37	Sub critical	56	8°	Gray I
F1401	39,5	Sub critical	55,5	9°	
R80011	62	3°	52	10°	Gray II
E8105	59	7°	<40	Subcritical	Gray III
R80012	51	10°	<40	Subcritical	
E8001A	49	12°	<40	Subcritical	
FV81004	49	13°	<40	Subcritical	
E8002B	48	14°	<40	Subcritical	
D8102	47	15°	<40	Subcritical	

Table 10 CI variation after increasing the weight of the cost related factors

II year standard weight			II year increased weight of B1, G1, H1, M1, N1 factors		
Item	CI	Rank	Item	NEW CI	Delta rank
C8901	97	1	C8901	131	0
P8004A	77	2	P8004A	101	0
R8602	70	3	R8602	96	0
E8920	61	4	TK8101	91	From 5th to 4th
TK8101	61	5	D8902B	90	From 8th to 5th
F3751	58	6	E8920	87	From 4th to 6th
C8903	58	7	R80011	86	From 10th to 7th
D8902B	56	8	F3751	84	From 6th to 8th
F1401	55.5	9	F1401	83.5	0
R80011	52	10	C8903	82	From 7th to 10th

Applying the CI method to the turnaround consists of performing the maximum number of tasks, allowed by the established TAM duration, starting from the items with highest CI value. However, all the maintenance operations related to items in the iper critical and critical class should be performed.

The results of the study demonstrate the effectiveness of such method that can be used as planning tools by the company management, together with the maintenance and reliability team, for selecting the items that need major maintenance and scheduling the turnaround activities. In this way, obtaining an optimized use of human and economical resources, as well as the highest level of reliability and safety.

Simulating the application of this decision tool on the same TAM, it was possible to estimate a reduction of the equipment of about 25 %, with a consequent costs decrease of 11 % and of resources usage of about 7 %.

7 Conclusions

Maintenance engineering represents the technical and organizational answer to promote the continuous improvement and the excellence of the performances, together with the costs reduction. The application of the maintenance engineering in the TAM allows to reduce the number of equipments that need to be included in the shutdown, decreasing the total time of the outage and the production loss, and increasing the interval between two shutdowns.

Two methods have been tested for defining the items to include in the programmed turnaround and discard those being potentially managed with current maintenance.

The decision-making one has been used as a tool to quantify the risk and define the priority actions in terms of "action/event risk—associated risks—possible improvements—resources used—result." This approach allowed to identify the really critical events and the items to include in TAM in terms of safety, environment, plant availability, quality of the product and maintenance costs. The saving generated by the risk-based method, in terms of costs, has been evaluated in the order of 1.8 million € and in terms of resources usage led to a reduction of about 20,000 working hours. Moreover it also brought to the development of a common technical language facilitating the communication between all the people interacting in the plants.

The criticality index, developed and demonstrated for the first time in the analyzed refinery, was applied in the simulation of the same turnaround management. The results showed that also with this method it was possible to improve the performance of the process in terms of equipment reduction, operations, and costs. A different result was obtained for the resources; more personnel were indeed involved in the management, due to the great amount of data needed to calculate the index and the effort made to create and update the database.

Despite this drawback the criticality index tool provides a complete overview of the current equipment state of maintenance; the major efforts requested are paid off by the minor complexity in the management of the reliability and maintenance programs that for a refinery constitute one of the main issues.

Acknowledgments The authors would like to acknowledge the reviewers for their constructive and helpful comments and suggestions that helped in improving the paper value.

References

Ahuja, I. P. S., & Khamba, J. S. (2008). Total productive maintenance implementation in a manufacturing organization. *International Journal of Productivity and Quality Management, 3* (3), 360–381.

Ahuja, I. P. S., & Khamba, J. S. (2009). Investigation of manufacturing performance achievements through strategic total productive maintenance initiatives. *International Journal of Productivity and Quality Management, 4*(2), 129–152.

Aissani, N., Beldjilali, B., & Trentesaux, D. (2009). Dynamic scheduling of maintenance tasks in the petroleum industry: A reinforcement approach. *Engineering Applications of Artificial Intelligence, 22,* 1089–1103.

American Petroleum Institute. (2002a). API Recommended practice 580 (1st Edn.). Risk-based Inspection.

American Petroleum Institute. (2002b). API Publication 581 (1st Edn.). Risk-based Inspection. Base Resource Document.

Arunraj, N. S., & Maiti, J. (2007). Risk-based maintenance—techniques and applications. *Journal of Hazardous Materials, 142*(3), 653–661.

Bahrami, G K., & Price, J. (2000). Mantenimiento basado en el riesgo. Gestion de activos industriales, ano III, numero 9, enero/2000. Alcion. Madrid.

Bevilacqua, M., Braglia, M., & Gabbrielli, R. (2000). Monte Carlo simulation approach for a modified FMECA in a power plant. *Quality and Reliability Engineering International, 16,* 313–324. doi:10.1002/1099-1638(200007/08).

Bevilacqua, M., Ciarapica, F., Giacchetta, G., & Marchetti, B. (2012). Development of an innovative criticality index for turnaround management in an oil refinery. *International Journal of Productivity and Quality Management, 9*(4), 519–544.

Bhangu, N. S., Singh, R., & Pahuja, G. L. (2011). Reliability centered maintenance in a thermal power plant: a case study. *International Journal of Productivity and Quality Management, 7*(2), 209–228.

Brissaud, F., Lanternier, B., & Charpentier, D. (2011). Modeling failure rates according to time and influencing factors. *International Journal of Reliability and Safety, 5*(2), 95–109.

Casal, J., Montiel, H., Planas, E., & Vilchez, J. (1999). *Analisis del riesgo en instalaciones industriales.* Barcelona: Ediciones UPC.

Cepin, M. (2002). Optimization of safety equipment outages improves safety. *Reliability Engineering System Safety, 80,* 77–71.

Christen, A. J., Ruggeri, F., & Villa, E. (2011). Utility based maintenance analysis using a Random Sign censoring model. *Reliability Engineering and System Safety, 96,* 425–431.

Conn, A. R., Deleris, L. A., Hosking, J. R. M., & Thorstensen, T. A. (2010). A simulation model for improving the maintenance of high cost systems, with application to an offshore oil installation. *Quality and Reliability Engineering International, 26,* 733–748. doi:10.1002/qre. 1136.

Dossier Special Algerie. (2002). Special Issue on the Petroleum Industry in Algery (2nd part). Petrole et techniques, no. 440.

de Leon, Gomez, Hijes, F., & Cartagena, J. J. R. (2006). Maintenance strategy based on a multicriterion classification of equipments. *Reliability Engineering and System Safety, 91*, 444–451.

Duffuaa, S. O., & Daya, B. (2004). Turnaround maintenance in petrochemical industry: practices and suggested improvements. *Journal of Quality and Maintenance Engineering, 10*(3), 184–190. doi:10.1108/13552510410553235.

Garg, S., Singh, J., & Singh, D. V. (2010). Availability and maintenance scheduling of a repairable block-board manufacturing system. *International Journal of Reliability and Safety, 4*(1), 104–118.

Ghosh, D., & Roy, S. (2010). A decision-making framework for process plant maintenance. *European Journal of Industrial Engineering, 4*(1), 78–98.

Hokstad, P., Flotten, P., Holmstrom, S., McKenna, F., & Onshus, T. (1995). A reliability model for optimisation of test schemes for fire and gas detectors. *Reliability Engineering System Safety, 47*, 15–25.

Khanna, V. K. (2008). Total productive maintenance experience: a case study. *International Journal of Productivity and Quality Management, 3*(1), 12–32.

Kister, T. C., & Hawkins, B. (2006). *Maintenance planning and scheduling: Streamline your organization for a lean environment.* Elsevier Inc. ISBN: 978-0-7506-7832-2.

Kosta, A. L., & Keshav, K. (2013). *Refinery Inspection and Maintenance, Petroleum refining and natural gas processing, ASTM manual series MNL 58.* West Conshohocken, PA: ASTM International.

Laggoune, R., Chateauneuf, A., & Aissani, D. (2009). Opportunistic policy for optimal preventive maintenance of a multi-component system in continuous operating units. *Computers & Chemical Engineering, 33*, 1499–1510.

Pérès, F., & Noyes, D. (2003). Evaluation of a maintenance strategy by the analysis of the rate of repair. *Quality and Reliability Engineering International, 19*, 129–148. doi:10.1002/qre.515.

Pham, H. (2013). A software reliability model with Vtub-shaped fault-detection rate subject to operating environment, *Proceedings of the 19th International Conference on Reliability and Quality in Design.* Honolulu. (August 5–7).

Prabhakar, P., Jagathy Raj, V. P. (2013). A new model for reliability centered maintenance in petroleum refineries, *International Journal Of Scientific & Technology Research, 2*(5), 56–64.

Rosmaini, A., Shahrul, K., Ishak, A. A., & Indra, P. A. (2011). Preventive replacement schedule: a case study at a processing industry. *International Journal of Industrial and Systems Engineering, 8*(3), 386–406.

Sachdeva, A., Kumar, D., & Kumar, P. (2008). A methodology to determine maintenance criticality using AHP. *International Journal of Productivity and Quality Management, 3*(4), 396–412.

Savino, M. M., Brun, A., & Riccio, C. (2011). Integrated system for maintenance and safety management through FMECA principles and fuzzy inference engine. *European J. of Industrial Engineering, 5*(2), 132–169.

S-RCM Training Guide. (2000). *Shell-reliability centered maintenance.* Shell Global Solution International.

Singh, A. B. (2000). *World-Class Turnaround Management.* Houston, USA: Everest Press.

Telford, S., Mazhar, M. I., & Howard, I. (2011). Condition based maintenance (CBM) in the oil and gas industry: An overview of methods and techniques. *Proceedings of the 2011 International Conference on Industrial Engineering and Operations Management.* Kuala Lumpur. (January 22–24).

Tsakatikas, D., Diplaris, S., & Sfantsikopoulos, M. (2008). Spare parts criticality for unplanned maintenance of industrial systems. *European Journal of Industrial Engineering, 2*(1), 94–107.

You-Tern, T., Kuo-Shong, W., & Lin-Chang, T. (2004). A study of availability centered preventive maintenance for multi-component systems. *Reliability Engineering & System Safety, 84*(3), 261–70.

Zhang, C., & Mostashari, A. (2011). Influence of component uncertainty on reliability assessment of systems with continuous states. *International Journal of Industrial and Systems Engineering, 7*(4), 542–552.

Zhaoyang, T., Jianfeng, L., Zongzhi, W., Jianhu, Z., & Weifeng, H. (2011). An evaluation of maintenance strategy using risk based inspection. *Safety Science, 49*, 852–860.

Zhou, X., Lu, Z., & Xi, L. (2010). A dynamic opportunistic preventive maintenance policy for multi-unit series systems with intermediate buffers. *International Journal of Industrial and Systems Engineering, 6*(3), 276–288.

Age Replacement Models with Random Works

Xufeng Zhao and Toshio Nakagawa

1 Introduction

Most operating units are repaired or replaced when they have failed. It may require much time and suffer higher cost to repair a failed unit or to replace it with a new one, so that it is necessary to replace it preventively before failures. Replacements after failure and before failure are called corrective replacement (CR) and preventive replacement (PR), respectively. The recently published books (Osaki 2002; Pham 2003; Nakagawa 2005, 2011; Wang and Pham 2007; Kobbacy and Murthy 2008; Manzini et al. 2010) collected many PR models in theory and their applications in industrial systems.

Some units in offices and industries execute jobs or computer procedures successively. For such units, it would be impossible or impractical to maintain or replace them in a strict regular fashion, e.g., a planned time T, even though the planned maintenance or replacement time comes. This is because the sudden suspension of the job may suffer losses of production in different degrees if there is no sufficient preparation in advance (Barlow and Proschan 1965, p. 72; Nakagawa 2005, p. 245). On the other hand, we sometimes are interested in certain quantities of units in which to analyze their reliabilities and maintenances that are usually observed by a unique time scale such as age or operating time, but they are often measured by some combined scales in reliability applications. Alternative time

Manuscript chapter submitted to Quality and Reliability Management and Its Applications, H. Pham (Ed.), Springer.

X. Zhao (✉)
Qatar University, Doha 2713, Qatar
e-mail: kyokuh@qu.edu.qa

X. Zhao · T. Nakagawa
Aichi Institute of Technology, Toyota 470-0392, Japan

© Springer-Verlag London 2016
H. Pham (ed.), *Quality and Reliability Management and Its Applications*,
Springer Series in Reliability Engineering, DOI 10.1007/978-1-4471-6778-5_8

scales were investigated and how to select effective ones to analyze mean time to failure (MTTF) was discussed (Duchesne and Lawless 2000). Replacement policies with two time scales, such as age and number of uses, were proposed (Nakagawa 2005, p. 83). Taking parts of an aircraft as an example, appropriate maintenance and replacement policies have been usually scheduled at a total hours of operations or at a specified number of flights since the last major overhaul (Nakagawa 2008, p. 149).

By considering the factors of working times in operations, the reliability quantities of a random age replacement policy were obtained (Barlow and Proschan 1965, p. 72). Several schedules of jobs that have random processing times were summarized (Pinedo 2002). The properties of replacement policies between two successive failed units, where the unit is replaced at random times, were investigated (Stadje 2003). Under the assumptions of random failure and maintenance, replacement and inspection with planned and random policies were considered, and their comparisons were made (Zhao and Nakagawa 2011; Nakagawa et al. 2011). Combining a planned replacement with working times, the age and periodic replacement policies, where the unit is replaced at a planned time T and at the N th random working time, were discussed (Chen et al. 2010a, b). Such a notion of maintenance was applied to a parallel system with random number of units to satisfy the random jobs (Nakagawa and Zhao 2012). Furthermore, a summary of newly proposed age replacement policies, including some models with random working cycles, was made (Zhao and Nakagawa 2012a).

It has been assumed in all policies that units are maintained or replaced before failure at some amount of quantities, e.g., age, periodic time, usage number, damage level, etc., whichever occurs first. These policies would be reasonable if failures are serious and sometimes may incur heavy production losses. By considering the cases when replacement costs suffered for failures would be estimated to be not so high and the factor of working times, age and periodic replacement policies—where the unit is replaced at a planned operating time T or at a random working cycle Y, whichever occurs last—were proposed (Zhao and Nakagawa 2012b). This also indicated that policies proposed in (Chen et al. 2010a, b) would cause frequent and unnecessary replacement which may incur production losses under the assumption of "whichever occurs first". The notion "whichever occurs last" was applied in a cumulative damage model, where the unit is replaced before failure at a planned time T and at a damage level Z, and an optimal Z for a given T was discussed (Zhao et al. 2011).

On the other hand, when a job has a variable working cycle or processing time, it would be better to do maintenance or replacement after the job is just completed even though the maintenance time has arrived (Sugiura et al. 2004; Nakagawa 2005, p. 245). From such a viewpoint, by considering both planned time and working times, the age and periodic replacement policies which are done at the Nth completion of works and at the first completion of some working time over a planned time T was proposed (Chen et al. 2010a, b), and optimal age replacement with overtime policy was derived in (Zhao and Nakagawa 2012a). A representative practical example for such a policy is to maintain a database or to perform a backup of data when a transaction is processing its sequences of operations, because it is necessary to guarantee ACID (atomicity, consistency, isolation, and durability) properties of database transactions, especially for a distributed transaction across a distributed database. That is, if any part of transaction fails, the entire transaction fails and the database state is left unchanged (Haerder and Reuter 1983; Lewis et al. 2002).

From the above considerations, we make the following assumptions and summarize four age replacement policies: It is assumed that an operating unit works for jobs successively with random working cycles, and it deteriorates with its operating time. As preventive replacement, (1) the unit is replaced at a total operating time T, which is called *standard age replacement*; (2) the unit is replaced at a total operating time T or at a random working cycle Y, whichever occurs first, which is called *replacement first*; (3) the unit is replaced at a total operating time T or at a random working cycle Y, whichever occurs last, which is called *replacement last*; (4) the unit is replaced at the first completion of some working cycles over a planned time T, which is called *replacement overtime*. We first summarize optimal replacement policies for the above four replacement policies in the following sections. Second, we compare each policy with one another from the viewpoint of cost rate when the working cycle is distributed exponentially and the replacement costs at time T and Y or Y_j are the same. It is shown theoretically that standard replacement is the best among the four. We also discuss that either replacement first or last is better than the other according to the ratio of replacement costs. Third, we optimize random age replacement policy which is a particular case of Policies (2)–(4), and an optimal mean time to replacement is obtained and computed. Last, under the assumptions of replacement costs after the completion of working cycles might be lower than those at time T, we discuss theoretically and numerically that if how much the cost at Y or Y_j is lower than that of T, random age replacement and replacement overtime would be better than standard replacement.

2 Four Replacement Policies

2.1 Standard Replacement

It is assumed that an operating unit works for jobs successively, and it deteriorates with its operating time and has a variable failure time X with a general distribution $F(t) \equiv \Pr\{X \le t\}$ and a finite mean $\mu \equiv \int_0^\infty \overline{F}(t)\mathrm{d}t < \infty$. In addition, let $h(t) \equiv f(t)/\overline{F}(t)$ be the instant failure rate of $F(t)$, where $f(t)$ is a density function of $F(t)$, where $\overline{\Phi}(t) \equiv 1 - \Phi(t)$ for any function $\Phi(t)$. It is also assumed that $h(t)$ increases strictly to $h(\infty) \equiv \lim_{t \to \infty} h(t)$, which might be ∞.

Suppose that when the unit fails, its failure is immediately detected, and then CR is done. As PR policies, the unit is replaced before failure at a total operating time

$T(0 < T \leq \infty)$. We call such a policy as *standard age replacement* (SAR), and the expected cost rate is (Barlow and Proschan 1965, p. 87; Nakagawa 2005, p. 72)

$$C_S(T) = \frac{c_T + (c_F - c_T)F(T)}{\int_0^T \overline{F}(t)dt}, \tag{1}$$

where costs c_F and c_T are the respective replacement costs at failure and at time T with $c_F > c_T$.

If $h(\infty) > c_F / [\mu(c_F - c_T)]$, then an optimal $T_S^*(0 < T_S^* < \infty)$ which minimizes $C_S(T)$ is given by a finite and unique solution of the equation

$$h(T) \int_0^T \overline{F}(t)dt - F(T) = \frac{c_T}{c_F - c_T}, \tag{2}$$

and the resulting cost rate is

$$C_S(T_S^*) = (c_F - c_T)h(T_S^*). \tag{3}$$

2.2 Replacement First

It is assumed that each job has a variable working cycle Y with a general distribution $G(t) \equiv \Pr\{Y \leq t\}$ and a finite mean $1/\theta \equiv \int_0^\infty \overline{G}(t)dt < \infty$, which is independent of the failure time X. Suppose that the unit is replaced before failure at a total operating time $T(0 < T \leq \infty)$ or at a random working cycle Y, whichever occurs first (Chen et al. 2010a, b; Zhao and Nakagawa 2012a), which is called *replacement first* (RF).

Then, the probability that the unit is replaced at time T is

$$\Pr\{X > T, Y > T\} = \overline{F}(T)\overline{G}(T), \tag{4}$$

the probability that it is replaced at time Y is

$$\Pr\{Y < X, Y < T\} = \int_0^T \overline{F}(t)dG(t), \tag{5}$$

and the probability that it is replaced at failure is

$$\Pr\{X \leq T, X \leq Y\} = \int_0^T \overline{G}(t)dF(t), \tag{6}$$

Thus, the mean time to replacement is

$$T\overline{F}(T)\overline{G}(T) + \int_0^T t\overline{F}(t)dG(t) + \int_0^T t\overline{G}(t)dF(t) = \int_0^T \overline{F}(t)\overline{G}(t)dt. \qquad (7)$$

Therefore, the expected cost rate is, from (4)–(7),

$$C_F(T) = \frac{c_T + (c_F - c_T)\int_0^T \overline{G}(t)dF(t) + (c_Y - c_T)\int_0^T \overline{F}(t)dG(t)}{\int_0^T \overline{F}(t)\overline{G}(t)dt}, \qquad (8)$$

where c_F and c_T are given in (1), and c_Y is the replacement cost at time Y with $c_F > c_Y$. Clearly,

$$C_F(0) \equiv \lim_{T \to 0} C_F(T) = \infty,$$

$$C_F(\infty) \equiv \lim_{T \to \infty} C_F(T) = \frac{c_Y + (c_F - c_Y)\int_0^\infty \overline{G}(t)dF(t)}{\int_0^\infty \overline{F}(t)\overline{G}(t)dt}, \qquad (9)$$

which is the expected cost rate when PR is only done at a random working cycle Y (Barlow and Proschan 1965, p. 86).

We find an optimal T_F^* which minimizes $C_F(T)$ in (8). Differentiating $C_F(T)$ with respect to T and setting it equal to zero,

$$(c_F - c_T)\left[h(T)\int_0^T \overline{F}(t)\overline{G}(t)dt - \int_0^T \overline{G}(t)dF(t) \right]$$
$$+ (c_Y - c_T)\left[r(T)\int_0^T \overline{F}(t)\overline{G}(t)dt - \int_0^T \overline{F}(t)dG(t) \right] = c_T, \qquad (10)$$

where $r(t) \equiv g(t)/\overline{G}(t)$ and $g(t)$ is a density function of $G(t)$, and $r(\infty) \equiv \lim_{t \to \infty} r(t)$.

Denote the left-hand side of (10) as $L_F(T)$, and suppose that $c_Y \leq c_T$ and $r(t)$ decreases with t. Then, it can be easily shown that $L_F(T)$ increases strictly with T from 0 to

$$L_F(\infty) = (c_F - c_T)\int_0^\infty \overline{F}(t)\overline{G}(t)[h(\infty) - h(t)]dt$$

$$+ (c_Y - c_T)\int_0^\infty \overline{F}(t)\overline{G}(t)[r(\infty) - r(t)]dt.$$

Thus, if $L_F(\infty) \leq c_T$, then $T_F^* = \infty$, and the expected cost rate is given in (9). If $L_F(\infty) > c_T$, then there exists a finite and unique $T_F^*(0 < T_F^* < \infty)$ which satisfies (10). It is clear that if $h(\infty) = \infty$, then $L_F(\infty)$ goes to infinity, and the minimum expected cost rate is

$$C_F(T_F^*) = (c_F - c_T)h(T_F^*) + (c_Y - c_T)r(T_F^*). \tag{11}$$

When $G(t) = 1 - e^{-\theta t}$, i.e., $r(t) = \theta$, (10) becomes

$$h(T) \int_0^T e^{-\theta t}\overline{F}(t)\mathrm{d}t - \int_0^T e^{-\theta t}\mathrm{d}F(t) = \frac{c_T}{c_F - c_T}, \tag{12}$$

and the resulting cost rate is

$$C_F(T_F^*) = (c_F - c_T)h(T_F^*) + (c_Y - c_T)\theta. \tag{13}$$

In particular, when $c_T = c_Y$, (10) becomes

$$h(T) \int_0^T \overline{F}(t)\overline{G}(t)\mathrm{d}t - \int_0^T \overline{G}(t)\mathrm{d}F(t) = \frac{c_T}{c_F - c_T}, \tag{14}$$

whose left-hand side increases strictly with T, and the resulting cost rate becomes

$$C_F(T_F^*) = (c_F - c_T)h(T_F^*). \tag{15}$$

When $G(t) = 1 - e^{-\theta t}$, because the left-hand side of (12) decreases strictly with θ, T_F^* decreases with $1/\theta$ to T_S^* given in (2).

2.3   Replacement Last

Suppose that the unit is replaced before failure at a total operating time $T(0 \leq T \leq \infty)$ or at a random working cycle Y, whichever occurs last (Zhao and Nakagawa 2012a, b), which is called *replacement last* (RL).

The above policy can be analyzed as follows: (i) If the unit fails before the job is completed, we replace it immediately; (ii) or if the unit is operating properly for a time duration T but the job has not been completed yet, we continue to operate it until the job is completed, and then replace it; (iii) or if the unit finishes the job using a time duration that is shorter than T, we start the second job, and so on. When the total operating time of this unit has reached T, we shut it down and replace it.

Then, the probability that the unit is replaced at time T is

$$\Pr\{Y<T<X\} = \overline{F}(T)G(T), \tag{16}$$

the probability that it is replaced at time Y is

$$\Pr\{T<Y<X\} = \int_T^\infty \overline{F}(t)\mathrm{d}G(t). \tag{17}$$

Replacement done at failure is divided into three cases: (i) The unit fails both before T and one job completion, and the probability is

$$\Pr\{X<T, X<Y\} = \int_0^T \overline{G}(t)\mathrm{d}F(t), \tag{18}$$

(ii) the unit fails after at least one job is completed before T, and the probability is

$$\Pr\{Y<X<T\} = \int_0^T G(t)\mathrm{d}F(t), \tag{19}$$

(iii) the unit fails after T before one job completion, and the probability is

$$\Pr\{T \le X \le Y\} = \int_T^\infty \overline{G}(t)\mathrm{d}F(t), \tag{20}$$

Thus, the mean time to replacement is, from (16) to (20),

$$T\overline{F}(T)G(T) + \int_T^\infty t\overline{F}(t)\mathrm{d}G(t) + \int_0^T t\mathrm{d}F(t) + \int_T^\infty t\overline{G}(t)\mathrm{d}F(t)$$

$$= \int_0^T \overline{F}(t)\mathrm{d}t + \int_T^\infty \overline{G}(t)\overline{F}(t)\mathrm{d}t. \tag{21}$$

Therefore, the expected cost rate is

$$C_L(T) = \frac{c_F - (c_F - c_T)\overline{F}(T)G(T) - (c_F - c_Y)\int_T^\infty \overline{F}(t)\mathrm{d}G(t)}{\int_0^T \overline{F}(t)\mathrm{d}t + \int_T^\infty \overline{F}(t)\overline{G}(t)\mathrm{d}t}. \tag{22}$$

Clearly,

$$C_L(0) \equiv \lim_{T \to 0} C_L(T) = C_F(\infty),$$

$$C_L(\infty) \equiv \lim_{T \to \infty} C_L(T) = \frac{c_F}{\mu}, \tag{23}$$

which is the expected cost rate when the unit is replaced only at failure.

Differentiating $C_L(T)$ with respect to T and setting it equal to zero,

$$(c_F - c_T) \left\{ [h(T) - v(T)] \left[\int_0^T \overline{F}(t)dt + \int_T^\infty \overline{F}(t)\overline{G}(t)dt \right] + \overline{F}(T)G(T) \right\}$$

$$+ (c_F - c_Y) \left\{ v(T) \left[\int_0^T \overline{F}(t)dt + \int_T^\infty \overline{F}(t)\overline{G}(t)dt \right] + \int_T^\infty \overline{F}(t)dG(t) \right\} = c_F, \tag{24}$$

where $v(t) \equiv g(t)/G(t)$. Denote $L_L(T)$ be the left-hand side of (24),

$$L_L(0) = (c_F - c_T)[h(0) - v(0)] \int_0^\infty \overline{F}(t)\overline{G}(t)dt$$

$$+ (c_F - c_Y) \left[v(0) \int_0^\infty \overline{F}(t)\overline{G}(t)dt + \int_0^\infty \overline{F}(t)dG(t) \right],$$

$$L_L(\infty) = \mu\{(c_F - c_T)[h(\infty) - v(\infty)] + (c_F - c_Y)v(\infty)\},$$

$$L_L'(T) = [(c_F - c_T)h'(T) - (c_T - c_Y)v'(T)] \left[\int_0^T \overline{F}(t)dt + \int_T^\infty \overline{F}(t)\overline{G}(t)dt \right].$$

Suppose that $c_Y \le c_T$ and $v(t)$ decreases with t. Then, $L_L(T)$ increases strictly from $L_L(0)$ to $L_L(\infty)$, and hence, we have the following optimal policies:

1. If $L_L(0) \ge c_F$, then $T_L^* = 0$ and the expected cost rate is given in (9).
2. If $L_L(0) < c_F < L_L(\infty)$, then there exists a finite and unique $T_L^* (0 < T_L^* < \infty)$ which satisfies (24), and the resulting cost rate is

$$C_L(T_L^*) = (c_F - c_T)h(T_L^*) - (c_T - c_Y)v(T_L^*). \tag{25}$$

3. If $L_L(\infty) \le c_F$, then $T_L^* = \infty$ and the expected cost rate is given in (23).

When $G(t) = 1 - e^{-\theta t}$, note that $v(t) = \theta/(e^{\theta t} - 1)$ which decreases strictly from ∞ to 0, then an optimal T_L^* exists, and (25) becomes

$$C_L(T_L^*) = (c_F - c_T)h(T_L^*) - \frac{(c_T - c_Y)\theta}{e^{\theta T_L^*} - 1}. \tag{26}$$

In addition, when $c_T = c_Y$, (24) becomes

$$h(T)\left[\int_0^T \overline{F}(t)dt + \int_T^\infty \overline{F}(t)\overline{G}(t)dt \right] + \int_T^\infty G(t)dF(t) = \frac{c_F}{c_F - c_T}, \tag{27}$$

whose left-hand side increases strictly from

$$h(0)\int_0^\infty \overline{F}(t)\overline{G}(t)dt + \int_0^\infty G(t)dF(t)$$

to $\mu h(\infty)$, and the resulting cost rate is

$$C_L(T^*) = (c_F - c_T)h(T_L^*). \tag{28}$$

When $G(t) = 1 - e^{-\theta t}$, because the left-hand side of (27) increases strictly with θ, T_L^* increases with $1/\theta$ from T_S^* given in (2).

2.4 Replacement Overtime

Suppose that the unit is replaced before failure at the first completion of some working cycles over a planned time $T(0 \le T \le \infty)$ (Zhao and Nakagawa 2012a), which is called *replacement overtime* (RO).

Denote $G^{(j)}(t)(j = 1, 2, \ldots)$ as the j-fold Stieltjes convolution of $G(t)$ with itself and $G^{(0)}(t) \equiv 1$ for $t \ge 0$. Then, the probability that the unit finishes exactly j jobs in $[0, t]$ is $G^{(j)}(t) - G^{(j+1)}(t)$, and the probability that the units is replaced before failure at the first completion of working cycles over T is

$$\sum_{j=0}^\infty \int_0^T \left[\int_{T-t}^\infty \overline{F}(t+u)dG(u) \right] dG^{(j)}(t), \tag{29}$$

the probability that it is replaced at failure before time T is

$$\sum_{j=0}^{\infty} \int_0^T \left[G^{(j)}(t) - G^{(j+1)}(t) \right] dF(t) = F(T), \tag{30}$$

and the probability that it is replaced at failure after time T is

$$\sum_{j=0}^{\infty} \int_0^T \left\{ \int_{T-t}^{\infty} [F(t+u) - F(T)] dG(u) \right\} dG^{(j)}(t). \tag{31}$$

Thus, the mean time to replacement is, from (29) to (31),

$$\sum_{j=0}^{\infty} \int_0^T \left[\int_{T-t}^{\infty} (t+u) \overline{F}(t+u) dG(u) \right] dG^{(j)}(t)$$

$$+ \int_0^T t dF(t) + \sum_{j=0}^{\infty} \int_0^T \left\{ \int_{T-t}^{\infty} \left[\int_T^{T+u} v dF(v) \right] dG(u) \right\} dG^{(j)}(t) \tag{32}$$

$$= \int_0^T \overline{F}(t) dt + \sum_{j=0}^{\infty} \int_0^T \left[\int_T^{\infty} \overline{G}(u-t) \overline{F}(u) du \right] dG^{(j)}(t).$$

Therefore, the expected cost rate is

$$C_O(T) = \frac{c_F - (c_F - c_Y) \sum_{j=0}^{\infty} \int_0^T [\int_{T-t}^{\infty} \overline{F}(t+u) dG(u)] dG^{(j)}(t)}{\int_0^T \overline{F}(t) dt + \sum_{j=0}^{\infty} \int_0^T [\int_T^{\infty} \overline{G}(u-t) \overline{F}(u) du] dG^{(j)}(t)}. \tag{33}$$

Clearly,

$$C_O(0) \equiv \lim_{T \to 0} C_O(T) = C_F(\infty) = C_L(0),$$

$$C_O(\infty) \equiv \lim_{T \to \infty} C_O(T) = \frac{c_F}{\mu}. \tag{34}$$

Differentiating $C_O(T)$ with respect to T and setting it equal to zero,

$$z(T; G) \left\{ \int_0^T \overline{F}(t) dt + \sum_{j=0}^{\infty} \int_0^T \left[\int_T^{\infty} \overline{G}(u-t) \overline{F}(u) du \right] dG^{(j)}(t) \right\}$$

$$- \left\{ 1 - \sum_{j=0}^{\infty} \int_0^T \left[\int_{T-t}^{\infty} \overline{F}(t+u) dG(u) \right] dG^{(j)}(t) \right\} = \frac{c_Y}{c_F - c_Y}, \tag{35}$$

where

$$z(T; G) \equiv \frac{\int_0^\infty [\overline{F}(T) - \overline{F}(t+T)] dG(t)}{\int_0^\infty \overline{F}(t+T)\overline{G}(t) dt}.$$

Denote $L_O(T)$ be the left-hand side of (35),

$$L_O(0) = z(0; G) \int_0^\infty \overline{F}(t)\overline{G}(t) dt - \int_0^\infty F(t) dG(t),$$

$$L_O(\infty) = \mu z(\infty; G) - 1,$$

$$L_O'(T) = \frac{dz(T; G)}{dT} \left\{ \int_0^T \overline{F}(t) dt + \sum_{j=0}^\infty \int_0^T \left[\int_T^\infty \overline{G}(u - t)\overline{F}(u) du \right] dG^{(j)}(t) \right\}.$$

Thus, if $z(T; G)$ increases strictly, then $L_O(T)$ also increases strictly from $L_O(0)$ to $L_O(\infty)$. Therefore, we have the following optimal policies:

1. If $L_O(0) \geq c_Y/(c_F - c_Y)$, then $T_O^* = 0$ and the expected cost rate is given in (9).
2. If $L_O(0) < c_Y/(c_F - c_Y) < L_O(\infty)$, then there exists a finite and unique $T_O^*(0 < T_O^* < \infty)$ which satisfies (35), and the expected cost rate is

$$C_O(T_O^*) = (c_F - c_Y)z(T_O^*; G). \tag{36}$$

3. If $L_O(\infty) \leq c_Y/(c_F - c_Y)$, then $T_O^* = \infty$ and the expected cost rate is given in (34).

In particular, when $G(t) = 1 - e^{-\theta t}$,

$$z(T; \theta) \equiv \frac{\int_0^\infty [\overline{F}(T) - \overline{F}(t+T)\theta e^{-\theta t}] dt}{\int_0^\infty \overline{F}(t+T)e^{-\theta t} dt} = \frac{\int_T^\infty e^{-\theta t} dF(t)}{\int_T^\infty e^{-\theta t}\overline{F}(t) dt}.$$

By differentiating $z(T; \theta)$ with respect to T,

$$\frac{dz(T; \theta)}{dT} = \int_T^\infty e^{-\theta t}\overline{F}(t)[h(t) - h(T)] dt > 0,$$

which follows that $z(T; \theta)$ increases strictly to $h(\infty)$ and $z(T; \theta) \geq h(T)$. In this case, (35) is simplified as

$$z(T; \theta) \int_0^T \overline{F}(t)dt - F(T) = \frac{c_Y}{c_F - c_Y}, \tag{37}$$

whose left-hand side increases from 0 to $\mu h(\infty) - 1$. Therefore, if $h(\infty) > c_F / [\mu(c_F - c_Y)]$, a finite and unique $T_O^*(0 < T_O^* < \infty)$ which satisfies (37) exists. Furthermore, differentiating $z(T; \theta)$ with θ,

$$\frac{dz(T; \theta)}{d\theta} = \frac{1}{[\int_T^\infty e^{-\theta t}\overline{F}(t)dt]^2} \left[-\int_T^\infty te^{-\theta t}dF(t) \int_T^\infty e^{-\theta t}\overline{F}(t)dt \right.$$

$$\left. + \int_T^\infty e^{-\theta t}dF(t) \int_T^\infty te^{-\theta t}\overline{F}(t)dt \right].$$

Denote $L(T)$ as the numerator of the left-hand side, $L(\infty) = 0$ and

$$L'(T) = \overline{F}(T)e^{-\theta T} \int_T^\infty e^{-\theta t}\overline{F}(t)[h(t) - h(T)](T - t)dt > 0,$$

i.e., $L(T) < 0$ for $0 \leq T < \infty$. So that, $z(T; \theta)$ decreases with θ. In addition,

$$\lim_{\theta \to \infty} \frac{\int_T^\infty e^{-\theta t}dF(t)}{\int_T^\infty e^{-\theta t}\overline{F}(t)dt} = \lim_{\theta \to \infty} \frac{\int_T^\infty f(t)d(e^{-\theta t})}{\int_T^\infty \overline{F}(t)d(e^{-\theta t})}$$

$$= \lim_{\theta \to \infty} \frac{f(T) + \int_0^\infty e^{-\theta t}df(t + T)}{\overline{F}(T) \int_0^\infty e^{-\theta t}d\overline{F}(t + T)} = h(T).$$

So that $z(T; \theta)$ decreases strictly with θ from $z(T; \theta)$ to $h(T)$, and T_O^* decreases with $1/\theta$ from T_S^* given in (2).

3 Comparisons with the Same Preventive Replacement Costs

It is assumed that replacement done at time T is well-regulated or carefully planned by persons or computers, the job would be suspended by sufficient preparation in advance, so that it would cause trifling production losses or not cause losses just like replacement done at working cycle Y. In this case, we suppose that $c_T = c_Y$ for the above policies. Note that if the failure rate $h(t)$ increases strictly from 0 to $h(\infty) = \infty$, then all optimal and finite policies of the above four replacement exist uniquely, and we compare four optimal policies as follows:

3.1 Comparisons of T_S^* and T_F^*, T_L^*, T_O^*

First, the difference between the left-hand sides of (2) and (14) is

$$
h(T) \int_0^T \overline{F}(t)\mathrm{d}t - F(T) - \left[h(T) \int_0^T \overline{F}(t)\overline{G}(t)\mathrm{d}t - \int_0^T \overline{G}(t)\mathrm{d}F(t) \right]
$$

$$
= \int_0^T G(t)\overline{F}(t)[h(T) - h(t)]\mathrm{d}t > 0,
$$

(38)

which follows that $T_F^* > T_S^*$. From (3) and (15), SAR is better than RF.

Second, the difference between the left-hand sides of (2) and (27) is

$$
h(T) \int_0^T \overline{F}(t)\mathrm{d}t - F(T) - h(T) \left[\int_0^T \overline{F}(t)\mathrm{d}t + \int_T^\infty \overline{F}(t)\overline{G}(t)\mathrm{d}t \right] + 1 - \int_T^\infty G(t)\mathrm{d}F(t)
$$

$$
= \int_T^\infty \overline{F}(t)\overline{G}(t)[h(t) - h(T)] > 0,
$$

(39)

which follows that $T_L^* > T_S^*$. From (3) and (28), SAR is better than RL.

Third, when $G(t) = 1 - \mathrm{e}^{-\theta t}$, the expected cost rate in (33) is

$$
C_O(T) = \frac{c_F - (c_F - c_Y) \int_0^\infty \theta \mathrm{e}^{-\theta t}\overline{F}(t+T)\mathrm{d}t}{\int_0^T \overline{F}(t)\mathrm{d}t + \int_0^\infty \mathrm{e}^{-\theta t}\overline{F}(t+T)\mathrm{d}t}.
$$

(40)

An optimal T_O^* satisfies (37), and hence, the resulting cost rate also can be written as

$$
C_O(T_O^*) = \frac{c_F - (c_F - c_Y)\overline{F}(T_O^*)}{\int_0^{T_O^*} \overline{F}(t)\mathrm{d}t}.
$$

(41)

It has already been shown that $z(T; \theta) \geq h(T)$, and from (2) and (37), $T_O^* < T_S^*$. In addition, recalling that T_S^* is an optimal solution for minimizing $C_S(T)$ in (1), so that $C_O(T_O^*)$ is greater than $C_S(T_S^*)$ from (41), i.e., SAR is better than RO.

Table 1 Optimal T_F^*, T_L^*, and T_O^* when $F(t) = 1 - e^{-t^m}$ and $1/\theta = 1$

$\dfrac{c_T}{c_F - c_T}$	$m = 2.0$			$m = 3.0$		
	T_F^*	T_L^*	T_O^*	T_F^*	T_L^*	T_O^*
0.01	0.10	0.41	0.01	0.17	0.49	0.01
0.02	0.14	0.42	0.02	0.22	0.49	0.02
0.05	0.23	0.44	0.06	0.30	0.51	0.06
0.10	0.34	0.48	0.11	0.39	0.53	0.11
0.20	0.49	0.56	0.21	0.49	0.57	0.19
0.50	0.84	0.77	0.47	0.69	0.69	0.37
1.00	1.33	1.10	0.84	0.90	0.83	0.57
2.00	2.24	1.69	1.48	1.19	1.05	0.85
5.00	4.99	3.38	3.26	1.78	1.50	1.38

Suppose that failure time X has a Weibull distribution $F(t) = 1 - e^{-t^m}$ ($m = 2.0, 3.0$) and random working cycle Y has an exponential distribution $G(t) = 1 - e^{-\theta t}$. Table 1 presents optimal T_F^* in (14), T_L^* in (27) and T_O^* in (37) when $c_T = c_Y$ and $1/\theta = 1$. This indicates that when the cost ratio is small, $T_L^* > T_F^* > T_O^*$, and when it is large, $T_F^* > T_L^* > T_O^*$.

3.2 Comparisons of T_F^* and T_L^*

First, compare the left-hand sides of (14) and (27): Denote

$$
\begin{aligned}
Q(T) \equiv h(T) &\left[\int_0^T \overline{F}(t)\,\mathrm{d}t + \int_T^\infty \overline{F}(t)\overline{G}(t)\,\mathrm{d}t \right] - \left[1 - \int_T^\infty G(t)\,\mathrm{d}F(t) \right] \\
&- \left[h(T) \int_0^T \overline{F}(t)\overline{G}(t)\,\mathrm{d}t - \int_0^T \overline{G}(t)\,\mathrm{d}F(t) \right] \\
= h(T) &\left[\int_0^T \overline{F}(t)G(t)\,\mathrm{d}t + \int_T^\infty \overline{F}(t)\overline{G}(t)\,\mathrm{d}t \right] \\
&- \left[\int_T^\infty \overline{G}(t)\,\mathrm{d}F(t) + \int_0^T G(t)\,\mathrm{d}F(t) \right] \\
= &\int_0^T \overline{F}(t)G(t)[h(T) - h(t)]\,\mathrm{d}t - \int_T^\infty \overline{F}(t)\overline{G}(t)[h(t) - h(T)]\,\mathrm{d}t.
\end{aligned}
$$

(42)

Clearly,

$$Q(0) \equiv \lim_{T \to 0} Q(T) = - \int_0^\infty \overline{G}(t) dF(t) < 0,$$

$$Q(\infty) \equiv \lim_{T \to \infty} Q(T) = \infty,$$

$$Q'(T) \equiv h'(T) \left[\int_0^T \overline{F}(t) G(t) dt + \int_T^\infty \overline{F}(t) \overline{G}(t) dt \right] > 0.$$

Thus, there exists a finite and unique $T_A^* (0 < T_A^* < \infty)$ which satisfies $Q(T) = 0$. Second, denote that

$$L(T_A^*) \equiv h(T_A^*) \int_0^{T_A^*} \overline{F}(t) \overline{G}(t) dt - \int_0^{T_A^*} \overline{G}(t) dF(t). \tag{43}$$

Then, it is easily shown that if $L(T_A^*) \geq c_T/(c_F - c_T)$, i.e., $c_F \geq c_T[1 + 1/L(T_A^*)]$, then $T_F^* \leq T_L^*$, and hence, from (15) and (28), RF is better than RL. Conversely, if $L(T_A^*) < c_T/(c_F - c_T)$, then $T_F^* > T_L^*$, and RL is better than RF. So that, it can be determined which policy is better than the other according to the ratio c_T/c_F, by comparing it with $L(T_A^*)/[1 + L(T_A^*)]$.

Table 2 presents respective optimal T_F^*, T_L^*, T_A^*, $L(T_A^*)$, and T_S^*, which satisfy (14), (27), (42), (43), and (2) for $1/\theta$ and $c_T/(c_F - c_T)$ when $F(t) = 1 - e^{-t^2}$. Clearly, optimal T_F^* decreases to T_S^* and T_L^* increases from T_S^* with $1/\theta$. In addition, T_A^* increases with $1/\theta$ from (42) and $L(T_A^*)$ also increases with $1/\theta$ from (43).

Table 2 Optimal T_F^*, T_L^*, T_A^*, $L(T_A^*)$, and T_S^* when $F(t) = 1 - e^{-t^2}$

$\frac{c_T}{c_F - c_T}$	$1/\theta = 0.1$		$1/\theta = 0.3$		$1/\theta = 0.5$		$1/\theta = 0.7$		$1/\theta = 1.0$		
	T_F^*	T_L^*	T_F^*	T_L^*	T_F^*	T_L^*	T_F^*	T_L^*	T_F^*	T_L^*	T_S^*
0.01	0.12	0.13	0.11	0.23	0.10	0.31	0.10	0.36	0.10	0.41	0.10
0.03	0.24	0.18	0.19	0.26	0.18	0.33	0.18	0.38	0.18	0.42	0.17
0.05	0.35	0.23	0.26	0.29	0.24	0.35	0.24	0.40	0.23	0.44	0.22
0.07	0.45	0.27	0.31	0.32	0.29	0.37	0.28	0.42	0.28	0.46	0.26
0.10	0.61	0.32	0.39	0.36	0.36	0.41	0.35	0.44	0.34	0.48	0.32
0.30	1.62	0.56	0.80	0.57	0.69	0.59	0.65	0.61	0.62	0.63	0.56
0.50	2.64	0.74	1.19	0.74	0.97	0.75	0.89	0.76	0.84	0.77	0.74
0.70	3.66	0.89	1.57	0.89	1.24	0.89	1.12	0.90	1.04	0.91	0.89
1.00	5.19	1.09	2.14	1.09	1.64	1.09	1.45	1.10	1.33	1.10	1.09
T_A^*	0.13		0.33		0.46		0.55		0.65		
$L(T_A^*)$	0.011		0.077		0.155		0.228		0.325		

It also indicates that, for RL, at least one job would be completed, no matter whether T_L^* is less than $1/\theta$. For RF, the first job would always be suspended when $T_F^* < 1/\theta$, e.g., when $1/\theta = 1.0$ and $c_T/(c_F - c_T) \leq 0.5$. When $1/\theta = 0.1$ and $c_T/(c_F - c_T) = 1.0$, RF can finish at most one job, however, 10 jobs would be completed for RL. That is, when two time scales are used for preventive replacement, RL can let the unit operate for a longer time and avoid unnecessary replacements caused by RF.

4 Modified Preventive Replacement Costs

It has been shown that when $c_T = c_Y$, standard replacement is better than the others from the viewpoint of replacement cost rate. In general, PR cost at Y might be lower than that at T (Barlow and Proschan 1965, p. 72), when preparation for suspension at T is not sufficient and production losses incur. We compute the modified PR costs for two random policies from above discussions when $c_Y < c_T$ while their optimal expected cost rates are equal to that of SAR.

4.1 Random Age Replacement

It has been shown from above sections that $C_F(\infty) = C_L(0) = C_O(0)$. We call such a policy as *random age replacement* (Barlow and Proschan 1965, p. 86; Nakagawa 2005, p. 247), i.e., PR is made only by considering factors of the job completion and done at a random working cycle Y, and its expected cost rate $C_R(G)$ could be written in (9).

It can be shown from (Barlow and Proschan 1965, p. 87) that

$$C_R(G) \geq = C_S(G_T) = C_S(T),$$

where $G_T(t)$ is the degenerate distribution placing unit mass at T, i.e., $G_T(t) \equiv 0$ for $t < T$ and 1 for $t \geq T$. If $T = \infty$, then the units is replaced only at failure and the expected cost rate is

$$C_R \equiv \lim_{T \to \infty} C_S(T) = \frac{c_F}{\mu}. \tag{44}$$

Therefore, the optimal replacement policy is nonrandom and the expected cost rate is given in (1).

When $G(t) = 1 - e^{-\theta t}$, we find an optimal $1/\theta_R^*$ which minimizes the expected cost rate

$$C_R(\theta) = \frac{c_Y + (c_F - c_Y) \int_0^\infty e^{-\theta t} dF(t)}{\int_0^\infty e^{-\theta t} \overline{F}(t) dt}. \tag{45}$$

Differentiating $C_R(\theta)$ with respect to θ and setting it equal to zero,

$$q(\theta) \int_0^\infty e^{-\theta t} \overline{F}(t) dt - \int_0^\infty e^{-\theta t} dF(t) = \frac{c_Y}{c_F - c_Y}, \tag{46}$$

where $q(\theta) \equiv \lim_{T \to \infty} q(T, \theta)$, and for $0 < T \le \infty$,

$$q(T, \theta) \equiv \frac{\int_0^T t e^{-\theta t} dF(t)}{\int_0^T t e^{-\theta t} \overline{F}(t) dt}.$$

We investigate the properties of $q(T, \theta)$: It can be easily seen that because $h(t)$ increases strictly with t, $q(T, \theta) \le h(T)$ and increases strictly with T from $h(0)$ to $q(\theta)$. Furthermore, differentiating $q(T, \theta)$ with θ,

$$\frac{dq(T,\theta)}{d\theta} = \frac{1}{[\int_0^T \theta t e^{-\theta t} \overline{F}(t) dt]^2} \left[\int_0^T t^2 e^{-\theta t} dF(t) \int_0^T t e^{-\theta t} \overline{F}(t) dt \right.$$
$$\left. - \int_0^T t e^{-\theta t} dF(t) \int_0^T t^2 e^{-\theta t} \overline{F}(t) dt \right].$$

Denote $L(T)$ as the numerator of the right-hand side, $L(0) = 0$ and

$$L'(T) = T e^{-\theta T} \overline{F}(T) \int_0^T t e^{-\theta t} \overline{F}(t) [h(T) - h(t)] (T - t) dt > 0,$$

i.e., $L(T) > 0$ for $0 < T < \infty$. So that, $q(T, \theta)$ increases strictly with $1/\theta$ to $\int_0^T t dF(t) / \int_0^T t \overline{F}(t) dt$.

From the above result, the left-hand side of (46) also increases with θ from 0 to $q(\infty)\mu - 1$, where $q(\infty) \equiv \int_0^\infty t dF(t) / \int_0^\infty t \overline{F}(t) dt$. Therefore, if $q(\infty) > c_F / [\mu(c_F - c_Y)]$, then there exists an optimal $1/\theta_R^* (0 < 1/\theta_R^* < \infty)$ which satisfies (46), and the resulting cost rate is

$$C_R(\theta_R^*) = (c_F - c_Y) q(\theta_R^*). \tag{47}$$

Note that $q(\theta)$ plays the same role as the failure rate $h(t)$ in the standard age replacement.

We compare the expected cost rates $C_S(T_S^*)$ in (3) and $C_R(\theta_R^*)$ in (47) when $G(t) = 1 - \mathrm{e}^{-\theta t}$. We derive a modified replacement cost $\widehat{c}_Y$ in which two optimal cost rates $C_S(T_S^*)$ and $C_R(\theta_R^*)$ are the same. First, compute $T_S^*(0 < T_S^* < \infty)$ which satisfies (2) for c_T and c_F, and compute $C_S(T_S^*)$ in (3). Next, compute $\widehat{c}_Y$ which satisfies

$$q(\widehat{\theta}_R) \int_0^\infty \mathrm{e}^{-\widehat{\theta}_R t}\overline{F}(t)\mathrm{d}t + \int_0^\infty (1 - \mathrm{e}^{-\widehat{\theta}_R t})\mathrm{d}F(t) = \frac{c_F}{c_F - \widehat{c}_Y},$$

$$(c_F - c_T)h(T_S^*) = (c_F - \widehat{c}_Y)q(\widehat{\theta}_R),$$

i.e., we obtain $\widehat{\theta}_R$ for T_S^* which satisfies

$$\frac{1}{q(\widehat{\theta}_R)} \int_0^\infty (1 - \mathrm{e}^{-\widehat{\theta}_R t})\mathrm{d}F(t) + \int_0^\infty \mathrm{e}^{-\widehat{\theta}_R t}\overline{F}(t)\mathrm{d}t = \frac{1}{h(T_S^*)}\frac{c_F}{c_F - c_T}, \qquad (48)$$

and using $\widehat{\theta}_R$, we compute $\widehat{c}_R$ which satisfies

$$\frac{c_F - \widehat{c}_Y}{c_F - c_T} = \frac{h(T_S^*)}{q(\widehat{\theta}_R)}. \qquad (49)$$

Table 3 presents T_S^*, $1/\theta_R^*$, $\widehat{c}_Y/c_F$, and $\widehat{c}_Y/c_T$ for c_T/c_F or c_Y/c_F when $G(t) = 1 - \mathrm{e}^{-\theta t}$ and $F(t) = \sum_{j=k}^\infty (t^j/j!)\mathrm{e}^{-t}$ $(k = 2, 3, 4)$. From the comparison results of T_S^* and $1/\theta_R^*$, it is shown that $T_S^* > 1/\theta_R^*$ for small c_T/c_F or c_Y/c_F, however, $T_S^* < 1/\theta_R^*$ for large ones. It could also be shown clearly that $\widehat{c}_Y/c_F$ decreases with k and increases with c_T/c_F or c_Y/c_F. From the numerical values of $\widehat{c}_Y/c_T$, we can find that if how much $\widehat{c}_Y$ is less than c_T, the expected costs for standard and random age replacements are almost the same. Taking $k = 2$ for an example, when $\widehat{c}_Y$ is a little more than 60 % of c_T, we should adopt random policies.

4.2   Random Overtime Replacement

When the unit is replaced before failure at the first completion of some working cycles over a planned time T, we also call it as *random overtime replacement*, because it is made at random times and there is no cost c_T which is greater than c_Y.

We obtain a modified replacement cost $\widehat{c}_Y$ in which two optimal cost rates $C_S(T_S^*)$ and $C_O(T_O^*)$ are the same when $G(t) = 1 - \mathrm{e}^{-\theta t}$. First, compute $T_S^*(0 < T_S^* < \infty)$ which satisfies (2) for c_T and c_F, and $C_S(T_S^*)$ in (3). Using T_S^* and $C_S(T_S^*)$, we obtain $\widehat{c}_Y$ which satisfies

Table 3 Optimal T_S^*, $1/\theta_R^*$, $\widehat{c}_Y/c_F$, and $\widehat{c}_Y/c_T$ when $G(t) = 1 - e^{-\theta t}$ and $F(t) = \sum_{j=k}^{\infty}(t^j/j!)e^{-t}$

c_T/c_F or c_Y/c_F	$k = 2$				$k = 3$				$k = 4$			
	T_S^*	$1/\theta_R^*$	$\widehat{c}_Y/c_F$	$\widehat{c}_Y/c_T$	T_S^*	$1/\theta_R^*$	$\widehat{c}_Y/c_F$	$\widehat{c}_Y/c_T$	T_S^*	$1/\theta_R^*$	$\widehat{c}_Y/c_F$	$\widehat{c}_Y/c_T$
0.01	0.157	0.051	0.006	0.600	0.357	0.195	0.005	0.500	0.631	0.348	0.004	0.400
0.02	0.233	0.062	0.012	0.600	0.468	0.299	0.010	0.500	0.784	0.471	0.008	0.400
0.05	0.412	0.335	0.031	0.620	0.697	0.540	0.026	0.520	1.069	0.757	0.022	0.440
0.1	0.680	0.793	0.064	0.640	0.984	0.955	0.053	0.530	1.400	1.220	0.046	0.460
0.2	1.306	4.017	0.129	0.645	1.512	2.462	0.109	0.545	1.957	2.649	0.099	0.495

Table 4 Optimal T_S^*, T_O^*, $\widehat{c}_Y/c_F$, and $\widehat{c}_Y/c_T$ when $G(t) = 1 - e^{-\theta t}$ and $F(t) = 1 - e^{-t^2}$

$\frac{c_T}{c_F - c_T}$	$1/\theta = 0.1$			$1/\theta = 0.5$			
	T_O^*	$\widehat{c}_Y/c_F$	$\widehat{c}_Y/c_T$	T_O^*	$\widehat{c}_Y/c_F$	$\widehat{c}_Y/c_T$	T_S^*
0.01	0.043	0.005	0.497	0.015	–	–	0.100
0.02	0.075	0.009	0.466	0.030	–	–	0.142
0.05	0.149	0.039	0.821	0.073	–	–	0.225
0.10	0.239	0.084	0.921	0.137	–	–	0.319
0.20	0.373	0.160	0.961	0.249	0.068	0.410	0.455
0.50	0.656	0.329	0.986	0.521	0.268	0.805	0.738
1.00	1.011	0.495	0.991	0.885	0.459	0.918	1.091
2.00	1.615	0.663	0.994	1.515	0.654	0.981	1.689
5.00	3.315	0.829	0.995	3.274	0.825	0.990	3.385

$$z(\widehat{T}_O; \theta) \int_0^{\widehat{T}_O} \overline{F}(t)dt + \overline{F}(\widehat{T}_O) = \frac{c_F}{c_F - \widehat{c}_Y}, \tag{50}$$

$$(c_F - c_T)h(T_S^*) = (c_F - \widehat{c}_Y)z(\widehat{T}_O; \theta), \tag{51}$$

i.e., we compute $\widehat{T}_O$ which satisfies

$$\int_0^{\widehat{T}_O} \overline{F}(t)dt + \frac{\overline{F}(\widehat{T}_O)}{z(\widehat{T}_O; \theta)} = \frac{c_F}{(c_F - c_T)h(T_S^*)}, \tag{52}$$

and using T_S^* and T_O^*, we compute $\widehat{c}_Y$ from (50) or (51).

Table 4 presents optimal T_S^* and T_O^* for $c_T = c_Y$, and $\widehat{c}_Y/c_F$ and $\widehat{c}_Y/c_T$ for $c_Y < c_T$. It indicates that T_O^* decreases with c_F/c_T and $1/\theta$ and $T_O^* < T_S^* < T_O^* + 1/\theta$, and when replacement cost at failure becomes lower, $\widehat{c}_Y$ is close to c_T. In this case, replacement overtime equals to the standard policy. For example, when $c_T/(c_F - c_T) = 5.00$ and $1/\theta = 0.1$, $\widehat{c}_Y/c_T = 0.995$, i.e., if we set PR cost c_T is 1, when the actual PR cost at the end of work is less than 0.995, then replacement overtime is better than the standard policy. However, there exist some particular cases in which replacement overtime would never be better than the standard policy, e.g., the cases when $1/\theta = 0.5$ which is marked with "–".

5 Conclusions

We have selected a total operating time T and random working cycle Y or Y_j as two time scales to do the preventive replacement for the unit. Three combined age replacement policies, i.e., replacement first, last, and overtime, have been discussed and optimized. Comparisons between such three policies and standard replacement have been made in two cases when replacement costs c_T for T and c_Y for Y or Y_j are the same or not. It has been clearly shown that when $c_T = c_Y$, standard age replacement is the best among them. It is of great interest that replacement last is better than replacement first when the ratio of replacement costs c_T/c_F is less than some value. We have optimized random age replacement policy which is particular case of replacement first, last, and overtime, and discussed whether how much c_Y is lower than c_T, random age replacement and replacement overtime are better than standard policy. From such results, the age replacement with random working times or the policy which combines random replacement should be used more in practical fields from economical and environmental viewpoints.

Finally, as extended age replacement policies in Sects. 2.2–2.4, we suppose that the unit is replaced before failure at a total operating time T or at the N th working time $(N = 1, 2, \ldots)$, whichever occurs first. Then, by replacing $G(t)$ in (8) with $G^{(N)}(t)$ formally, the expected cost rate is

$$C_F(T, N) = \frac{c_T + (c_F - c_T) \int_0^T [1 - G^{(N)}(t)] \mathrm{d}F(t) + (c_Y - c_T) \int_0^T \overline{F}(t) \mathrm{d}G^{(N)}(t)}{\int_0^T \overline{F}(t)[1 - G^{(N)}(t)] \mathrm{d}t},$$

$$(53)$$

Clearly, when $N = \infty$, $C_F(T, \infty) = C_S(T)$ in (1).

Next, suppose that the unit is replaced before failure at a total operating time T or at the N th working time $(N = 0, 1, 2, \ldots)$, whichever occurs last. Then, by replacing $G(t)$ in (22) with $G^{(N)}(t)$ formally, the expected cost rate is

$$C_L(T, N) = \frac{c_F - (c_F - c_T)\overline{F}(T)G^{(N)}(T) - (c_F - c_Y) \int_T^\infty \overline{F}(t) \mathrm{d}G^{(N)}(t)}{\int_0^T \overline{F}(t) \mathrm{d}t + \int_T^\infty \overline{F}(t)[1 - G^{(N)}(t)] \mathrm{d}t}. \qquad (54)$$

Furthermore, suppose that the unit is replaced at the first completion some working cycles over a planned time T or at the Nth completion of working cycles, whichever occurs first. Then, the expected cost rate is

$$C_O(T, N) = \frac{\begin{aligned} & c_F - (c_F - c_T) \sum_{j=0}^{N-1} \int_0^T [\int_{T-t}^\infty \overline{F}(t + u) \mathrm{d}G(u)] \mathrm{d}G^{(j)}(t) \\ & \quad -(c_F - c_Y) \int_0^T \overline{F}(t) \mathrm{d}G^{(N)}(t) \end{aligned}}{\begin{aligned} & \sum_{j=0}^{N-1} \int_0^T [\int_T^\infty \overline{G}(u - t)\overline{F}(u) \mathrm{d}u] \mathrm{d}G^{(j)}(t) \\ & \quad + \int_0^T \overline{F}(t)[1 - G^{(N)}(t)] \mathrm{d}t \end{aligned}}. \qquad (55)$$

It would be interesting that problems as further studies to derive and compare optimal policies T^* and N^* analytically and numerically which minimize the expected cost rates $C_F(T, N)$ in (53), $C_L(T, N)$ in (54), and $C_O(T, N)$ in (55), by using similar methods proposed in this paper.

References

Barlow, R. E., & Proschan, F. (1965). *Mathematical theory of reliability*. New York: Wiley.
Chen, M., Mizutani, S., & Nakagawa, T. (2010a). Random and age replacement policies. *International Journal of Reliability, Quality and Safety Engineering, 17,* 27–39.
Chen, M., Nakamura, S., & Nakagawa, T. (2010b). Replacement and preventive maintenance models with random working times. *IEICE Transaction on Fundamentals, E93-A,* 500–507.
Duchesne, T., & Lawless, J. (2000). Alternative time scales and failure time models. *Lifetime Data Analysis, 6,* 157–179.
Haerder, T., & Reuter, A. (1983). Principles of transaction-oriented database recovery. *ACM Computing Surveys, 15,* 287–317.
Kobbacy, K. A. H., & Murthy, D. N. P. (2008). *Complex system maintenance handbook*. London: Springer.
Lewis, P. M., Bernstein, A. J., & Kifer, M. (2002). *Databases and transaction processing: An application-oriented approach*. Boston: Addison Wesley.
Manzini, R., Regattieri, A., Pham, H., & Ferrari, E. (2010). *Maintenance for industrial systems*. London: Springer.
Nakagawa, T. (2005). *Maintenance theory of reliability*. London: Springer.
Nakagawa, T. (2008). *Advanced reliability models and maintenance policies*. London: Springer.
Nakagawa, T. (2011). *Stochastic process with applications to reliability theory*. London: Springer.
Nakagawa, T., & Zhao, X. (2012). Optimization problems of a parallel system with a random number of units. *IEEE Transactions on Reliability, 61,* 543–548.
Nakagawa, T., Zhao, X., & Yun, W. (2011). Optimal age replacement and inspection policies with random failure and replacement times. *International Journal of Reliability, Quality and Safety Engineering, 18,* 405–416.
Osaki, S. (2002). *Stochastic models in reliability and maintenance*. Berlin: Springer.
Pham, H. (2003). *Handbook of reliability engineering*. London: Springer.
Pinedo, M. (2002). *Scheduling theory, algorithms and systems*. Englewood Cliffs, NJ: Prentice Hall.
Stadje, W. (2003). Renewal analysis of a replacement process. *Operations Research Letters, 31,* 1–6.
Sugiura, T., Mizutani, S., & Nakagawa, T. (2004). Optimal random replacement policies. In H. Pham & S. Yamada (Eds.), *Proceedings of the Tenth ISSAT International Conference on Reliability and Quality in Design* (pp. 99–103). Nevada: Las Vegas.
Wang, H., & Pham, H. (2007). *Reliability and optimal maintenance*. London: Springer.
Zhao, X., & Nakagawa, T. (2011). Optimal age replacement and inspection policies with random failure and replacement times. In H. Pham (Ed.), *Proceedings of the 17th ISSAT International Conference on Reliability and Quality in Design* (pp. 200–204). B.C.: Vancouver.
Zhao, X., Nakayama, K., & Nakamura, S. (2011). Cumulative damage models with replacement last. In T. Kim, et al. (Eds.), *Communications in computer and information science* (vol. 257, pp. 338–345). Berlin: Springer.
Zhao, X., & Nakagawa, T. (2012a). A summary of newly proposed age replacement policies. In H. Pham (Ed.), *Proceedings of the 18th ISSAT International Conference on Reliability and Quality in Design* (pp. 106–110). Boston: Massachusetts.
Zhao, X., & Nakagawa, T. (2012b). Optimization problems of replacement first or last in reliability theory. *European Journal of Operational Research, 223,* 141–149.

Availability of Systems with or Without Inspections

Kai Huang and Jie Mi

1 Introduction

Let S be a system that is designed for implementing certain function and is put into field operation at time $t = 0$. It is desired that the system performs excellently in field use. To measure its performance naturally, we can use its reliability characteristics such as survival probability, mean lifetime, mean residual life, or hazard rate etc., as criteria. If, however, we look at this process dynamically, then we will have to consider whether the system will be still functioning at any given time $t > 0$.

No matter how reliable the system is, it will fail sooner or later. So a problem is how to deal with failed system. Commonly, there are two ways to restore the failed system to operation. One is to repair the failed system if the system is repairable. Two types of repair methods are commonly studied in reliability literature. One method is called perfect or complete repair which repairs the failed system as good as new, i.e., the lifetime of the repaired system has the same distribution function as the original system S. Of course, it is implicitly assumed that the lifetime of the repaired system is independent of that of the original system S. The other type of repair method is named imperfect or incomplete repair with which the distribution function of the repaired system is not exactly the same as that of the original system. There is a special imperfect repair-minimal repair that repairs the failed system as good as it was prior to the failure of the original system. The other way to restore the failed system to operation is to replace it by a system that is iid as the original S. Both

K. Huang · J. Mi (✉)
Department of Mathematics and Statistics,
Florida International University, Miami, FL 33199, USA
e-mail: mi@fiu.edu

K. Huang
e-mail: khuang@fiu.edu

© Springer-Verlag London 2016   249
H. Pham (ed.), *Quality and Reliability Management and Its Applications*,
Springer Series in Reliability Engineering, DOI 10.1007/978-1-4471-6778-5_9

perfect repair and replacement change the failed system to be as good as new, they are equivalent in this regard and so we will use them alternatively later.

In addition to different types of repair methods, another related problem is how we can know whether a system is failed or unfailed. With respect to this, systems can be classified into two categories. Systems in one category can be under continuously monitoring, and consequently their failures are self-announcing, whereas systems in the other category cannot be continuously monitored due to either technical difficulties or expensive costs and therefore their failures are not self-announcing. In this case, system failure can only be detected by applying inspections. Two inspection policies, calendar-based and age-based policy will be described in Sect. 3.

Maintenance policies of both repair/replacement and inspection will be considered. With the help of these maintenance policies, it becomes more meaningful to measure the likelihood for a system to be functioning at any given time $t > 0$. To this purpose, we consider a system which can be in one of two states, namely 'unfailed' (or 'up') and 'failed' (or 'down'). By 'up,' we mean the system is still functioning and by 'down,' we mean the system is not working. Suppose that the status of system in field use can be revealed through certain way. Depending on the status of the system, the unfailed system may be upgraded or modified, and the failed system will be repaired or replaced. Let a system (the original system) starts operation at time $t = 0$ and works until certain maintenance measures, which includes but not limited to minimal repair, perfect repair, replacement, and many others, are going to take place. At this time, the first up period is over and the first down period begins. The first up and down periods constitute the first cycle of the system. At the end of each subsequent down period, a new cycle of the system will be completed and the system will resume operating, and so on and so forth. Let U_j and D_j denote the duration of the jth up and down periods, respectively. Basically U_j is the lifetime of the system after the $(j - 1)$th down period, while D_j is the length of time required to finish the planned maintenances like repairing or replacement. For any time $t \geq 0$, we can use binary random variable $\xi(t)$ to indicate the status of the system, namely $\xi(t) = 1$ meaning the system is unfailed or still working, and $\xi(t) = 0$ meaning the system failed or is not working. The probability that the system is still working at time t called instantaneous availability is denoted as $A(t) = P(\xi(t) = 1)$. This review will focus on the recent research works on $A(t)$ and some other related quantities such as the steady-state availability $A(\infty)$, the limiting average availability $A_{av}(\infty)$ defined later. The focus of Sect. 2 is the availability of systems whose failures are self-announcing, and so there is no need of applying inspections. Section 3 reviews major works on availability of systems whose failure are not self-announcing and hence inspections are necessary. The last section will mention some other works on system availability. There is a great vast of papers contributed to the topic of system availability, so it is inevitable that our review may miss some meaningful works or even very significant ones. But the authors hope this chapter can provide readers an overview about the progresses made in recent years toward the very important topic of system availability.

2 Availability of System Under Continuously Monitoring

It is assumed in this section that the failure of the system is self-announcing and thus inspection policy will not be involved.

Usually it is a formidable work to give explicit formula of $A(t)$ except for a few simple cases, so other measures have been proposed, and more attention is being paid to the limiting behavior of these quantities, i.e., engineers are more interested in the extent to which the system will be available after it has been run for a long time.

In the case, when $\{(U_j, D_j), j \geq 1\}$ consists of a sequence of iid random variables and U_j is independent of D_j for each $j \geq 1$ (for convenience we will call this as the IID Model in the following), some desirable properties have been obtained using results from alternative renewal processes. For instance, it has been proved that $A(t)$ is the unique solution of the renewal equation

$$A(t) = \overline{F}(t) + \int_0^t A(t - s)\mathrm{d}H(s),$$

where $H(t)$ is the convolution of $F(t)$ and $G(t)$ due to the assumed independence of U_i and D_i, i.e., $H(t) = \int_0^t F(t - x)\mathrm{d}G(x)$ for any $t \geq 0$. The solution can actually be expressed explicitly as

$$A(t) = \left(\overline{F} * \sum_{n=0}^{\infty} H^{(n)}\right)(t),$$

where $H^{(n)}$ is the n fold convolution of H. However, in the most cases this equation does not help much. In the case, when both F and G have density functions f and g the function H also has density given be $h(t) = H'(t) = \int_0^t g(t - x)f(x)\mathrm{d}x$ and consequently $A(t)$ is the unique solution of the renewal equation

$$A(t) = \overline{F}(t) + \int_0^t A(t - s)h(s)\mathrm{d}s.$$

Moreover, as $t \to \infty$ both the instantaneous and the average availability

$$\bar{A}(t) = \frac{1}{t} \int_0^t A(u)\mathrm{d}u \tag{1}$$

converge to a common limit $\mathbb{E}(U)/[\mathbb{E}(U) + \mathbb{E}(D)]$ where (U, D) is iid as (U_1, D_1). More details can be found in Barlow and Proschan (1975).

In addition, Takács (1957), Rényi (1957), Rise (1979), and Gut and Janson (1983) discussed the asymptotic normality property of $A(t)$ for the IID Model.

Mi (1995) studies the case when $\{(U_j, D_j), j \geq 1\}$ are independent but not necessarily identically distributed. The concepts that a sequence of random variables or their CDFs are dominated by a function and the average availability in the first n cycles defined by the ratio of accumulated up time in the first n cycles to the total length of time in the n cycles

$$\bar{A}_n = \frac{\sum_{j=1}^{n} U_j}{\sum_{j=1}^{n} U_j \sum + \sum_{j=1}^{n} D_j} \tag{2}$$

were introduced there. Assuming that $\{(U_j, D_j), j \geq 1\}$ are dominated by a function and that

$$\frac{1}{n} \sum_{j=1}^{n} \mathbb{E}(U_j) \to \mu, \quad \frac{1}{n} \sum_{j=1}^{h} \mathbb{E}(D_j) \to v \quad \text{as } n \to \infty \tag{3}$$

it was shown that

$$\lim_{n \to \infty} \bar{A}_n = \frac{\mu}{\mu + v} \quad a.s. \text{ and } L_p$$

and

$$A_{av}(\infty) = \lim_{t \to \infty} \bar{A}(t) = \frac{\mu}{\mu + v}, \quad a.s. \text{ and } L_p, \tag{4}$$

where $A_{av}(\infty)$ is called the limiting average availability.

Furthermore, under some additional mild conditions both $\bar{A}_n$ and $\bar{A}(t)$ are asymptotically normal as $n \to \infty$ or $t \to \infty$.

Assuming the IID Model, Sarkar and Chaudhuri (1999) found the Fourier transform $\tilde{b}(z)$ of the derivative $b(t)$ of unavailability $B(t) = 1 - A(t)$ defined by

$$\tilde{b}(z) = \int_{-\infty}^{\infty} e^{iz\ddot{u}} b(u) du, \tag{5}$$

where $i = \sqrt{-1}$ is the imaginary unit. Then they defined function $c_u(z) = e^{-iuz} \tilde{b}(z)$ for any $u > 0$. The function $c_u(z)$ is analytic except at finite number of isolated singularities, say $z_j, 1 \leq j \leq k$, and the authors further expressed $b(u)$ as a sum of residues

$$b(u) = -i \sum_{j:Im(z_j) < 0} \text{Res}(c_u, z_j), \tag{6}$$

where $Im(z_j)$ is the imaginary part of the complex number z_j, and $Im(z_j) < 0$ means z_j locates in lower half of the complex plane. Finally the instantaneous availability $A(t)$ was expressed in terms of the integral of $b(u)$

$$A(t) = 1 - \int_0^t b(u)du, \quad \forall t \geq 0 \tag{7}$$

In that paper, Fourier transformation is applied instead of Laplace transformation in order to avoid problem with inverting the Laplace transform of $A(t)$. As an example, let the lifetime of the system have gamma distribution with density

$$f(t) = \frac{\lambda^\alpha}{\Gamma(\alpha)} e^{-\lambda t} t^{\alpha-1}, \quad t > 0, \tag{8}$$

where α is a positive integer, and let the repair time have exponential distribution with density

$$g(t) = \lambda e^{-\lambda t}, \quad t > 0. \tag{9}$$

Then $A(t)$ is obtained as

$$A(t) = \frac{\alpha}{\alpha+1} - \frac{1}{\alpha+1} \sum_{j=1}^\alpha \theta_j e^{-\lambda(1-\theta_j)t}, \tag{10}$$

where $\theta_0 = 1, \theta_1, \ldots, \theta_\alpha$ are the $(\alpha+1)$-th roots of 1. That is, $\theta_j = [\exp\{i2\pi/(\alpha+1)\}]^j$.

Example 2.1 Suppose that $T \sim Gamma(4, \alpha)$ and $D \sim Gamma(2, \alpha)$.
In this case, we have

$$f(t) = \frac{\alpha^4}{\Gamma(4)} e^{-\alpha t} t^3, \quad t > 0,$$

$$\tilde{f}(s) = (1 - is/\alpha)^{-4}, \quad -\infty < s < \infty,$$

$$g(t) = \frac{\alpha^2}{\Gamma(2)} e^{-\alpha t} t, \quad t > 0,$$

$$\tilde{g}(s) = (1 - is/\alpha)^{-2}, \quad -\infty < s < \infty,$$

$$c_u(z) = e^{-iuz} \tilde{f}(z) \frac{1 - \tilde{g}(z)}{1 - \tilde{f}(z)\tilde{g}(z)}$$

$$= \frac{e^{-itz} \alpha^4}{7\alpha^2 z^2 + 6i\alpha^3 z - 4i\alpha z^3 - 3\alpha^4 - z^4}.$$

There are 4 singularities of $c_u(z)$:

$$z_1 = \frac{-3i + \sqrt{3}}{2}\alpha, \quad z_2 = \frac{-3i - \sqrt{3}}{2}\alpha \quad \text{and} \quad z_3 = \frac{-i + \sqrt{3}}{2}\alpha, \quad z_4 = \frac{-i - \sqrt{3}}{2}\alpha.$$

We can calculate the residue at $z = z_1$:

$$\text{Res}(c_u, z_1) = \lim_{z \to z_1}(z - z_1)c_u(z) = \frac{\alpha e^{-3u\alpha/2} e^{iu\alpha\sqrt{3}/2}}{\sqrt{3} - 3i}.$$

The residues at $z = z_2, z_3$ and z_4 can be calculated similarly. Thus, we have

$$b(u) = -i(\text{Res}(c_u, z_1) + \text{Res}(c_u, z_2) + \text{Res}](cu,z3) + \text{Res}(cu,z4))$$

and

$$
\begin{aligned}
A(t) &= 1 - \int_0^t b(u)du \\
&= \frac{2}{3} + \frac{\sqrt{3}e^{-\alpha t/2}\sin(\sqrt{3}\alpha t/2)}{3} + \frac{e^{-3\alpha t/2}\cos(\sqrt{3}\alpha t/2)}{3}.
\end{aligned}
$$

Without using Fourier or Laplace transformation, it seems not likely to obtain this expression through directly solving the renewal equation mentioned above.

Keeping the assumption of independence of all U_j and D_j, $j \geq 1$, Biswas and Sarkar (2000) modified the IID model as follows. A positive integer k is fixed in advance. At the $(k+1)$ th failure of the system, either it is replaced by a new system that is iid to the original one and the replacement is finished instantly without taking any time (Model A), or it is perfectly repaired that takes time D_{k+1} (Model B). Obviously, in either case, the system is brought back to a condition as good as new and so the time when D_{k+1} ends is the renewal point. Afterward, the process will evolve in the same pattern. In other words, the two models allow k imperfect repairs before a complete repair or replacement that will bring the process to a renewal point. It is natural to further assume that

$$F_1 \overset{st}{\geq} F_2 \overset{st}{\geq} \cdots \overset{st}{\geq} F_{k+1} \tag{11}$$

and

$$G_1 \overset{st}{\leq} G_2 \overset{st}{\leq} \cdots \overset{st}{\leq} G_{k+1} \tag{12}$$

This paper employed the same Fourier transformation approach in Sarkar and Chaudhuri (1999). Denote the instantaneous system availability as $A_j(t)$ when at time $t = 0$ the system with lifetime U_j and then again take the ending time of D_{k+1}

as the renewal point. The equations satisfied by the Fourier transforms of the derivatives $b_j(t)$ of unavailability $B_j(t) = 1 - A_j(t), 1 \leq j \leq k+1$ were derived for both Model A and Model B. Upon determination of $\tilde{b}(u) \equiv \tilde{b}_1(u)$, the desired availability $A(t) \equiv A_1(t)$ then can be obtained by (7). The explicit expression of $A(t)$ were shown for the case of exponential lifetimes and repair times.

In the above studies at each system failure, it is deterministic that the failed system undergoes either perfect repair or imperfect repair. Brown and Proschan (1983) considered a model according to which a perfect repair is implemented with probability p and an imperfect repair, which is actually a minimal repair restoring the failed system to its condition just prior to failure, is performed with probability $1 - p$ at each system failure. Their model has been generalized by Block et al. (1985) to the case in which the probability of perfect repair is state dependent. Lim et al. (1998) proposed the Bayesian imperfect repair model, according to which the probability of performing a perfect repair is a random variable P with distribution function $\Pi(p)$ on $(0, 1]$, and the probability of applying minimal repair is $1 - P$ at each system failure. Cha and Kim (2001) examined the same model under the assumptions that the perfect repair times are iid, the minimal repair times are iid, and these times are independent of each other. Under these assumptions the steady-state system availability $A(\infty)$ was derived as

$$A(\infty) = \frac{\int \int_0^\infty \exp\{-p\Lambda(t)\}\mathrm{d}t\mathrm{d}\Pi(p)}{\int \int_0^\infty \exp\{-p\Lambda(t)\}\mathrm{d}t\mathrm{d}\Pi(p) + v_1 \int_0^1 \frac{1-p}{p}\mathrm{d}\Pi(p) + v_2}, \tag{13}$$

where $\Lambda(t) = \int_0^t \lambda(x)\mathrm{d}x$, $\lambda(x)$ is the failure rate function of the system, v_1 is the mean perfect repair time, and v_2 is the mean minimal repair time.

In the special case of $P = 1$ with probability one, that is, only perfect repair is performed at each system failure, this model is reduced to the IID one. Certainly in this case $v_2 = 0$ and so $A(\infty)$ becomes

$$A(\infty) = \frac{\int_0^\infty \bar{F}(t)\mathrm{d}t}{\int_0^\infty \bar{F}(t)\mathrm{d}t + v_1} \tag{14}$$

which is exactly the same as in the classic IID model since

$$\bar{F}(t) = \int_0^t \Lambda(x)\mathrm{d}x. \tag{15}$$

In the previous works on availability, only one type failure was taken into consideration, Cha et al. (2004) generalizes the study of Mi (1994) and considered repairable system with two types of failures: one is Type I failure (minor failure) that occurs with probability $1 - p(t)$, where t is the age of the system at failure, the other is Type II failure (catastrophic failure, i.e., the usual failure) that occurs with probability $p(t)$. The failed system with Type I failure can be restored to operation by a minimal repair, whereas the failed system with Type II failure can be restored to operation only by a perfect repair (or a replacement). This model is called the general failure model. The study on availability in Cha et al. (2004) combined burn-in policy b and age replacement T together and obtained the expression of the steady-state availability $A(\infty)$ as follows:

Suppose that a new system is burned-in for time b, and it will be put in field operation if it survives the burn-in. In the field use, the system is replaced by another system, which has also survived the same burn-in time b, either at the use "age" T or at the time of the first Type II failure, whichever occurs first. However, for each Type I failure occurring during field use, only minimal repair will be performed.

It is further assumed that the repair times are not negligible. Let v_1, v_2, and v_3 be the means of a minimal repair time, time for an unplanned replacement caused by the Type II failure, and time for a replacement done at the system field use age T by planned preventive maintenance policy, respectively. For technique reason, it is required that $\int_0^\infty p(t)r(t)dt = \infty$ where $r(t)$ is the hazard rate function of the lifetime of a new system. Under these assumptions, then by similar arguments described in Cha and Kim (2002), it can be shown that the steady-state availability of the system under the policy (b, T) is given by

$$A(\infty) = \frac{\int_0^T \bar{G}_b(t)dt}{\int_0^T \bar{G}_b(t)dt + \left[\int_0^T r(b+t)\bar{G}_b(t)dt - G_b(T)\right]v_1 + G_b(T)v_2 + \bar{G}_b(T)v_3}$$

(16)

where

$$\bar{G}_b(t) = \exp\{-\int_0^t p(b+x)r(b+x)dx\}$$ (17)

Letting $b = 0$ and $p(t) = 1, \forall t \geq 0$, we see that $\bar{G}_b(t) = \bar{F}(t)$. It also implies that there is only perfect repair but no minimal repair and so $v_1 = 0$, and $v_2 = v_3 \equiv v$. Thus $A(\infty)$ is reduced to

$$A(\infty) = \frac{\int_0^T \bar{F}(t)dt}{\int_0^T \bar{F}(t)dt + v}.$$ (18)

If further let the age replacement policy $T = \infty$, that is replacement can only take place at system failure, then finally $A(\infty)$ is obtained as

$$A(\infty) = \frac{\int_0^\infty \bar{F}(t)dt}{\int_0^\infty \bar{F}(t)dt + v} = \frac{\mu}{\mu + v}$$ (19)

which is exactly the result in the case of the IID Model.

Mi (2006a, b) reconsidered the system with nonidentical lifetime distributions and nonidentical repair time distributions studied in Mi (1995). Let U_j and D_j have distribution functions F_j and G_j, respectively, for each $j \geq 1$. Assuming that both sequences $\{U_j\}, j \geq 1$ and $\{D_j, j \geq 1\}$ are dominated, there exist two CDFs F and G such that $F_j \rightarrow F$ and $G_j \rightarrow G$ in distribution as $j \rightarrow \infty$, and some other technical requirements, Mi (2006a, b) gave three sets of conditions under which the steady-state availability $A(\infty)$ exists and is given by

$$A(\infty) = \frac{\mu}{\mu + \nu}$$

where

$$\mu = \int_0^\infty \bar{F}(t)dt, \quad \nu = \int_0^\infty \bar{G}(t)dt \tag{20}$$

Moreover, it was shown there that if there exists an integer $k \geq 0$ such that $F_{nk+j}(t) = F_j(t)$, $G_{nk+j}(t) = G_j(t)$, for any $1 \leq j \leq k$, $t \geq 0$, then it holds that

$$A(\infty) = \frac{\sum_{j=1}^k \mu_j}{\sum_{j=1}^k \mu_j + \sum_{j=1}^k \nu_j}, \tag{21}$$

where μ_j and ν_j are the means associated with F_j and G_j, $1 \leq j \leq k$. Clearly, the results of both Model A and Model B discussed in Biswas and Sarkar (2000) can be obtained as special cases of this result in Mi (2006a, b).

In the models reviewed above, there is no spare system on cold standby and there is only one repair facility so failed system can be placed for repairing without any waiting time. However, in the model considered in Sarkar and Li (2006) in addition to the original system, there are $s \geq 1$ identical spares remain on cold standby, and there are $r \geq 1$ repair facilities which serves the failed systems in the order in which they join the repair queue. The lifetimes of the original system and the s spares are iid; the repair times of the r repair facilities are also iid; further, these lifetimes and repair times are independent of each other.

At time $t = 0$, the original system is put on operation and at its failure one spare is placed on operation immediately without taking any time and the failed system is sent for repairing. In general, at the instant of failure of an operating system, the failed system always joins the repair queue and its repair starts as soon as one of the repair facilities is free, in the mean time one spare, if available, is placed to operation immediately without taking any time. If, however, at the failure of an operating system, there is no any spare available, that is all the $(s+1)$ systems are either undergoing or awaiting repair, then the entire system enter the down state. It is obvious that $r \leq s+1$ since otherwise at any time, there are always some repair facilities remain idle.

Let the original system be supported by $r \geq 1$ repair facilities and $s \geq r - 1$ spare systems. Assuming that the lifetime distribution is exponential with mean α^{-1} and repair time distribution is exponential with mean β^{-1}, the authors derived the limiting average availability as

$$A_{av}(\infty) = \frac{r\rho \sum_{j=0}^{s} \gamma_j \rho^{s-j}}{1 + r\rho \sum_{j=0}^{s} \gamma_j \rho^{s-j}}, \tag{22}$$

where $\rho = \beta/\alpha$ and

$$\gamma_j = \begin{cases} \frac{r! r^{s-r}}{j!}, & j = 0, 1, \ldots, r-1 \\ r^{s-j}, & j = r, r+1, \ldots, s. \end{cases} \tag{23}$$

In a more general case, if again there are at least one repair facilities ($r \geq 1$), repair time has exponential distribution with mean β^{-1}, but the number of spare systems satisfies $s \geq \max\{1, r-1\}$, and the lifetime distribution of systems has density and is arbitrary other than this. Based on these assumptions, the limiting average availability was obtained as

$$A_{av}(\infty) = \frac{\mu(0, \ldots, 0, 1)(I - Q)^{-1}(1, \ldots, 1)'}{\mu(0, \ldots, 0, 1)(I - Q)^{-1}(1, \ldots, 1)' + (r\beta)^{-1}}, \tag{24}$$

where μ denotes the mean system lifetime, $(1, \ldots, 1)'$ is a $s \times 1$ column vector with all components of 1, and the $s \times s$ matrix Q can be determined by some equations given in Sarkar and Li (2006).

Sarkar and Biswas (2010) employed the same Fourier transformation approach proposed in Sarkar and Chaudhuri (1999) to the model studied in Sarkar and Li (2006). Keeping the same assumption of the exponential system lifetimes and repair times, the authors expressed the instantaneous availability $A(t)$ as

$$A(t) = 1 - \int_0^t b_0(u) \mathrm{d}u$$

for the case of $s \geq 1$ and $r = 1$ and $r = 2$, where the function $b_0(u)$ is the derivative of $B_0(u)$, and $B_0(u)$ denotes the unavailability of the system at time $u > 0$ when there is no failure of spares. Actually, $A(t) = 1 - B_0(t)$. It turns out that $b_0(u)$ is the sum of residues of a complex-valued function that is analytic except finite number of isolated singularities. For details, the readers are referred to the Appendix of Sarkar and Biswas (2010).

At the end of this section recall that usually it is difficult to obtain a closed-form expression for $A(t)$ as mentioned before. As a matter of fact, the behavior of $A(t)$ can also be very complicate as shown in the following example.

Example 2.2 Consider a system that has $U \sim Gamma(p, \alpha)$ and $D \sim \ln \mathcal{N}(\mu, \sigma)$ with density functions

$$f(t) = \frac{\alpha^p}{\Gamma(p)} e^{-\alpha t} t^{p-1} \quad \text{and} \quad g(t) = \frac{1}{t\sigma\sqrt{2\pi}} e^{-\frac{(\ln t - \mu)^2}{2\sigma^2}}.$$

In the following figure, the left panel shows the system availability functions $A(t)$ corresponding to different parameters $p = 2, 4, 8$ and $p = 10$. The right panel shows the availability functions with different parameters $\sigma = 0.25, 0.50$ and $\sigma = 0.75$. The function $A(t)$ for all these cases does not have closed form and thus are obtained numerically.

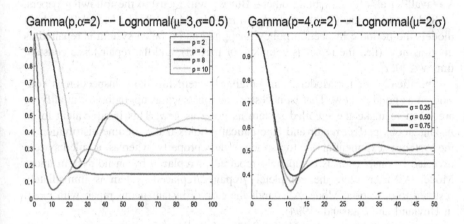

3 Availability of System with Inspections

In this section, we will review research works on availability of systems that can be maintained through inspections. Inspection policy was proposed in Barlow and Proschan (1975) or even earlier. Inspections are important for systems whose failures are not self-announcing. This type of systems is common in industries. For instance, some industrial safety and protection system such as circuit breakers, fire detectors, gas detectors, pressure detectors, and safety valves are installed to prevent various specific risks. Depending on the status of the system being inspected the system will be repaired, replaced, upgraded, or modified. The system then will be restored to operation upon completion of these maintenances.

Two types of inspection policies are widely applied in practice. The first type called calendar-based inspection policy schedules inspections at fixed calendar intervals, say at times $\tau, 2\tau, \ldots$, where $\tau > 0$ is a predetermined constant. This policy is also named as periodic inspection policy. According to the calendar-based inspection policy, a system starting its operation at time $t = 0$ is inspected at time $t = \tau$, then at time $t = 2\tau$ and so on.

The second type of inspection polity, the age-based inspection policy schedules inspections at fixed age intervals. Suppose that constant $\tau > 0$ is determined in advance. Let the system be inspected at time $t = \tau$ and resume operation at time $\tau + m$, where m represents the required time to complete the above-mentioned maintenances. According to the age-based inspection policy, the system will be inspected at time $t = \tau + m$ and this pattern will be continued in the same way.

Much has been done in studying availability of systems that are maintained through inspection. For example, Wortman et al. (1994), Wortman and Klutke (1994), Yeh (1995), Klutke et al. (1996), Dieulle (1999), Vaurio (1999), Ito and Nakagawa (2000), Chelbi and Ait-Kadi (2000), Yang and Klutke (2000, 2001) and Yadavalli et al. (2002), among others. But we will focus on the following papers.

Sarkar and Sarkar (2000) studied two models: Model A and Model B. In both models the periodic inspection policy is applied and a failed system is repaired as good as new (i.e., the repair is complete or perfect), and the repair takes constant time $v \in [0, \tau]$.

Specifically, under Model A an unfailed system found by inspection is considered as good as new. That is, necessary actions such as upgrading or modifying are taken to make the unfailed system as good as new. This is equivalent to an instantaneous perfect repair and automatically holds if the lifetime distribution of the system is exponential due to its memoryless property; whereas, a failed system revealed by inspection is completely repaired or replaced by an iid system under Model A. Thereafter, the completely repaired/replaced system is immediately restored to operation. Model A extends the case of instantaneous repair with $v = 0$ in Høyland and Rausand (1994).

On the other hand, under Model B an unfailed system continues its operation without any intervention, i.e., the system remains as good as it is; a failed system will undergo perfect repair or replacement as under Model A, but the operation of the repaired system will start at the next scheduled inspection time after the repair/replacement, not immediately which is different from Model A.

To display the results in Sarkar and Sarkar (2000) we denote the life time of a given system starting operation at time $t = 0$ as U, the distribution function of U as $F(\cdot)$. This notation will be kept in the rest of this paper.

For Model A with constant repair/replacement time $0 \leq v \leq \tau$ the availability $A(k\tau)$ is given as

$$A(k\tau) = \frac{[\bar{F}(\tau) - \bar{F}(\tau - v)]^k F(\tau) + \bar{F}(\tau - v)}{F(\tau) + \bar{F}(\tau - v)} \qquad (25)$$

Based on it the instantaneous availability $A(t)$ is given as

$$A(t) = \begin{cases} \bar{F}(t), & \text{if } 0 \leq t \leq \tau; \\ \bar{F}(\tau)\bar{F}(t-\tau), & \text{if } \tau < t < \tau + v; \\ A(k\tau)\bar{F}(t-k\tau) & \text{if } k\tau+v \leq t < (k+1)\tau+v, \\ +[1-A(k\tau)]\bar{F}(t-k\tau-v)' & k = 1,2,\ldots \end{cases} \qquad (26)$$

It is easy to see that when $v = 0$ the expression of $A(t)$ has the form

$$A(t) = \begin{cases} \bar{F}(\tau), & \text{if } 0 \leq t \leq \tau \\ \bar{F}(t-k\tau) = \bar{F}\left(t - \lfloor \frac{t}{\tau} \rfloor \tau \right) & \text{if } k\tau \leq t < (k+1)\tau, \\ & k = 1,2,\ldots \end{cases} \qquad (27)$$

where $\lfloor x \rfloor$ is the largest integer part of x. This is exactly the result in Høyland and Rausand (1994). For the same Model A, the limiting average availability of the system is

$$A_{av}(\infty) = \tau^{-1}\{\phi[H(\tau+v) - H(v)] + (1 - \phi)H(\tau)\} \qquad (28)$$

where

$$\phi = \frac{\bar{F}(\tau - v)}{\bar{F}(\tau - v) + F(\tau)} \qquad (29)$$

and

$$H(t) = \int_0^t \bar{F}(x)dx \qquad (30)$$

For Model B, in the case of $v = 0$, the instantaneous availability $A(t)$ is given as

$$A(t) = \sum_{j=0}^{k} c_j \bar{F}(t - j\tau), \quad \text{if } k\tau \leq t < (k+1)\tau, \ k = 0,1,\ldots \qquad (31)$$

where $c_0 = 1$ and $c_j, j \geq 1$ are determined recursively by

$$c_j = \sum_{i=0}^{j} p_i c_{j-i}$$

with $p_i = F(i\tau) - F((i-1)\tau)$.

In addition, the limiting average availability of the system is

$$A_{av}(\infty) = \frac{\mathbb{E}(U)}{\tau \mathbb{E}\left(\left\lceil \frac{U}{\tau} \right\rceil\right)} \tag{32}$$

where $\lceil x \rceil$ is the smallest integer satisfying $\lceil x \rceil \geq x$.

In the case of $v > 0$, without loss of generality it can be assumed that $v = m\tau$ for some integer $m \geq 1$. This holds because under Model B the failed system is restored to operation only at the next scheduled inspection time following its perfect repair or replacement; that is, if repair/replacement is completed during the time interval $((m-1)\tau, m\tau]$, then the repaired/replaced system is restored to operation at time $m\tau$.

In the case of $v > 0$ suppose $v = m\tau$ for an integer $m \geq 1$, it wan shown that

$$A(t) = \begin{cases} \bar{F}(t) & \text{if } 0 \leq t < (m+1)\tau \\ \sum_{j=0}^{k+m} d_j \bar{F}(t - j\tau), & \text{if } (k+m)\tau \leq t < (k+m+1)\tau, \quad k = 1, 2, \ldots \end{cases} \tag{33}$$

where $d_0 = 1$, $d_1 = d_2 = \cdots = d_m = 0$, and d_j, $\quad j \geq m$ is determined by

$$d_j = \sum_{i=1}^{j} q_i d_{j-i}, \quad \forall j \geq m+1$$

with $\quad q_1 = q_2 = \cdots = q_m = 0 \quad$ and $\quad q_{m+i} = F(i\tau) - F((i-1)\tau), \quad \forall i \geq 1$.
Moreover, the limiting average availability of the system is

$$A_{av}(\infty) = \frac{\mathbb{E}(U)}{\tau \mathbb{E}\left(m + \left\lceil \frac{U}{\tau} \right\rceil\right)} = \frac{\mathbb{E}(U)}{v + \tau \mathbb{E}\left(\left\lceil \frac{U}{\tau} \right\rceil\right)} \tag{34}$$

Example 3.1 Consider Model A in Sarkar and Sarkar (2000). The system availability $A(t)$ is determined by (25) and (26).

Specifically, let $\bar{F}(t) = e^{-t}$ for all $t \geq 0$, $v = \ln 2$ and $\tau = 2\ln 2$. According to (25)

$$A_k \equiv A(2k \ln 2) = \frac{[e^{-2\ln 2} - e^{-\ln 2}]^k (1 - e^{-2\ln 2}) + e^{-\ln 2}}{(1 - e^{-2\ln 2}) + e^{-\ln 2}}$$

$$= \frac{4}{5}\left[(-1)^k \frac{3}{4^{k+1}} + \frac{1}{2}\right] = \frac{1}{5}\left[(-1)^k \frac{3}{4^k} + 2\right], \quad k = 1, 2, \ldots.$$

From (26) it follows that

$$A(t) = \begin{cases} e^{-t} & \text{if } \ 0 \le t \le 3\ln 2 \\ 2^{2k}e^{-t}(2 - A_k), & \text{if } \ (2k+1)\ln 2 \le t < (2k+3)\ln 2, \\ & \quad k = 1, 2, \ldots \end{cases}$$

Obviously $A(t)$ does not converge as $t \to \infty$ but the limiting average availability $A_{av}(\infty)$ exists and is given by (28). We have

$$\phi = \frac{e^{-\ln 2}}{e^{-\ln 2} + (1 - e^{-2\ln 2})} = \frac{2}{5}$$

$$H(t) = \int_0^t e^{-x}dx = 1 - e^{-t}$$

and

$$A_{av}(\infty) = \tau^{-1}[\phi[H(\tau+v) - H(v)] + (1 - \phi)H(\tau)]$$
$$= (2\ln 2)^{-1}\left[\frac{2}{5}[e^{-\ln 2} - e^{-3\ln 2}] + \frac{3}{5}(1 - e^{-\ln 4})\right]$$
$$= = \frac{3}{5}(2\ln 2)^{-1} \approx 0.4328.$$

Mi (2002) discussed a model which is similar to Model B with $v = 0$ in Sarkar and Sarkar (2000) except that the system will undergo complete repair or replacement at time $\eta\tau$ regardless whether the system is failed or unfailed from the result of inspection, where η is either an integer or $\eta = \infty$. Under this assumption, the limiting average availability of the system was derived as

$$A_{av}(\infty) = \frac{\int_0^{\eta\tau} \bar{F}(x)dx}{\tau \sum_{k=0}^{\eta-1} \bar{F}(k\tau)}. \tag{35}$$

Note that assuming $v = 0$ then the limiting average availability of the system under Model B in Sarkar and Sarkar (2000) is the special case of $\eta = \infty$ in Mi (2002). As a matter of fact, we have

$$\mathbb{E}\left(\left\lceil\frac{U}{\tau}\right\rceil\right) = \sum_{k=1}^{\infty}\int_{(k-1)\tau}^{k\tau}\left\lceil\frac{x}{\tau}\right\rceil dF(x) = \sum_{k-1}^{\infty}\int_{(k-1)\tau}^{k\tau} k dF(x)$$

$$= \sum_{k-1}^{\infty} k\{\bar{F}(k-1)\tau - \bar{F}(k\tau)\} = \sum_{k-1}^{\infty}\bar{F}(k\tau) \tag{36}$$

Example   3.2 Let   the   system   lifetime   distribution   be   exponential $F(t) = 1 - exp(-\lambda t)$. The following figures show the limiting average availabilities obtained from (35) at fixed λ (left panel) and at fixed τ (right panel).

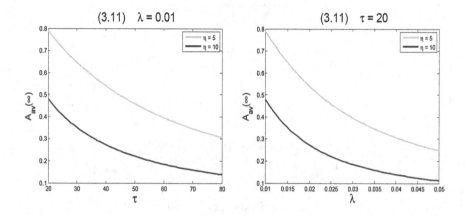

In the model studied by Sarkar and Sarkar (2001), the system in application is periodically inspected with inspection interval $\tau > 0$ and is supported by an iid spare system which is in cold standby. At time $t = \tau$, the spare system takes over the operation no matter whether the status of the inspected system is failed or unfailed. If the inspection found the inspected system failed then it is sent for repair/replacement; otherwise, it is upgraded. Both the repair and upgrade are perfect meaning that repaired or upgraded system becomes as good as new. At time $t = 2\tau$, the inspection is performed again. The system being repaired/replaced or upgraded at time $t = \tau$ will take over operation if the repair/replacement or upgrade has been completed before $t = 2\tau$, and the system inspected at $t = 2\tau$ will undergo either repair/replacement or upgrade; otherwise, the inspection will be suspended and only after completion of repair/replacement or upgrade the repaired or upgraded system will take over the operation at the next scheduled inspection time. Denote the random time needed for repair/replacement and upgrade as Y_r and Y_w, respectively. It can be assumed that the random times Y_r and Y_w have $\{\tau, 2\tau, \ldots\}$ as their support sets because of the assumption about the inspection policy. Furthermore, it is assumed naturally that $Y_w \overset{st}{\le} Y_r$.

Let $P(Y_w = i\tau) = p_i$ and $P(Y_r = i\tau) = q_i$ for $i \geq 1$. Under their model, the authors obtain the instantaneous system availability $A(t)$ as follows:

$$A(t) = \begin{cases} \bar{F}(t), & \text{if } 0 \leq t < \tau; \\ \bar{F}(t - \tau), & \text{if } \tau \leq t < 2\tau; \\ A_1(t - \tau), & \text{if } 2\tau \leq t < \infty. \end{cases} \tag{37}$$

Where $A_1(s) = \mathbb{P}(\xi(s + \tau) = 1)$ and has the form

$$A_1(t) = \sum_{j=0}^{k} w_{kj}\bar{F}(s - j\tau), \quad \text{for } k\tau \leq s < (k+1)\tau, \quad k = 0, 1, \cdots \tag{38}$$

with w_{kj} determined by (2.1a), (2.1b) of Sarkar and Sarkar (2001). Based on the expression of $A(t)$, the limiting average availability is also obtained as

$$A_{av}(\infty) = \frac{\sum_{i=0}^{\infty}(1 - R_i)[H((i+1)\tau) - H(i\tau)]}{\tau \sum_{i=0}^{\infty}(1 - R_i)} \tag{39}$$

where $R_0 = 0$,

$$R_i = \sum_{k=1}^{i}[\bar{F}(\tau)p_k + F(\tau)q_k], \quad \forall i \geq 1. \tag{40}$$

and

$$H(t) = \int_0^t \bar{F}(x)dx \tag{41}$$

Cui and Xie (2001) investigated two models: Model A and Model B similar to those of Sarkar and Sarkar (2000). But the models in Cui and Xie (2001) allow a perfect repair or a replacement that takes a random time Y with distribution $G(y)$ and density function $g(y)$. Special case when Y is a constant v that was assumed in Sarkar and Sarkar (2000) is also discussed in Cui and Xie (2001).

Using a random walk approach, Cui and Xie (2001) established the relationship between the random walk in a plane and the periodically inspected system. Based on this relationship, both the explicit and recursive formulas of $A(t)$ for the case of constant perfect repair time or replacement time were displayed for the two models. In the case of random time Y, the recursive formula of $A(t)$ for the two models were obtained too. All these expressions and formulas of $A(t)$ are complicate, so the readers are referred to their paper for details.

The model proposed in Biswas et al. (2003) assumed that for $1 \leq i \leq h$, where $h \geq 1$ is a given integer, at the time the ith system failure is detected by inspection, the failed system will undergo an incomplete repair and the lifetime distribution of the repaired system may not be the same as F, the lifetime distribution of the original system. At the time, when the $(h+1)$th system failure is detected, the failed system will be repaired perfectly or replaced by one whose lifetime is iid as that of the original system. Here, the times needed for performing incomplete repair, perfect repair, and replacement can be either constants or random variables. It is also assumed that the repaired system will be restored to operation at the next scheduled inspection time but not immediately. As a consequence of this assumption, the support set of the random perfect repair time, replacement time, incomplete repair time can be limited on the set $\{\tau, 2\tau, \ldots\}$, and constant repair/replacement times can be limited to multiples of τ.

Under these assumptions, both the instantaneous system availability $A(t)$ and limiting average availability are obtained when there is only a single incomplete repair, i.e., $h = 1$. The expression of $A(t)$ is complicate, so we display only the limiting average availability $A_{av}(\infty)$. In the deterministic case, that is repair/replacement times are constant, $A_{av}(\infty)$ is given as

$$A_{av}(\infty) = \frac{\mathbb{E}(U_1 + U_2)}{\tau \left[\mathbb{E}\left(\left\lceil \frac{U_1}{\tau} \right\rceil\right) + \mathbb{E}\left(\left\lceil \frac{U_2}{\tau} \right\rceil\right) + m_1 + m_2 \right]}, \tag{42}$$

where U_1 is the lifetime of the original system and U_2 is the lifetime of the system upon completion of the first incomplete repair, constants $m_1\tau$ and $m_2\tau$ are the required times for perfect repair (or replacement) and incomplete repair, respectively.

In the stochastic case, denote the required times of perfect repair and incomplete repair as random variables D_1 and D_2, respectively. Then the limiting average availability is

$$A_{av}(\infty) = \frac{\mathbb{E}(U_1 + U_2)}{\tau \left[\mathbb{E}\left(\left\lceil \frac{U_1}{\tau} \right\rceil\right) + \mathbb{E}\left(\left\lceil \frac{U_2}{\tau} \right\rceil\right) + \mathbb{E}(D_1) + \mathbb{E}(D_2) \right]}. \tag{43}$$

Obviously, if $\mathbb{P}(D_i = m_i\tau) = 1$, $i = 1, 2$ for two integers m_1 and m_2, then Eq. (43) becomes (42).

Note that the expression of $A_{av}(\infty)$ shown in (43) becomes that one in Sarkar and Sarkar (2000) if $U_1 \overset{st}{=} U_2$ and $m_1 = m_2$ which is equivalent to the case $h = 0$.

Consider the case of single incomplete repair ($h = 1$) but the repair/replacement times are random. This time the expressions for $A(t)$ and $A_{av}(\infty)$ were also derived in Biswas et al. (2003). For example,

$$A_{av}(\infty) = \frac{\mathbb{E}(U_1 + U_2)}{\left[\mathbb{E}\left(\left\lceil \frac{U_1}{\tau} \right\rceil\right) + \mathbb{E}\left(\left\lceil \frac{U_2}{\tau} \right\rceil\right) + \mathbb{E}(D_1) + \mathbb{E}(D_2) \right]},$$

where D_1 represents the random time required for an incomplete repair and D_2 represents the random time required for a perfect repair. Clearly, if $P(D_i = m_i\tau) = 1$, $i = 1, 2$, then (43) is reduced to (42).

Cui and Xie (2005) assumed the following: failed system found by inspection is completely repaired or replaced; the times required for a perfect repair or replacement is either a constant $v \geq 0$ or a random variable Y which has distribution $G(y)$ and density function $g(y)$, the repaired or replaced system is restored to operation immediately, i.e., it does not take any time, and treating the time at the completion of the perfect repair or replacement as a new starting point then the periodic inspection will be resumed. In their Model A, it is also assumed that unfailed system determined by inspection is upgraded as good as new, whereas in Model B unfailed system continues operation without any intervention.

Under Model A with constant repair/replacement time v, the instantaneous system availability $A(t)$ is determined recursively as follows

$$A(t) = \begin{cases} \bar{F}(t), & \text{if } 0 \leq t \leq \tau \\ \bar{F}(t - \tau), & \text{if } \tau < t < \tau + v \\ \bar{F}(t)A(t - \tau) + F(\tau)A(t - \tau - v), & \text{if } t \geq \tau + v \end{cases} \quad (44)$$

From this recursive equation, it was shown there that the limit of $A(t)$ as $t \to \infty$ does not exist if $v = 0$ or τ/v is a rational number when $v > 0$, i.e., the steady-state availability does not exist. If the repair/replacement time is random, $A(t)$ can be determined by two different recursive equations:

$$A(t) = (\bar{F}(\tau))^{\lceil t/\tau \rceil - 1} \bar{F}\left(t - \left(\left\lceil \frac{t}{\tau} \right\rceil - 1\right)\tau\right)$$
$$+ \sum_{i=1}^{\lfloor t/\tau \rfloor} (\bar{F}(\tau))^{i-1} F(\tau) \int_0^{t - i\tau} A(t - y - i\tau) g(y) dy \quad (45)$$

and

$$A(t) = \bar{F}(\tau)A(t - \tau) + F(\tau) \int_0^{t-\tau} A(t - y - \tau) g(y) dy \quad (46)$$

Moreover, in this case the steady-state availability and consequently the limiting average availability exist and are given as

$$A(\infty) = A_{av}(\infty) = \frac{\tau - \int_0^\tau F(x) dx}{\tau + F(\tau) \int_0^\infty \bar{G}(y) dy}. \quad (47)$$

For Model B with constant repair/replacement time v, the instantaneous availability $A(t)$ is given in a recursive way

$$A(t) = \bar{F}(t) + \sum_{i=1}^{\lfloor (t-v)/\tau \rfloor} A(t - i\tau - v)[F(i\tau) - F((i-1)\tau)] \qquad (48)$$

On the other hand when the repair/replacement time Y has density function $g(y)$ then $A(t)$ satisfies equation

$$A(t) = \bar{F}(t) + \sum_{i=1}^{\lfloor t/\tau \rfloor}[F(i\tau) - F((i-1)\tau)] \int_0^{t-i\tau} A(t - y - i\tau)g(y)dy \qquad (49)$$

The steady-state availability and consequently the limiting average availability exist and are given by

$$A(\infty) = A_{av}(\infty) = \frac{\int_0^\infty \bar{F}(x)dx}{\tau \sum_{i=1}^\infty i[F(i\tau) - F((i-1)\tau)] + \int_0^\infty \bar{G}(y)dy} \qquad (50)$$

It is interesting that if we let $\eta = \infty$ in Mi (2002), $Y = 0$ in Cui and Xie (2005), and notice that

$$\sum_{i=1}^\infty i[F(i\tau) - F((i-1)\tau)] = \sum_{i=1}^\infty i[\bar{F}((i-1)\tau) - \bar{F}(i\tau)] = \sum_{k=0}^\infty \bar{F}(k\tau) \qquad (51)$$

then the results in Mi (2002) and Cui and Xie (2005) are the same as that one given in (50).

Example 3.3 Assume that system lifetime has Weibull distribution $F(t) = 1 - \exp(-(x/\lambda)^k)$ with mean $\lambda\Gamma(1 + 1/k)$ and $G(t) = 1 - \exp(-\beta t)$. The following figures show the limiting average availabilities given by Eqs. (47) and (50) with different parameters τ, λ and β when $k = 2$.

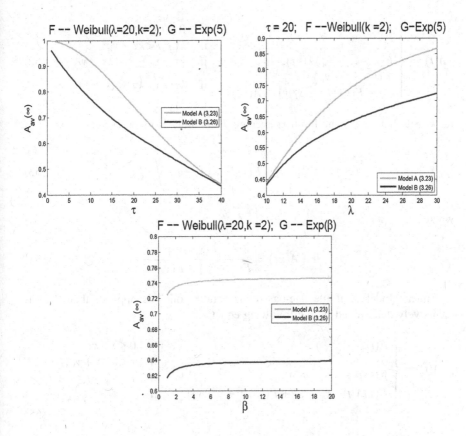

Different from most of previous research work on system availability, Tang et al. (2013) considered both calendar-based and age-based inspection policy. In their study, not only the downtime due to repair or replacement but also the downtime due to inspection are taken into consideration. In the past, only few works considered both nonnegligible times, for instance, Barroeta (2005), Jardine and Tsang (2006), Jiang and Jardine (2006), and Pak et al. (2006).

In the following, we use v_w to denote the constant downtime due to inspection for unfailed system and constant v_f to denote the total time when a failed system is detected including the downtime due to repair/replacement for failed system.

Model A studied in Tang et al. (2013) assumes that unfailed system found by inspection is upgraded or modified to be as good as new. But Model B assumes that unfailed system is put back to operation, i.e., the unfailed system remains as good as it is.

For Model A, the instantaneous system availability $A(t)$ with a calendar-based inspection polity is given recursively by

$$A(t) = \begin{cases} \bar{F}(t), & \text{if } 0 \le t \le \tau; \\ 0, & \text{if } k\tau < t < k\tau + v_w; \\ \bar{F}(t - k\tau - v_w)A(k\tau), & \text{if } k\tau + v_w \le t < k\tau + v_f; \\ \bar{F}(t - k\tau - v_w)A(k\tau) \\ \quad + \bar{F}(t - k\tau - v_f)(1 - A(k\tau)), & \text{if } k\tau + v_f \le t < (k+1)\tau. \end{cases} \quad (52)$$

for $k = 1, 2, \ldots$. And the limiting average availability is given as

$$A_{av}(\infty) = \tau^{-1}[\phi \int_0^{\tau - v_w} \bar{F}(x)dx + (1 - \phi) \int_0^{\tau - v_f} \bar{F}(x)dx], \quad (53)$$

where

$$\phi = \lim_{k \to \infty} A(k\tau) = \frac{\bar{F}(\tau - v_f)}{F(\tau - v_w) + \bar{F}(\tau - v_f)}. \quad (54)$$

Under Model A if the age-based inspection policy is applied, then $A(t)$ is recursively determined by the following equation:

$$A(t) = \begin{cases} \bar{F}(t), & \text{if } 0 \le t \le \tau; \\ 0, & \text{if } \tau < t < \tau + v_w; \\ \bar{F}(\tau)\bar{F}(t - \tau - v_w), & \text{if } \tau + v_w \le t < \tau + v_f; \\ \bar{F}(\tau)A(t - \tau - v_w) + F(t)A(t - \tau - v_f), & \text{if } t \ge \tau + v_f. \end{cases}$$

$$(55)$$

and the limiting average availability is

$$A_{av}(\infty) = \frac{\int_0^\tau \bar{F}(x)dx}{\tau + v_w \bar{F}(\tau) + v_f F(\tau)} \quad (56)$$

For Model B, the instantaneous system availability $A(t)$ with a calendar-based inspection policy is determined recursively as follows:

$$A(t) = \begin{cases} \bar{F}(t), & \text{if } 0 \le t < \tau; \\ 0, & \text{if } k\tau < t < k\tau + v_w; \\ \bar{F}(t - kv_w) + \sum_{i=1}^{k-1} B(t - i\tau - v_f)p_i, & \text{if } k\tau + v_w \le t < k\tau + v_f; \\ \bar{F}(t - kv_w) + \sum_{i=1}^{k} B(t - i\tau - v_f)p_i, & \text{if } k\tau + v_f \le t \le (k+1)\tau. \end{cases} \quad (57)$$

The equations for determining $B(s)$ appearing in the above expression can be found in Tang et al. (2013) and is omitted here because it is tedious to display them. The limiting average availability in this case is obtained as

$$A_{av}(\infty) = \frac{\int_0^\infty \bar{F}(x)dx}{\tau \sum_{k=0}^\infty \bar{F}(s_k)}, \tag{58}$$

where $s_0 = 0$, and $s_k = k(\tau - v_w) + v_w - v_f, \forall k \geq 1$.

In the age-based inspection policy is applied under Model B, then $A(t)$ is determined recursively by

$$A(t) = \begin{cases} \bar{F}(t - n(t)v_w) + \sum_{k=1}^{m(t)} A(t - t_i - v_f)p_i, & \text{if } 0 \leq t \leq n(t)(\tau + v_w) + \tau; \\ \sum_{k=1}^{m(t)} A(t - t_i - v_f)p_i, & \text{if } t > n(t)(\tau + v_w) + \tau \end{cases} \tag{59}$$

where

$$m(t) = \left\lfloor \frac{t - v_f + v_w}{\tau + v_w} \right\rfloor, \quad n(t) = \left\lfloor \frac{t}{\tau + v_w} \right\rfloor \tag{60}$$

and

$$t_i = (i - 1)(\tau + v_w) + \tau, \quad p_i = \bar{F}((i - 1)\tau) - \bar{F}(i\tau). \tag{61}$$

Under the same assumptions the limiting average availability is obtained as

$$A_{av}(\infty) = \frac{\int_0^\infty \bar{F}(x)dx}{(\tau + v_w) \sum_{k=0}^\infty \bar{F}(k\tau) + v_f - v_w} \tag{62}$$

Note that when $v_w = 0$ it holds that

$$A_{av}(\infty) = \frac{\int_0^\infty \bar{F}(x)dx}{\tau \sum_{k=0}^\infty \bar{F}(k\tau) + v_f}$$

and it turns out to be the same as the expression of $A_{av}(\infty)$ in Sarkar and Sarkar (2000).

Example 3.4 Suppose that system lifetime has exponential distribution $F(t) = 1 - exp(-\alpha t)$, the following figures show the limiting average availabilities determined by Eqs. (53), (58) and (56), (62), respectively, when $\alpha = 0.05, v_w = 1$ and $v_f = 2$. The left panel corresponds to the calendar-base inspection policy, and the right panel corresponds to the age-base inspection policy.

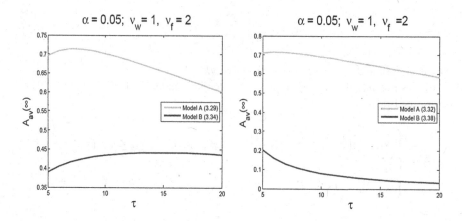

4 Other Works on System Availability

The previous two sections address availability of systems without specifying their configurations. In the field of reliability, the k-out-of-n system has particular importance since it is widely used in practice. Recent research works on the availability of k-out-of-n system include, for example, Fawzi (1991), Frostig (2002), De Smidt-Destombes et al. (2004, 2006, 2007, 2009), Li et al. (2006), Yam et al. (2003), and Zhang et al. (2000, 2006) among others.

There are also lots of studies on availability of various other systems appearing in industry. For instance, Berrade (2012), Chung (1994), Dhillon (1993), Dhillon and Yang (1992), Klutke et al. (1996, 2002), Lau et al. (2004), Mishra (2013), Pascual (2011), Pham-Gia and Turkkan (1999), Vaurio (1997, 1999), Wang et al. (2006), Wang and Chen (2009), and works referred therein. In these works warm standby, imperfect switch, reboot delay, common cause failures, and random deterioration were considered. Specifically, Mi (2006a, b) introduced the concept of pseudo availability which differs from the traditional availabilities in that once the system is in 'up' state, it will remain there forever without change.

It is worthy of mentioning that in addition to $A(t)$, the interval availability and the steady-state interval availability are probably as important as the instantaneous system availability or even more important for certain situations. For any $w \geq 0$ the interval availability is defined as $A_w(t) = P(\xi(s) = 1, \quad t \leq s \leq t+w)$. It is the probability that the system is functioning during the interval $[t, t+w]$. Of course, if $w = 0$ then $A_w(t)$ becomes the instantaneous system availability $A(t)$. Mi (1999) and Huang and Mi (2013) discussed interval availability. We think it would be worthwhile studying those models mentioned above to derive more results about interval availability.

References

Barlow, R. E., & Proschan, F. (1975). *Statistical theory of reliability and life testing*. New York: Holt, Rinehart & Winston.

Barroeta, C. E. (2005). Risk and economic estimation of inspection policy for periodically tested repairable components, MS thesis.

Berrade, M. D. (2012). A two-phase inspection policy with imperfect testing. *Applied Mathematical Modelling, 36*, 108–114.

Biswas, J., & Sarkar, J. (2000). Availability of a system maintained through several imperfect repairs before a replacement or a perfect repair. *Statistics & Probability Letters, 50*, 105–114.

Biswas, A., Sarkar, J., & Sarkar, S. (2003). Availability of a periodically inspected system, maintained under an imperfect-repair policy. *IEEE Transactions on Reliability, 52*(3).

Block, H. W., Borges, W. S., & Savits, T. H. (1985). Age-dependent minimal repair. *Journal of Applied Probability, 22*, 370–385.

Brown, M., & Proschan, F. (1983). Imperfect repair. *Journal of Applied Probability, 20*, 851–859.

Cha, J. H., & Kim, J. J. (2001). On availability of Bayesian imperfect repair model. *Statistics & Probability Letters, 53*, 181–187.

Cha, J. H., & Kim, J. J. (2002). On the existence of the steady state availability of imperfect repair model. *Sankhya Series B, 64*, 76–81.

Cha, J. H., Lee, S., & Mi, J. (2004). Bounding the optimal burn-in time for a system with two types of failure. *Naval Research Logistics: An International Journal, 51*, 1090–1101.

Chelbi, A., & Ait-Kadi, D. (2000). Generalized inspection strategy for randomly failing systems subjected to random shocks. *International Journal of Production Economics, 64*, 379–384.

Chung, W. K. (1994). Reliability and availability analysis of warm standby systems with repair and multiple critical errors. *Microelectronics Reliability, 34*(1), 153–155.

Cui, L. R. & Xie, M. (2001). Availability analysis of periodically inspected systems with random walk model *38*, 860–871.

Cui, L. R., & Xie, M. (2005). Availability of a periodically inspected system with random repair or replacement times. *Journal of Statistical Planning and Inference, 131*, 89–100.

De Smidt-Destombes, K. S., van der Heijdenb, M. C., & van Harten, A. (2004). On the availability of a k-out-of-N system given limited spares and repair capacity under a condition based maintenance strategy. *Reliability Engineering and System Safety, 83*, 287–300.

De Smidt-Destombes, K. S., van der Heijden, M. C., & van Harten, A. (2006). On the interaction between maintenance, spare part inventories and repair capacity for a k-out-of-N system with wear-out. *European Journal of Operational Research, 174*, 182–200.

De Smidt-Destombes, K. S., van der Heijdenb, M. C., & van Harten, A. (2007). Availability of k-out-of-N systems under block replacement sharing limited spares and repair capacity. *International Journal of Production Economics, 107*, 404–421 (2007)

De Smidt-Destombes, K. S., vander Heijden, M. C., & van Harten, A. (2009). Joint optimization of spare part inventory, maintenance frequency and repair capacity for k-out-of-N systems. *International Journal of Production Economics, 118*, 260–268.

Dhillon, B. S., & Yang, N. (1992). Reliability and availability analysis of warm standby systems with common-cause failures and human errors. *Microelectron Reliability, 32*(4), 561–575.

Dhillon, B. S. (1993). Reliability and availability analysis of a system with warm standby and common cause failures. *Microelectron Reliability, 33*(9), 1343–1349.

Dieulle, L. (1999). Reliability of a system with Poisson inspection times. *Journal of Applied Probability, 36*, 1140–1154.

Fawzi, B. B., & Hawkes, A. G. (1991). Availability of an R-out-of-N system with spares and repairs. *Journal of Applied Probability, 28*, 397–408.

Frostig, E., & Levikson, B. (2002). On the availability of R out of N repairable systems. *Naval Research Logistics, 49*(5), 483–498.

Gut, A., & Jansons, S. (1983). The limiting behaviour of certain stopped sums and some applications. *Scandinavian Journal of Statistics, 10*, 281–292.

Høyland, A., & Rausand, M. (1994). *System reliability theory*. New York: Wiley.

Huang, K., & Mi, J. (2013). Properties and computation of interval availability of system. *Statistics and Probability Letter, 83*, 1388–1396.

Ito, K., & Nakagawa, T. (2000). Optimal inspection policies for a storage system with degradation at periodic tests. *Mathematical and Computer Modelling, 31*, 191–195.

Jardine, A. K. S., & Tsang, A. H. C. (2006). *Maintenance, replacement, and reliability: Theory and applications*. Boca Raton: CRC Press.

Jiang, R. Y., & Jardine, A. K. S. (2006). Optimal failure-finding interval through maximizing availability. *International Journal of Plant Engineering and Management, 11*, 174–178.

Klutke, G. A., Wortman, M. A., & Ayhan, H. (1996). The availability of inspected systems subject to random deterioration. *Probability in the Engineering and Informational Sciences, 10*, 109–118.

Klutke, G. A., & Yang, Y. J. (2002). The availability of inspected systems subject to shocks and graceful degradation. *IEEE Transactions on Reliability, 51*(3), 371–374.

Lau, H. C., Song, H. W., See, C. T., & Cheng, S. Y. (2004). Evaluation of time-varying availability in multi-echelon spare parts systems with passivation. *European Journal of Operational Research, 170*(1), 91–105.

Li, X., Zuo, M. J., & Yam, R. C. M. (2006). Reliability analysis of a repairable k-out-of-n system with some components being suspended when the system is down. *Reliability Engineering and System Safety, 91*, 305–310.

Lim, J. H., Lu, K. L., & Park, D. H. (1998). Bayesian imperfect repair model. *Communication in Statistics Theory, 27*(4), 965–984.

Mi, J. (1994). Burn-in and maintenance policies. *Advanced Applied Probability, 26*, 207–221.

Mi, J. (1995). Limiting behavior of some measures of system availability. *Journal of Applied Probability, 32*, 482–493.

Mi, J. (1999). On measure of system interval availability. *Probability in the Engineering and Informational Sciences, 13*, 359–375.

Mi, J. (2002). On bounds to some optimal policies in reliability. *Journal of Applied Probability, 39*, 491–502.

Mi, J. (2006a). Limiting availability of system with non-identical lifetime distributions and non-identical repair time distributions. *Statistics & Probability Letter, 76*, 729–736.

Mi, J. (2006b). Pseudo availability of repairable system. *Methodology and Computing in Applied Probability, 8*, 93–103.

Mishra, A., & Jain, M. (2013). Maintainability policy for deteriorating system with inspection and common cause failure. *International Journal of Engineering Transactions C: Aspects, 26*(6), 631–640.

Pak, A., Pascual, R., & Jarding, A. K. S. (2006). Maintenance and replacement policies for protective devices with imperfect repairs, Technical report.

Pascual, R., Louit, D., & Jardine, A. K. S. (2011). Optimal inspection intervals for safety systems with partial inspections. *Journal of the Operational Research Society, 62*, 2051–2062.

Pham-Gia, T., & Turkkan, N. (1999) System availability in a gamma alternating renewal process. *Naval Research Logistics, 46*

Rényi, A. (1957). On the asymptotic distribution of the sum of a random number of independent random variables. *Acta Mathematica Academiae Scientiarum Hungaricae, 8*, 193–199.

Rise, J. (1979). Compliance test plans for reliability. In *Proceedings of the 1979 annual reliability and maintenance symposium*.

Sarkar, J., & Biswas, A. (2010). Availability of a one-unit system supported by several spares and repair facilities. *Journal of the Korean Statistical Society, 39*, 165–176.

Sarkar, J., & Chaudhuri, G. (1999). Availability of a system with gamma life and exponential repair time under a perfect repair policy. *Statistics & Probability Letters, 43*, 189–196.

Sarkar, J., & Li, F. (2006). Limiting average availability of a system supported by several spares and several repair facilities. *Statistics & Probability Letters, 76*, 1965–1974.

Sarkar, J., & Sarkar, S. (2000). Availability of a periodically inspected system under perfect repair. *Journal of Statistical Planning and Inference, 91*, 77–90.

Sarkar, J., & Sarkar, S. (2001). Availability of a periodically inspected system supported by a spare unit, under perfect repair or perfect upgrade. *Statistics & Probability Letters, 53*, 207–217.

Takács, L. (1957). On certain sojourn time problems in the theory of stochastic processes. *Acta Mathematica Academiae Scientiarum Hungaricae, 8*, 169–191.

Tang, T. Q., Lin, D. M., Banjevic, D., & Andrew, K. S. J. (2013). Availability of a system subject to hidden failure inspected at constant intervals with non-negligible downtime due to inspection and downtime due to repair/replacement. *Journal of Statistical Planning and Inference, 143*, 176–185.

Vaurio, J. K. (1997). On time-dependent availability and maintenance optimization of standby units under various maintenance policies. *Reliability Engineering and System Safety, 56*, 79–89.

Vaurio, J. K. (1999). Availability and cost functions for periodically inspected preventively maintained units. *Reliability Engineering and Systems Safety, 63*, 133–140.

Wang, K. H., Dong, W. L., & Jyh-Bin Ke, J. B. (2006). Comparison of reliability and the availability between four systems with warm standby components and standby switching failures. *Applied Mathematics and Computation, 183*, 1310–1322.

Wang, K. H., & Chen, Y. J. (2009). Comparative analysis of availability between three systems with general repair times, reboot delay and switching failures. *Applied Mathematics and Computation, 215*, 384–394.

Wortman, M. A., & Klutke, G. A. (1994). On maintained systems operating in a random environment. *Journal of Applied Probability, 31*, 589–594.

Wortman, M. A., Klutke, G. A., & Ayhan, H. (1994). A maintenance strategy for systems subjected to deterioration governed by random shocks. *IEEE Transaction on Reliability, 43*, 439–445.

Yadavalli, V. S. S., Botha, M., & Bekker, A. (2002). Asymptotic confidence limits for the steady state availability of a two-unit parallel system with preparation time for the repair facility. *Asia-Pacific Journal of Operational Research, 19*, 249–256.

Yam, R. C. M., Zuo, M. J., & Zhang, Y. L. (2003). A method for evaluation of reliability indices for repairable circular consecutive-k-out-of-n: F systems. *Reliability Engineering and System Safety, 79*(1), 1–9.

Yang, Y. J., & Klutke, G. A. (2000). Improved inspection schemes for deteriorating equipment. *Probability Engineering Information Sciences, 14*(4), 445–460.

Yang, Y. J., & Klutke, G. A. (2001). A distribution-free lower bound for availability of quantile-based inspection schemes. *IEEE Transactions on Reliability, 50*(4), 419–421.

Yeh, L. (1995). An optimal inspection-repair-replacement policy for standby systems. *Journal of Applied Probability, 32*, 212–223.

Zhang, T. L., & Horigome, M. (2000). Availability of 3-out-of-4: G warm standby system. *IEEE Transactions on Fundamentals of Electronics Communications and Computer, E83-A*(5), 857–862.

Zhang, T. L., Xie, M., & Horigome, M. (2006). Availability and reliability of k-out-of-(M+N): G warm standby systems. *Reliability Engineering and System Safety, 91*, 381–387.

Reliability and Maintenance of the Surveillance Systems Considering Two Dependent Processes

Yao Zhang and Hoang Pham

1 Introduction

The application of surveillance systems is a great enhancement of security level to the monitored area by providing important reference for the security teams to make prompt actions against threats or incidents. The widespread implementation of the surveillance cameras significantly deters criminal behavior and reduces vandalism to agency property. With the rapid progress in automated control, image processing, and high-performance computing, the surveillance systems become more and more capable of providing comprehensive information on the monitored area (Singh et al. 2008).

Because of the critical role of the surveillance systems, the design, and modeling receives wide attention by multiple areas of researchers, including people from computer science, electrical engineering, operational research, statistical, and more (Valera and Velastin 2005; Kim et al. 2010). Many questions arise from the effort to build functional and cost-effective intelligent surveillance systems. Among a variety of available sensors, how to select the best combination and where to place the sensors to provide optimal coverage of any arbitrary-shaped area, with the consideration of overlap and occlusion? How do we automate the surveillance system to detect events/incidents/intrusions with the information collected from the sensors? All these problems lead to sophisticated and innovative modeling, some of which will be discussed in this chapter.

Since the wide implementation of the surveillance systems and the fact that either failure or deterioration in performance of the system may result in severe damage to the protected facility, the reliability estimation of the system, and inspection or maintenance scheduling is worth receiving serious attention (Zio 2009). Two incidents are discussed briefly here just to emphasize the importance of

Y. Zhang · H. Pham (✉)
Rutgers University, New Brunswick, USA
e-mail: hoang84pham@gmail.com

© Springer-Verlag London 2016
H. Pham (ed.), *Quality and Reliability Management and Its Applications*,
Springer Series in Reliability Engineering, DOI 10.1007/978-1-4471-6778-5_10

modeling and scheduling for the surveillance system. On January 3rd, 2010, the Newark Liberty International Airport had a security breach that one man reached the secure sterile area through a checkpoint exit without being screened by airport security (Katersky 2010). Due to the breakdown of the surveillance recording system, the airport authority failed to identify the inadvertent intruder until they got the footage from the redundant cameras 2 h later. The incident caused hours of delay in flights and thousands of passengers to be rescreened before boarding. The second example is the incident occurred on August 13, 2012. A man ran out of gas of his jet ski at Jamaica Bay in New York. He climbed the 8-foot-high perimeter fence and walked across the two runways seeking for help, without being detected by the perimeter intrusion detection system, which should be given out series of warnings under the circumstance (Ibarra 2012). Those lessons raise the requirement of more comprehensive models for assessing the reliability of such critical systems.

The characteristics of the surveillance system can help shape some basic requirements of the reliability and maintenance models. A complex surveillance system often includes multiple sensors that coordinate with each other to enhance the performance. Each sensor may have different failure modes and/or degradation levels. Thus, it is reasonable to consider multiunit multistate systems in the surveillance modeling. Accordingly, the maintenance of such type of systems should take consideration of the dependency between units. Moreover, the safety of the protected area does not only rely on the functioning of the surveillance system, but also highly related to the attack/incident arrival process. Thus multiprocess modeling is also a key factor to include in a comprehensive surveillance system modeling. Hence, the reliability and maintenance model for surveillance system should involve at least the following factors: multiunit, multistate, multiprocess, and maintenance schedules with consideration of dependency between units.

This chapter discusses the existing works related to surveillance systems modeling, including sensor deployment, intelligent surveillance system design involving data mining and computer automation techniques, and the attack-defense model that quantifies the interaction behavior between the defender and adversary. It also points out the needs of reliability modeling and maintenance scheduling on surveillance systems. It then further discusses some recent works in the field of reliability modeling that can be applied for complex surveillance systems. The chapter concludes with several examples on surveillance system reliability modeling with the consideration of two stochastic processes.

2 Surveillance System Design and Modeling

The modeling of the surveillance systems receives wide attention by multiple areas of researchers over the years. Many efforts are dedicated in searching ways to build functional, cost-effective, automated surveillance systems with consideration of interactions between the systems and their adversaries (either making effort to avoid detection or sabotage the system units). Three distinguish categories are discussed in this section. They are sensor placement and coverage models, intelligent video surveillance systems, and attack-defense models.

2.1 Sensor Placement and Coverage Models

One of the designs of surveillance system problems that receive the most attention from the research communities is the sensor deployment problem. Given the geometric layout of the facility that needs surveillance coverage, a designer of the system needs to determine the types, number of sensors required, and the locations of the sensors to meet the safety specification. This subsection reviews selected works discussing about ways to quantify the performance of the surveillance system and models for the sensor deployment problem. Generally, these models are computationally complex to solve. Many approximation models and heuristic search algorithms are developed to solve the optimal deployment problem more efficiently.

Bai et al. (2010) design a surveillance system detecting intruders of an empty area using two types of sensors to enhance the performance. The first type of sensor applied is the ultrasonic sensor detects a moving object when the signal of the ultrasonic from the transmitter to the receiver is cut off. The second type of sensor is the Pyroelectric Infrared sensor that is used to detect the environment temperature change. A majority voting algorithm is used to interpret the conflict signals between multiple sensors.

Zhao et al. (2009) propose a general visibility model as a flexible sensor planning framework. The designed model takes the self and mutual occlusion of the objects in the surveillance area into consideration. It can be used to search optimal sensor placement in an arbitrary-shaped 3D structure. The optimizer in the model tries to maximize the system performance and minimize the cost of the system at the same time using a greedy search via binary integer programming.

The above proposed binary integer programming model has a high computational complexity. Zhao extends the research in his dissertation (Zhao 2011) by comparing multiple approximation algorithms such as simulated annealing and semi-definite program to simplify the sensor planning model. The author further investigates the geometric fusion of the object information observed from different sensors in the surveillance network in order to improve the human body segmentation accuracy in the surveillance scene and generate better views of the object with the collected information in real time.

Dhillon and Chakrabarty (2003) propose a probabilistic optimization framework for sensor placement under the constraint of sufficient coverage. The optimizer minimizes the number of sensors deployed while maintaining desired coverage level of the monitored area. To reduce the number of sensors in the network also indicates the reduction of the transmission of data and power consumptions, along with low initial investment of the system. The model considers the nature of the terrain, such as the obstacles blocking the sight of the cameras and the preferential coverage of different locations. It also considers the imprecise detection of each sensor and different sensor capabilities.

Wang et al. (2003) present a wireless sensor network configuration model that is flexible to provide different degrees of coverage options based on system requirements. The applicable system of the model should have high node density so that many sensors can be scheduled with sleep intervals and are not required to work continuously to conserve energy. For wireless sensor network, connectivity means that all the sensors are able to communicate with every other sensor in the network. In other words, the graph of the working nodes should not be broken into isolated pieces. The model needs to maintain the connectivity of the network in addition to satisfying the coverage requirements when deciding which sensors are selected to provide continuous service.

Zou and Chakrabarty (2003) develop a virtual force metric to describe the geometric relationship between multiple sensors. The metric is applied to aim the deployment adjustment to enhance coverage after initial random deployment of multiple sensors that is practical in military applications (throwing the sensors in the field). The virtual force metric defines an attractive force if the distance between the two sensors is longer than twice the radius of the sensor coverage. A repulsive force is defined by contraries. The total force on each sensor provides the adjustment direction and distance. Hence, the final deployment is more uniform and provides better coverage compared to the initial random scattering.

Krishnamachari and Iyengar (2004) develop two distributed Bayesian algorithms to distinguish false alarms from real event detection for a wireless sensor network. Intuitively, if a real event occurs in a region, the sensor detections are likely to have agreements with neighbors. On the contrary, a false alarm appears more randomly. The designed approaches use randomized decision scheme and threshold decision scheme of the Bayesian algorithms to determine the correlation between event detection of the sensors.

Gupta et al. (2006) present three algorithms that select a subset of sensors to execute a given query in a large-scale sensor network. The centralized approximation algorithm provides the most near-optimal solution of the minimal set of censors providing the desired coverage level. The two distributed algorithms saves communication data transfer between sensors so that they extend the service life of the battery driven sensors. As a trade-off, the performance of the solutions by the distributed algorithms is degraded compared to the centralized one.

Ram et al. (2006) develop a metric calculating average probability of target discovery to evaluate the performance of a surveillance system consisted of video cameras and motion sensors. When s desired performance level is given, the model is able to find the optimal solution with information of the locations of the cameras, minimal number of motion sensor required, and the minimal field of view of the camera. The field of vision represents an important characteristic of the camera such that the wider field of vision requirement often indicates a more sophisticated type of camera thus significantly increase the total cost to set up the entire system.

Yao et al. (2010) propose a sensor positioning algorithm for persistent surveillance considering the handoff safety margin between adjacent cameras. The handoff of the target between cameras can be achieved smoothly only if the two cameras have sufficient overlap of effective coverage range, as shown in Fig. 1. In addition, the excessiveness of the overlap is considered as a waste of resource. The authors define an observation metric as the combination of camera resolution and the distance to edge of the field of view. A max coverage and min-cost problem is formulated to balance the overall coverage, appropriate level of overlap margins, and total cost of the system by optimizing the deployment of the sensors. Experimental results on a real-world implementation is carried out and compared with the work proposed by Erdem and Sclaroff (2006). With a minimal sacrifice of the overall coverage, the handoff successful rate of the target increases dramatically for the proposed model.

Herrera et al. (2011) develop a coverage strength model that takes many camera intrinsic parameters into consideration when evaluating the coverage performance of the sensors. The intrinsic parameters considered in this work include camera visibility, pixel resolution, depth of field, and angle of view. An example that only involves one camera with one laser line projector is studied to demonstrate the use of the proposed model.

Liang et al. (2011) propose a localization-oriented coverage (L-coverage) model based on Bayesian estimation to measure the overall performance of random deployment sensor networks. The random deployment is modeled as a two-dimension stationary Poisson point process. At any discretized point in the monitored field, it is defined to be L-covered if at least k cameras are able to estimate the target location at that point within an acceptable estimation error range. Then, the total L-coverage probability is calculated by the ratio of L-covered points over all points in the field. The relationship between the L-coverage probability and the random deployment intensity parameter λ is further investigated so that one can find the minimum Poisson intensity of the random deployment for a desired level of L-coverage probability.

Nam and Hong (2012) present an agent space trajectory model to simulate the trajectories of people traveled in an arbitrary-shaped monitoring field. Within the simulation one can estimate the different weight of the importance for each spot in the field and develop the camera placement algorithm in order to cover the most

Fig. 1 Graphical demonstration of the overlap concept introduced in (Yi et al. 2010)

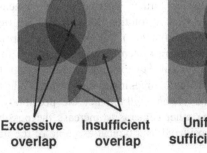

Excessive overlap Insufficient overlap Uniform and sufficient overlap

significant spots. Thus, the optimizer can be understood as a min-cost max-weighted-coverage probabilistic model. In the experiments, the authors demonstrate the selection between three different types of cameras and their optimal deployment plan (optimal number of cameras and layouts) under different budget constraints.

Rashmi and Latha (2013) develop a surveillance network using IP cameras that can transmit signals via network. The operator can directly control the camera network via his smart phone devices remotely to realize facial recognition, object identification, and other tasks. The development enhances the mobility and ease of access to the surveillance system control. However, the security issues such as hackers to the remote control system are also raised by the development.

Junbin et al. (2014) proposed a trans-dimensional simulated annealing algorithm to efficiently search near-optimal solutions to camera placement problem subject to different system design requirements. Four different constraints have been discussed for the camera placement model. The first constraint is the common 100 % floor coverage. The second one is the 100 % floor coverage with important targets covered by multiple cameras (critical coverage redundancy). The third constraint is to guarantee 100 % facial recognition success rate. The last one is to replanning the existing camera network so that the total number of cameras is given. The placement plan to satisfy the floor coverage is compared with some existed algorithms to show its effectiveness.

Wang (2011) discusses the classification of the sensor network coverage problem based on different types of coverage model assumptions. The first type is the point coverage problems, in which the area under surveillance are discretized into individual points or there only exists a finite number of targets to be monitored in the field. The second type is the area coverage problem, in which the whole surveillance area is treated equally and the percentage of the coverage is studied to estimate the effectiveness of the deployment. The last type is noted as the barrier coverage problem, where the goal of designing certain type of surveillance system is to form a protection barrier so that the intruder cannot find any uncovered path between possible entrances to the targeted locations. The author reviews many existing works with the focus on the computational complexity of the models and the different optimizing techniques that applied to solve the coverage problems.

Mavrinac and Chen (2013) separate the coverage models by distinguishing their coverage geometry, coverage overlap, and transition topology. A geometric coverage model may be further deferred by the dimension of the monitored field, camera's field of view, resolution, focus and angle, treatment of the occlusion (not considered, static or dynamic). A coverage overlap model describes the physical topology of the camera system, while a transition model covers more functional topology of the system. For example, a nonoverlap deployed surveillance system can perform a prediction tracking of the intruder. When the target leaves the view of one camera, the system will predict the possible movement of the target and coordinating the other cameras to increase the possibility to recapture the target. This can be considered as a typical example of the transition model.

2.2 Intelligent Video Surveillance Systems

The performance of the human operators is a limitation of the surveillance systems, as it is hard to conduct consistent and focused monitoring to the screens. Many incidents and events that captured by the optimally deployed surveillance cameras may be missed due to the lack of awareness of the officers. With the rapid progress of vision processing and data mining techniques, the new generation of intelligent surveillance systems can automate many detection tasks to improve the performance of the system.

Marcenaro et al. (2001) propose a decomposition model of surveillance functionalities, including video tracking, object classification, and behavior understanding, etc. The decomposed logical tasks are then optimally allocated to the physical distributed nodes subject to available bandwidth, processing power, and the dynamic loads of the logical task. Through demonstration examples, the authors show the convenience to concentrate the processing power to the central office when the system is composed of a small number of cameras. If the bandwidth cost is high, allocating the intelligence tasks to distributed processing unit is preferred to the centralized processing.

Marchesotti et al. (2002) present a semi-automatic alarm generation technique applied to a parking lot surveillance to draw the operator's attention by sending blink icons on screen and generating sound signals when detecting events, such as car parking in non-parking zones, pedestrian detection in car limited zones, and erratic trajectories inside the lot. The application also compares the accuracy of the auto-alarm under different environment conditions, such as bad weather or night illumination scenarios.

Trivedi et al. (2005) develop the distributed interactive video array (DIVA) system to track and identify vehicles and people, monitoring facilities, and interpreting activities. The authors demonstrate the vehicle tracking and identification in bridge and roadway surveillance via overlap of camera view coordination and car feature extraction (color, size, speed, etc.) techniques. In a long-term room watch example, the system successfully records and identifies nine different people entering the room multiple times. In some cases, two people are presented under the surveillance the same time.

Saini et al. (2009) propose a queuing model consisted of four levels of components: sensor, colocated processing elements (CoPE), aggregate processing element (APE) and the network. Each of the components for every level has a finite processing rate hence, the incidents are possible to be missed due to the resource limitation or unexpected delay of the system. The missing probability and response time are studied under different combinations of the surveillance system parameters.

Doblander et al. (2005) propose a multi-objective optimization algorithm to balance service availability, quality of service, and energy consumption of the intelligent video surveillance system. The reported system runs several video analysis algorithms on a network processor to enhance the surveillance performance. These algorithms consume a lot of calculating power if all running at full quality at the same time. The proposed algorithm determines which algorithm could be run at lower quality mode and how the algorithms are distributed on different processors in the network to minimize the energy consumption while maximizing the quality of service and system availability.

Malik et al. (2013) present a surveillance system design and implementation programming language Systemj. The program simulates the surveillance system in a highly abstract manner to evaluate the system performance based on the sensor distribution, sensor type selection, communication between sensor, controller, operating unit, and storage units. Then Systemj can also generate executable codes for computers serving as different roles in the system (camera controller, Systemj control box, operating unit, and storage servers, etc.) as shown in Fig. 2 in the deployment stage.

Riveiro et al. (2008) develop a maritime anomaly detection surveillance system that can interact with operators. The anomaly detection module applies Gaussian mixture model and self-organizing map to realize the detection of the anomaly ·behavior of the vessel movement. The system notices the operator for each detection and waits for feedback. If the detection is indeed an anomaly, it is then added to the training model. If the observation is considered false, the operator can adjust the weight vector in the model to prevent the false alarm in the future. Thus,

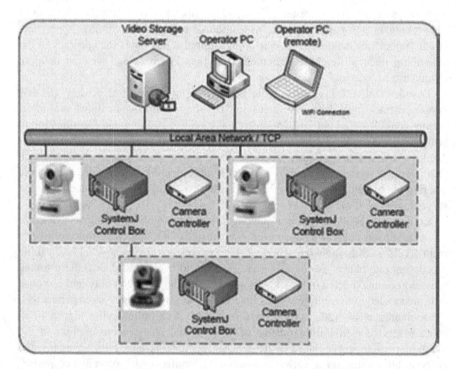

Fig. 2 The smart distributed surveillance system physical implementation from Malik et al. (2013)

the system updates itself with the human experience until the performance reaches satisfaction.

Stauffer and Grimson (1999) present a background tracking model by treating each pixel of the camera image as a mixture of Gaussians. The mixture of Gaussians for the whole image is then analyzed to identify the range of the background. The model is robust to deal with shadows, specularities, and swaying branches, and can be applied to different types of cameras, lightening conditions, and different objects being studied.

Szczodrak et al. (2010) borrow the idea of the three metrics introduced in Mariano et al. (2002) to evaluate the performance of video object tracking algorithms applied to 4 pieces of video recordings. The three metrics are known as fragmentation, average object area recall, and average detected box area precision, where these metrics are designed to evaluate the precision and processing speed of each object tracking algorithm.

Atrey et al. (2011) propose a human-centric approach to provide adaptive schedule of the best views of cameras for better observation of events. Based on the findings in Wallace and Diffley (1988), the human operator can typically monitor four screens effectively at a time. Thus, the model applies adaptive Gaussian type event detection and an operator eye tracking feedback method to select the four best views of camera screens when an incident occurs in the monitored area.

Anwar et al. (2012) present an anomaly event detection algorithm by monitoring the sequential pattern of a frequently occurred series of events. This method is effective in detecting unknown anomalous events that are not likely to follow a pattern of a frequent series of events. Various experiments are carried out to test the computational complexities of the proposed algorithm subject to variations of model parameters such as the number of input event and duration of the sequence of events.

Clapés et al. (2013) propose a facial identification and object recognition module for an intelligent surveillance system. The model estimates the environment and applies background subtraction to extract objects in the camera image. Then the extracted object is compared with a skeletal model in order to identify if a person is detected. Then facial recognition is conducted on the human tracking result. The extraction is updated online to realize robustness against partial occlusions and camera 3D rotation.

2.3 Attack-Defense Models of Surveillance Systems

Many existing models related to surveillance reliability are the attack-defense models where game theory is often applied for consideration of both intelligent attackers and defenders. Hausken and Levitin (2012) develop a table which categorizes the literature according to system structure, defense measures, attack tactics and circumstances. System structure is further divided into single element, series systems, parallel systems, series-parallel systems, networks, multiple elements, interdependent systems, and other types of systems. Defense measures are divided into separation of system elements, redundancy, protection, multilevel defense, false targets deployment and preventive strike. Attack tactics and circumstances are divided into attack against single element, attack against multiple elements, consecutive attacks, random attack, combination of intentional and unintentional impacts, incomplete information, and variable resources. The classification is intended to give an overview of the field and implicitly suggest future research trajectories, false targets, separation, redundancy, and number of attacks. Guikema (2009) points out three classical critiques of applying game theory on the intelligent actors in reliability analysis: assumption of instrumental rationality, common knowledge of rationality, and knowledge of the game rules. One should treat these assumptions carefully when modeling with game theory since the nature of uncertainty of the attackers. Thorough discussion of robustness of the parameters and even violation of assumptions would be useful to enhance the effectiveness and generalization of the modeling.

Most of the attack-defense models, if not all, consider protection on the potential targets. Protection is defined as the type of actions carried out by the defenders to reduce the probability of target destruction by attackers. Golany et al. (2007) compare optimal resource allocation plans under two types of risks, random, and strategic attack. Under the probabilistic risk assumption, the optimal allocation plan tends to fully protect the states that have the largest population but remains some low population states unprotected. The optimal plan under strategic attack balances the risk and achieves an average expected loss for every state despite of the population size. Paté-Cornell and Guikema (2002) present a probabilistic model for determining priorities among different types of threats attempted by different terrorist groups. This model considers multiple scenarios, the objectives of both the attackers and the defenders and the dependency between them. The model can help achieving a rational balance between enhancing the defense on previously occurred types of attacks and overinvestment on prevention of repeated attacks. It can also help the decision makers to avoid inaccurate intuition on priorities of threats. Dighe et al. (2009) state that the secrecy in the allocation plan can be beneficial to the defender. They study a two-node system by enumerating all possible attack-defense strategy combinations to support the statement. Both centralized and decentralized defending structures are considered. Siqueira and Sandler (2010) analyze the allocation of agents by general terrorist organizations based on different types of governments and local terrorist supporters. The findings further provide information to the government on how to effectively alter the positive attitude to terrorists of the local supporters. Against the intuition, the higher cost of investment for government actions may not reduce terrorism. Bandyopadhyay and Sandler (2011) compare the preemptive and defensive measures. The preemptive actions weaken terrorist assets or ability to attack, while the defensive actions reduce the damage after an attack occurred. The proposed model analyzes the interaction between the preemptive and defensive actions by the nation, considering the dependency between the two types of actions (the effective preemptive action is likely to reduce the need of the defensive action) and the interaction between the decision makers for different nations. The allocation of resources are also related to the terrorist preference of attack and both domestic and oversea assets. Hausken and Levitin (2011) present a model that both actors can invest in offensive and defensive resources. Each actor can either maximize its own survivability or minimize the other's. The result includes the optimal solutions on how to allocate the resources on offensive and defensive actions for all four combinations of the possible actors' objectives. Nikoofal and Zhuang (2012) derive a model including extreme bounds of the estimation of attacker's target valuation in the decision of optimal defensive resource allocation. Azaiez and Bier (2007) propose an optimal resource allocation algorithm for a general series-parallel system configuration. When under a constraint budget, the defender tends to protect the most attractive component, but can also consider some less attractive components if the cost of enhancing the security level on these components is minor. The algorithm is under the assumption of perfect knowledge for the attackers that they fully aware the improvement the defender can make to the system prior to the attack. Thus, the result of expected attack cost is a lower bound since it is more realistic that the attacker can only achieve imperfect information on the defensive plan. Hausken et al. (2009) develop a defense model including both terrorism and natural disaster threat. The defender has the options to invest in protection against both threats simultaneously, or in protection against either threat separately. In the modeling, three different scenarios of two-step games are considered: both attacker and defender move simultaneously, the attacker moves first and the defender moves first. It shows that the player that has the lower cost per unit tends to make the first move. The selection of the defensive policy depends on the relationship of the costs for different plans (either joint protection or separate protection). Hausken (2011) proposes a two-period game model with consideration of a multistate two-unit system. In each period, both actors make one investment decision on each non-failed components. Detailed discussions have been made for various parameter selections. Levitin and Hausken (2012) consider a situation that the attacker can make repeat attempts of attack to ensure destruction of the target. The attacker has an imperfect observation of the attack outcomes with a probability to falsely identify a destroyed target as undestroyed and vise versa. The error rates of the wrong outcome observation by the attacker have a great impact on the strategy of the attacker. For example, it is suggested for the attacker to alter the favor from multiple attacks to single attack when the false probability to identify an undestroyed target as destroyed rises above a threshold.

Besides the protection on individual units, some action can benefit to multiple or all targets at the same time, as the consideration of dependency between units. Golalikhani and Zhuang (2011) develop an attack-defense model with consideration of joint protection based on similarities of the threats or the protected targets. For example, the chemical and biological threats can be both monitored through the public-health surveillance program, but not the explosion threats. The investment on the arbitrary layer protection is compared with individual target protection and traditional boarder hardening which is a technique to enhance the security of all targets together (on the contrary of the arbitrary cluster of protected targets).

Haphuriwat and Bier (2011) compare the protection effect of target hardening and overarching protection where hardening represents the enhancement of the security level of individual targets and overarching stands for the protection over all targets available. The result shows that when the total number of the significant valuable targets is small, hardening of individual dominates the protective action, vise versa.

As alternatives to protection, other defensive actions are available for the defenders to form a more complex defense strategy to compete against the attackers. Creating false targets also receives some attention in the literature. Levitin and Hausken (2009) study the effectiveness of deploying false target comparing with investment in protection of the genuine target. The assumptions include that the attacker is not able to distinguish the real target from the false distraction. The model also requires that both the attacker and the defender are rational players for the game. The optimal solution of that how many false targets the defender decides to create and how many targets the attacker decides to attack is obtained for the cases that the Nash equilibrium of the described game existed. The conclusion shows that the optimal solution relies on the resources both players have, the cost of each false target and the intensity of the contest. Levitin and Hausken (2009) also study the effect of the random and strategic attacks on the defense planning mentioned in Golany et al. (2007). Moreover, the different attacks can be mixed in the modeling. Both redundancy and protection (resources to put on the unit to reduce the risk of successful attacks) are considered in the defensive strategy. In Levitin and Hausken (2009), they expand the comparison of defensive strategies to three types: redundancy, protection and false elements. Redundancy requires genuine units placed in the system more than needed, while false units cannot provide the function of genuine units. It is only replaced to confuse the attackers for selecting the correct targets thus normally is cheaper than distribution of a genuine unit. With a limited resource, the optimal allocation is dependent on the total resources (defensive and attacking), attacking intensity and the relative cost for each type of defensive strategies. Levitin and Hausken (2009) further extend the noncooperative game to a more generalized model, in which the defender considers all possible actions of protection, redundancy and false targets to enhance the survival rate of the protected system. To be specific, the attacker and defender compete in an intelligence contest prior to the attack-defense game. If the attacker wins the intelligence contest, he can identify all the false targets and take down all genuine targets unprotected with minimal effort. Then, he can further attack the protected real targets with the resources left from the intelligence contest. If the defender wins the intelligence contest, the attacker has to attack all targets (both real and false ones) randomly with the left resource. Peng et al. (2010) propose an attack-defense model considering imperfect false targets that have some probability to be identified by the attacker. In the modeling, it is first considered of the identical imperfect false targets with the same probability of being detected. Then, the model is further extended to one that the probability of being detected is a decision variable that the defender can choose, along with the number of false targets he wants to create. The numerical examples show that the flexibility of choosing different types of false targets with different probability of being detected for the defender is beneficial especially when the contest intensity in uncertain.

Many of the above examples on attack-defense model are based on general system models, while others are directly related to real life applications (networks, power system, transportation, etc.). Lin et al. (2007) derive a mixed nonlinear integer programming model to study the allocation of resources to protect the network. The objective of the defender is to either maximize the total cost of the attacker or minimize the probability of the core node under attack. Wang et al. (2013) discuss a dynamic voting system of networks, in which all the available units can either be selected as redundancy or used to create false targets. This is special compared to the false targets discussed earlier since all the false targets created in this work are capable to function as a working unit. They are simply not selected in the voting cluster. In this way, the defender can focus the resource on protecting the small set of voting clusters, meanwhile distracting the attackers with hard-to-detect false targets. This is proven in the paper effective especially when the defender is limited with sparse resource. Singh and Kankanhalli (2009) address the concern of the adversary in the surveillance scheme. They propose a zero-sum game for an ATM lobby defense scenario and a nonzero-sum game for a traffic control surveillance scenario. A generic treatment for enhancing the surveillance performance against rational adversary is provided with discussion on how to change the factors such as spatial, temporal and external, etc. Bier et al. (2007) propose an algorithm of interdiction strategies for a transmission system. The power systems are highly interconnected systems. One of the advantages is that the spare can be shared over the grid to enhance the capacity against shock loads such as a failure of a single generating site. However, if the shock is strong enough, the chain reaction of one site failing down after the other will cause wide-area black outs. Terrorist attacks can be one of the reasons to cause certain shocks. In the paper, an algorithm is developed to identify the critical transmission lines for interdiction. The algorithm is two-staged that it first chooses candidate lines then has them strengthened. The process is repeated until the desired resources run out. Bier and Haphuriwat (2011) discuss a model of determining the number of containers for inspection at the US ports to protect against terrorist attacks. The optimal portion that requires inspection should minimize the loss of the defender while the attackers are trying to maximize their rewards. It is found out that it is easier to deter an attack risk when the terrorists invest high attack costs. Thus lowering down the portion to be inspected will increase the chance of small threats (assault rifles) but not likely to impact the risk of huge threat much (nuclear). The model also considers the effect of retaliation on deterring terrorist attacks. Kanturska et al. (2009) study the safety of transportation networks against random incidents and terrorist attacks by using an attacker-defender model proposed by Bell (2000). The model can help identify the critical routs in the network and shows the advantages of applying mixed route strategies. Visible, invisible, and announced but not specified protections are compared as possible defensive measures.

3   Traditional Reliability Modeling

With all the discussion of the coverage optimization model and the new technologies to automate the event/threat/intrusion detection, there are little attention on
the reliability and maintenance of the surveillance system. How much a failure of
component will affect the overall performance of the surveillance system? Is
redundancy required to enhance the system reliability? How soon the first failure is
expected and how often the maintenance action should be carried out? Without the
proper answers to these questions, the designed system may perform well at the
beginning but soon deteriorates to fail the safety requirements of the protected area.
Although there are only few papers directly address the reliability and maintenance
issues of surveillance systems, many traditional reliability models can be borrowed
according to the characteristics of the surveillance systems. Three categories of
reliability modeling will be discussed in this section.

3.1   Model for Multiunit Multistate Systems

Adding redundancy is one of the most effective techniques to improve the system
reliability level through the use of replicated units (Kobbacy et al. 2008). Among all
possible structures of the system configuration, k-out-of-n system receives a lot of
attention in the reliability modeling. The parameter k represents the minimal set of
components to maintain a functional state of the system. Once there are $n - k + 1$
components down in total, the system is considered as failure. The reliability of
such a system configuration is easy to estimate as in Pham (2010):

$$R(t) = \sum_{i=k}^{n} \binom{n}{i} R_0(t)^i (1 - R_0(t))^{n-i} \qquad (1)$$

where $R_0(t)$ is the reliability function of the individual component.

    In the literature, models are developed with more complexity than the k-out-of-
n system with identical units. Mathur (1971) presents an N-modular redundancy
(NMR) system operated in simplex mode with spares as a majority voting system.
The system uses $N = (2n + 1)$ modules to form a majority voting system such that if
at least $(n + 1)$ units make the correct decision, the system outcome will be correct.
It equivalents to the system with $(n + 1)$-out-of-$(2n + 1)$ configuration plus S spare
units. The simplex mode is worked as that the failure in $(2n + 1)$ modules is simply
replaced when spares are available. If no more spare units left, the further failure in
the $(2n + 1)$ modules is discarded, along with a good unit so that the system reduces
to $(2n - 1)$ units. This process repeats until a single working unit is left in the
system. The reliability function of such a system with triple modules and S spares is
developed in the paper. Mathur and de Sousa (1975) modify this model with
multistate units. The units are used to identify binary input thus they are functional

if they can read 1 when the input is 1 and 0 otherwise. Traditionally, the failure is only considered as stuck-at-x, which means the failed module randomly gives value despite the input. Two more failure modes, stuck-at-0 and stuck-at-1, are considered in the modeling. For the voting system, when a pair of stuck-at-0 and stuck-at-1 failures occurred, they compensate each other so that the system outcome is not affected. The reliability model for NMR system considering this type of compensation is developed for comparison with the NMR simplex model to determine if discarding of good units along with failures are beneficial to the system reliability. Mathur and de Sousa (1975) further generalize the model to k-out-of-n configuration (k can be any value between 1 and n) with S spares, in which each unit still has three failure modes (stuck-at-a, $a = 0, 1, x$). They show in the work that with careful selection of parameters, many existed models can be summarized as special cases of the general modular redundant system.

Pham (1997) applies the similar idea of the multi-unit voting system with 2 failure mode (stuck-at-0, stuck-at-1) for each unit. Instead of majority voting, the model uses a variable threshold k to determine the output of the system. When less than k units transmit signal 1, the system has output 0. When at least k units transmit signal 1, the system decides to transmit 1. The majority voting is a special case of this model with odd number of n and $k = (n + 1)/2$. Selection of k to maximize the system reliability is also discussed in the paper. Nordmann and Pham (1997) implement the model in decision making of human organizations where the probability of stuck-at-a ($a = 0, 1$) differs from individual decision makers. The outcome of each voter is further weighted to represent more realistic modeling. A recursion algorithm to simplify the evaluation of the reliability of the weighted voting system is reported in Nordmann and Pham (1999) by constraining the weight parameters for only integers and the threshold k as a rational number. Pham (1999) also explores the effect of varying the total number of units in the system on the reliability function of the system with three failure modes.

The above models on multiunit multistate systems have the following limitations. First, the probability of each failure mode for individual component was considered time invariant. Second, although multiple failure modes for each component are taken into account, on the system level, there is either functional or failure state existed. Take a deeper look at the voting systems that is majorly focused in the above works. Based on the voters' observation, the system can either misread an input "1" as "0," or the opposite. It is acceptable to consider both types of failures as the same when making decisions of whether accepting or rejecting a project, or either transmitting a "1" or "0" bit in the computer. But for a safety-related system, the outcome of misreporting a threat that does not exist is essentially different from one of misdetection of a real threat. A safety-related system is the type of system that the failure of which will result in significant increment of risk to human lives and/or the environment (IEC 61508 2010). Knight (2002) discusses the definition, types, challenges of development and made prognosis on the technology and applications of the future safety-critical systems. Bukowski (2001) develops a Markov based reliability model for a 1-out-of-2 safety-shutdown controller. On the component level, each unit has five possible states: working; fail-safe recognized; fail-safe unrecognized; fail-danger recognized; fail-danger unrecognized. Fail-safe mode for each component means that the component falsely shuts down a process that is operating properly. Thus, for a 2-unit system there are in total of 25 combined elementary states, which can be further combined to system level states for reliability and mean time to failure (MTTF) estimations. Zhang et al. (2008) repeats the work with similar assumptions on the component (5 states) and system structure (1-out-of-2). The development of the Markov model for MTTF calculation is revised from the previous study. Both of the examples were only considered system with 2 units. As pointed out in Guo and Yang (2008), the complexity of the model grows exponentially with the increment of the number of units considered for the system structure. They develop a framework of automatic generation of Markov model for k-out-of-n structured safety-instrumented system. The model also includes the common cause failure into consideration, which can cause two or more failure occurred at the same time. As a demonstration of the proposed framework, a numerical estimation of the reliability for a 2-out-of-3 architecture is given at the end of the work. Bukowski and van Beurden (2009) further take proof test completeness and correctness into the loop. If the undetectable failures are failed to be identified by the inspection point (the proof test is not 100 % complete and/or cannot correct all the errors), they will remain as undetectable failures and degrade the performance of the safety-instrumented system. The new assumption adds more possible states on the system level. Torres-Echeverría et al. (2011) add a testing reconfiguration that if a component is under test, the system is downgraded to the state that the tested item is treated as known failure, hence further extends the complexity of the system level outcomes. Levitin et al. (2006) apply the same 5-level component assumption on a series-parallel structure of a fuel supply system. A recursive method is derived to obtain the system state distribution.

Another significant portion of the research on modeling of multistate components is the competing risk model where the component can have failures due to either degradation or fatal shocks. The competing risk model also involves the modeling with multiple processes, thus will be discussed in the next section.

3.2 Maintenance Model for Multiunit System

Let us take a brief look at the maintenance schedules for single-unit system first. As summarized in Wang (2002), thousands of papers and models are published on the maintenance schedule for single-unit systems. In our opinion, the categories of the maintenance schedule can be determined when the maintenance is carried out and how complete the maintenance has been performed. To be specific, if the maintenance is only conducted upon failure then it is a corrective maintenance; if the maintenance is scheduled before failure occurred to the unit from time to time, it is preventive maintenance. Judging by the completeness, if the maintenance restores the unit to "as good as new," it is a perfect maintenance; if it only recovers partially of a unit to a state somewhere between "as good as new" and "as bad as old," it is imperfect maintenance; if the maintenance intentionally to restore the system to a stage with the same failure rate before it fails, it is called minimal repair. Imperfect preventive maintenance, possible with combinations of the minimal repair for early period of operation, receives the most attention in the literature, as that the modeling assumptions are more realistic and the models are more complicated in the forms. Renewal theory is a common choice for application of optimal maintenance policy, as there is often a time point that the system deteriorates significantly and has to be repaired fully back to state "as good as new", which is clearly a renewal point. The optimal policy is obtained by either maximizing the availability of the system or minimizing the cost per unit time to operate the system, which is evaluated by

$$Cost\,per\,unit\,time = \frac{E[\text{total cost in one renewal cycle}]}{E[\text{renewal cycle length}]} \qquad (2)$$

For multiunit systems, if there is no dependency between components, the development of the maintenance schedule is similar to the ones for single-unit systems. However, it is typical that there exists dependency between components, such as economic dependence, failure dependence and structural dependence (Wang and Pham 2011). The maintenance strategy then defers from the ones for single-unit systems, as that the failure of one component provides the opportunity to maintenance other components in the system as well. Wang et al. (2006) highlight two main categories of the maintenance schedule for multi-unit systems in their survey. The first type is group maintenance that the maintenance actions are always carried out for multiple units at the same time, either upon each failure and preventively maintaining a group of other working components, or holding some failures and performing corrective maintenance until accumulation of some predetermined number of failures. The other type is opportunistic maintenance that the group maintenance is only carried out when some criteria has been met in the system. To our understanding, the only difference between these two types of policies for multiunit systems is that whether it is possible that the maintenance action on single units is performed. For group maintenance, maintenance actions are never carried out along for a single unit. One can argue that the opportunistic maintenance is a generalization of the group maintenance. Examples of these works can be found in Sheu and Jhang (1997), Barros et al. (2002), Tsai et al. (2004), Li and Pham (2005), Vaughan (2005), de Smidt-Destombes et al. (2006), Wang and Pham (2006), Nicolai and Dekker (2008), Sung and Schrage (2009), Nowakowski and Werbińka (2009), Liu and Huang (2010), Park and Pham (2012), Almgren et al. (2012), Cheng et al. (2012), Golmakani and Moakedi (2012), Hou and Jiang (2012), Koochaki et al. (2012), Liu (2012), Patriksson et al. (2012), Sarkara et al. (2012), Vu et al. (2012), Zhou et al. (2012), Martorell et al. (1999).

4   Multiprocess Modeling

For a single process modeling, the reliability function $R(t)$ is defined as the probability that the process is in working status by the time t. As the modeling becoming more complicated, events in one process often represent the triggers of consequences in other processes. Thus, the reliability estimation has to consider the dependencies between multiple processes. Three categories of the existed works are summarized in this section as examples of multiprocess reliability modeling.

4.1   Modeling of the Standby Systems

Standby redundancy has been widely applied in industry such as computer fault-tolerant systems (Mathur and de Sousa 1975), power plants (Singh and Mitra 1997) and space exploration systems (Sinaki 1994). From the relationship between the failure rate of the standby units and the active units, the standby systems can be categorized into three groups. The first type is that the standby units have the same failure rate as the active units, referred to as the hot-standby systems, or the active-standby systems. If the switching mechanism is perfect, this type of system can be easily modeled as the k-out-of-n system. In the second type of the standby systems, the standby units have 0 failure rate until being switched into use. This type of the systems is referred to as the cold-standby systems. The third category lies in between the first two types that the failure rate of the standby units is less than the active ones, but not 0. When modeling the reliability of the standby systems, the moment of the process that modeling the active units working status altering from success to failure triggers the working status of the standby process, if there are still spare units available at the time. The key of modeling the reliability of the system is the analysis of the sequence of events in the time line. Coit (2001) proposes the reliability estimation of a cold-standby system with one primary unit and $(n - 1)$ cold-standby units.

$$R(t) = r(t) + \int_0^t \Pr[T_2 > t - u] f(u) du + \int_0^t \Pr[T_3 > t - u] f_{S_2}(u) du$$

$$+ \cdots + \int_0^t \Pr[T_n > t - u] f_{S_{n-1}}(u) du \tag{3}$$

where components in the system are identical with reliability function $r(t)$ and failure density $f(t)$. $f_{Si}(t)$ is the density function of the failure time of the sum of the total i components.

Many other researches on standby systems can be found considering different system scenarios. She and Pecht (1992) derive a reliability model to study a k-out-of-n warm-standby system. In the modeling, k units are in active status and the other n-k units are spares thus have a lower failure rate than the active ones. Once a working component fails, one spare item is being activated and starts to fail at the active failure rate. This procedure repeats until no more spare unit is available. The system fails at the time point that the $(n - k + 1)$th failure occurred. Levitin and Amari (2010) estimated the reliability with the similar system assumptions, but using a universal generating function approach. In the numerical example, it is demonstrated that the reliability distribution for each component in this modeling has not to be identical. The example also shows that the sequence of activating the spares has an impact on the system reliability, given that the failure rates of the spare units vary from one to another. Yun and Cha (2010) develop a two-unit hybrid model that the standby component first serves as cold standby then shift to warm standby after some time of the successful operation of the active unit. If the active unit failure before the standby unit switches to warm mode, the system fails. Otherwise the warm-standby unit can be activated immediately to replace the failed one. Given the failure densities of both components, and the switching mechanism (either perfect switch or imperfect), there exists an optimal switching time, which is carefully studied with several numerical cases. Amari (2012) presents a k-out-of-n cold-standby system with components following Erlang distribution. Amari et al. (2012) explore the effects of changing the number of spares on the reliability improvement factor. Adding more redundant units to the system will always improve the reliability of the system, but the improvement is not linear. They find out that with the number of spares increases from 0, the reliability improvement first increases and then decreases. Moreover, the reliability improvement factor follows the probability mass function of the negative binomial distribution. These findings can be considered as factors to determine the optimal number of spares for k-out-of-n warm standby systems.

4.2 Competing Risk Models

The reliability of the components in the complex system is usually estimated using life testing techniques (Elsayed 1996). For some cases, it is not necessary to complete the test until failure of the tested component. Instead, there are some measurements that can provide enough information on how fast the unit wears out. Those types of measurements can be used to model the degradation path of each component. Compared to the reliability function estimation, the degradation analysis is more related to the physical representation of the failure mechanism. Both the reliability and degradation analysis are used to describe the normal wear process of a component. In reality, the components not only endure normal usage, but also suffer random shocks from the environment. Some of the shocks are strong enough to be fatal to the component. The competing risk model takes both degradation analysis and shock model under consideration. The processes for degradation and random shocks are competing with each other. The earlier arrival of failure in either process will cause failure of the component. Thus, the competing risk models are also categorized as multiprocess modeling. If one achieves the cumulative distributions of both the degradation $F_d(t)$ and the shock $F_s(t)$ separately (without the consideration of the influence of the other), then the competing risk of one hazard arrives earlier than the other can be calculated as

$$F_{T_s}(t) = \int_0^t (1 - F_d(u))dF_s(u) \tag{4}$$

$$F_{T_d}(t) = \int_0^t (1 - F_s(u))dF_d(u) \tag{5}$$

where $F_{T_s}(t)$ represents the cumulative distribution that the shock causes the system to failure earlier than the degradation, vise versa for $F_{T_d}(t)$.

Wang and Pham (2011) give a comprehensive structural review on dependent competing risk models with degradation and random shocks. The shock models can be categorized into cumulative shock model (Qian et al. 2003), extreme shock model (Chen and Li 2008) and δ-shock model (Rangan and Tansu 2008). The degradation model includes general path model (Yuan and Pandey 2009), stochastic model (Kharoufeh and Cox 2005), parametric (Bae and Kvam 2004), and nonparametric statistical model (Zuo et al. 1999). The combination of the two risk models can be either independent with simpler representation (Chiang and Yuan 2001), or dependent that is more realistic and complex (Fan et al. 2000). In this type of modeling, both the degradation and shock model are not limited to 1 process only, where examples can be found in Wang and Coit (2004). Some summaries on additional literature of the competing risk models are provided as a supplement to the review. Li and Pham (2005) present a multiple competing risk model considering two degradation processes and one cumulative random shock process. There is no interaction between these processes in the modeling. On the system level, the states are combination of states for individual process. The probability function of each system outcome is developed but no maintenance model based on the probability analysis is performed. Wang and Zhang (2005) consider a model that two types of failures can be generated by the shock process. One denotes for the type caused by short interval between arrivals of consecutive shocks and the other stands for the type caused by the magnitude of the shock strength. Since the two types of failures are due to the same shock process, there exists some dependency between the failure modes. Cui and Li (2006) apply the shock model on a multi-unit system that each shock has the same damage cumulated for different components of the system so that their failure rates become dependent. They further derive an opportunistic maintenance schedule to lower down the maintenance cost based on the dependency between components. Liu et al. (2008) bring both degradation and shock process into the modeling of a series-parallel system. Peng et al. (2010) consider the dependency by assuming that the shocks contribute as a step increment in the degradation process, if they are not strong enough to fail the component. Jiang et al.

(2011) push the dependency in the model one step further by considering not only the raising of degradation by the shocks, but also the dependency of thresholds by the shocks. To be more specific, each shock may lower the threshold for other processes and drive the system faster to failures than the case without the shock, besides the accumulation in the degradation process. Wang and Pham (2011a, b, c, 2012) develop several competing risk models considering dependency between processes. In Wang and Pham (2011), they develop an imperfect maintenance scheduling for a system with only one degradation process and one shock process. They modified this model in Wang and Pham (2011) by considering hidden failures that the system status is only available by each inspection point. The optimal scheduling is achieved by multi-objective optimization instead of optimization on the single cost per unit function. In Wang and Pham (2012), they derive a model with multiple dependent degradation processes and multiple shock processes using Copulas, a statistical method to estimate the joint distribution based on marginal distributions.

4.3 Modeling of the Surveillance System Considering the Demand Process

There are only a few works directly linked multiple processes modeling with surveillance systems. Pham and Xie (2002) propose a two-process model to determine the unfavorable ratings of airplane repair stations. The agents from Federal Aviation Administration have had inspection records for different repair stations. Due to the resource limitation, they have to wait for a certain period of time between paying each visit to a selected station for inspection. Hence, it is important to choose the station with the worst favorable rating based on the historical service record in order to maximize the overall performance of all the repair stations. The two processes under consideration in the model are the frequency of inspection to each station and the occurrence of unfavorable rating of each station. Both processes are modeled using NHPP with parametric time dependent models, which can also be found in Lewis and Shedler (1976), Muralidharan (2008). The inspection process determines the time point that the individual repair station restores to favorable status, while the performance of the station affects the frequency of the inspection. Thus, the two processes are dependent and have impacts on each other. In the modeling, the arrival rate of surveillance (the first process) at station k is represented as

$$\lambda_I^{(k)}(t) = \lambda_0 e^{G^{(k)}(t,\gamma)} \tag{6}$$

where λ_0 is a baseline visit rate and $G^{(k)}(t, \gamma)$ contains all the factors (such as time from last inspection and the number of unfavorable inspections over a period of time) that influence the surveillance rate to station k, weighted by scalar vector γ. The intensity of the occurrence of the unfavorable rating at a station k (the second process) is modeled as

$$\lambda_Z^{(k)}(t) = \alpha r_k^{i-1} t^{\beta-1} e^{bx^{(k)}} \tag{7}$$

where α, β are global parameters (same for every station) and r_k is a scalar for individual station k. $x^{(k)}$ includes all the information of each individual station to affect the performance, such as number of different types of employees. b is the corresponding coefficient vector of $x^{(k)}$.

Recently, some research focus on the safety-related modeling considering the two processes modeling with demand. Xu et al. (2012) purpose a model for a multi-unit multi-state safety-related system. The system is used to monitor the status of the production line and to shut down the production system to reduce damage when dangerous situation occurs. Thus, if the production system is safe but only the safety-related system fails, the damage is much smaller than the situation that the safety-related system fails to respond to a failure in the production line. A universal generating function approach (Ushakov 1986) is applied to obtain the probability outcomes of the system. Although multiple components are considered for the safety-related system, maintenance actions can only be applied on the whole system (either replace the whole system or do nothing). The component status is not available to the maintenance team also. For the surveillance system, it is possible to conduct different types of maintenance on individual subsystems. Each subsystem constantly transmits video for the central officer thus the interruption of working status has the chance to be discovered during operation. These assumptions are not compatible with the modeling in Xu et al. (2012) thus a lot of modifications are needed for a more realistic surveillance modeling.

5   Recent Studies

As this review has been pointed out, the modeling of the surveillance systems has drawn considerable attention from several areas of researchers. Nevertheless, the reliability and maintenance modeling of the surveillance systems has been somehow neglected. Recently, some research efforts have been made to address the problem with the consideration of multiple processes and different failure modes.

Zhang and Pham first introduce the two-process reliability modeling of the surveillance system in Zhang and Pham (2013) The model includes both the incident arrival process and the system failure process. It also takes into consideration of the randomness of the operational environments. One can adopt different system configurations, consider different environmental effects based on the collected data and evaluate intruder's effort to avoid being detected using different mechanisms. The quantitative evaluation of the reliability and soft-failure and hard-failure probabilities with the variation of the inspection interval length is derived and illustrated with numerical examples and several sensitivity analyses.

Zhang and Pham (2013) develop a k-out-of-n surveillance system in which the subsystems are subject to two competing failure modes—detectable and undetectable failure. The two stochastic-process reliability of the surveillance system is derived with the consideration of the intrusion process and a (m, T) opportunistic maintenance policy. The probability of two-level system outcomes has been developed in the modeling. Several numerical examples are given to demonstrate the validity of the modeling and the sensitivity of several important parameters. They further extend the model into a cost based structure to determine the near-optimal maintenance schedule and study the sensitivity of several key parameters in the model (Zhang and Pham 2015a).

For individual sensor, it can not only miss the detection of real incidents in the field, but also falsely report incident detection while there is absolutely nothing happened in the surveillance area. Once a single sensor constantly reports incident detection regardless the true state of the protected region, it is defined as the "false alarm" state for that individual sensor. The k-out-of-n surveillance system has the ability to overcome several sensor failures, either the fail-dangerous type of missing the real failures or the false alarm type. With the assumption that the sensors can fail both ways, and each failure can be either detectable or undetectable, the complexity of the system level outcome structure is elevated. In Zhang and Pham (2015b), Zhang and Pham discuss the modeling with the assumption of the complete failure modes while considering the incident arrival process and a detailed numerical analysis on the TMR (2-out-of-3) model as a special case.

6 Conclusion

In this chapter, three topics of design and implementing the surveillance system including the optimal sensor placement, intelligent video surveillance system design, and surveillance attack/defense model have been reviewed. The lack of the reliability and maintenance models for surveillance systems is pointed out. Based on the general characteristics of the surveillance systems, we suggest that the reliability and maintenance modeling should at least include the following aspects: multiunit, multistate, multiprocess, and maintenance schedule considering the dependency between units. Selected works of the reliability modeling that involves all the above keywords have been discussed. Several recent works that address the reliability modeling of the surveillance systems have been marked.

References

Alarcon Herrera, J. L., Mavrinac, A., & Xiang, C. (2011). Sensor planning for range cameras via a coverage strength model. In *2011 IEEE/ASME International Conference on Advanced Intelligent Mechatronics (AIM)* (pp. 838–843).

Almgren,T., Andréasson, N., Palmgren, M., Patriksson, M., Strömberg, A.-B., Wojciechowski, A., & Önnheim, M. (2012). Optimization models for improving periodic maintenance schedules by utilizing opportunities. In *Proceedings of 4th Production and Operations Management World Conference*.

Amari, S. V. (2012). Reliability analysis of k-out-of-n cold standby systems with Erlang distributions. *International Journal of Performability Engineering, 8,* 417.

Amari, S. V., Pham, H., & Misra, R. B. (2012). Reliability characteristics of k-out-of-n warm standby systems. *IEEE Transactions on Reliability, 61,* 1007–1018.

Anwar, F., Petrounias, I., Morris, T., & Kodogiannis, V. (2012). Mining anomalous events against frequent sequences in surveillance videos from commercial environments. *Expert Systems with Applications, 39,* 4511–4531.

Atrey, P. K., El Saddik, A., & Kankanhalli, M. S. (2011). Effective multimedia surveillance using a human-centric approach. *Multimedia Tools and Applications, 51,* 697–721.

Azaiez, M. N., & Bier, V. M. (2007). Optimal resource allocation for security in reliability systems. *European Journal of Operational Research, 181,* 773–786.

Bae, S. J., & Kvam, P. H. (2004). A nonlinear random-coefficients model for degradation testing. *Technometrics, 46,* 460–469.

Bandyopadhyay, S., & Sandler, T. (2011). The interplay between preemptive and defensive counterterrorism measures: A two-stage game. *Economica, 78,* 546–564.

Barros, A., Grall, A., & Berenguer, C. (2002). Maintenance policies for a two-units system: A comparative study. *International Journal of Reliability, Quality and Safety Engineering, 9,* 127–149.

Bell, M. G. H. (2000). A game theory approach to measuring the performance reliability of transport networks. *Transportation Research Part B: Methodological, 34,* 533–545.

Bier, V. M., Gratz, E. R., Haphuriwat, N. J., Magua, W., & Wierzbicki, K. R. (2007). Methodology for identifying near-optimal interdiction strategies for a power transmission system. *Reliability Engineering & System Safety, 92,* 1155–1161.

Bier, V., & Haphuriwat, N. (2011). Analytical method to identify the number of containers to inspect at U.S. ports to deter terrorist attacks. *Annals of Operations Research, 187,* 137–158. (2011/07/01).

Bukowski, J. V. (2001). Modeling and analyzing the effects of periodic inspection on the performance of safety-critical systems. *IEEE Transactions on Reliability, 50,* 321–329.

Bukowski, J. V., & van Beurden, I. (2009). Impact of proof test effectiveness on safety instrumented system performance. In *Annual Reliability and Maintainability Symposium* (pp. 157–163).

Chen, J., & Li, Z. (2008). An extended extreme shock maintenance model for a deteriorating system. *Reliability Engineering & System Safety, 93,* 1123–1129.

Cheng, Z., Yang, Z., & Guo, B. (2012). Opportunistic maintenance optimization of a two-unit system with different unit failure patterns. In *International Conference on Quality, Reliability, Risk, Maintenance, and Safety Engineering* (pp. 409–413).

Chiang, J. H., & Yuan, J. (2001). Optimal maintenance policy for a Markovian system under periodic inspection. *Reliability Engineering & System Safety, 71,* 165–172.

Clapés, A., Reyes, M., & Escalera, S. (2013). Multi-modal user identification and object recognition surveillance system. *Pattern Recognition Letters, 34,* 799–808.

Coit, D. W. (2001). Cold-standby redundancy optimization for nonrepairable systems. *IIE Transactions, 33,* 471–478.

Cui, L., & Li, H. (2006). Opportunistic maintenance for multi-component shock models. *Mathematical Methods of Operations Research, 63,* 493–511.

de Smidt-Destombes, K. S., van der Heijden, M. C., & van Harten, A. (2006). On the interaction between maintenance, spare part inventories and repair capacity for a k-out-of-N system with wear-out. *European Journal of Operational Research, 174,* 182–200.

Dhillon, S. S., & Chakrabarty, K. (2003). Sensor placement for effective coverage and surveillance in distributed sensor networks. In *Wireless Communications and Networking, 2003. WCNC 2003. 2003 IEEE* (vol. 3, pp. 1609–1614).

Dighe, N. S., Zhuang, J., & Bier, V. M. (2009). Secrecy in defensive allocations as a strategy for achieving more cost-effective attacker detterrence.

Doblander, A., Maier, A., Rinner, B., & Schwabach, H. (2005). Improving fault-tolerance in intelligent video surveillance by monitoring, diagnosis and dynamic reconfiguration. In *Third International Workshop on Intelligent Solutions in Embedded Systems, 2005* (pp. 194–201).

Elsayed, E. A. (1996). *Reliability engineering*. Prentice Hall.

Erdem, U. M., & Sclaroff, S. (2006). Automated camera layout to satisfy task-specific and floor plan-specific coverage requirements. *Computer Vision and Image Understanding, 103*, 156–169.

Fan, J., Ghurye, S. G., & Levine, R. A. (2000). Multicomponent lifetime distributions in the presence of ageing. *Journal of Applied Probability, 37*, 521–533.

Golalikhani, M., & Zhuang, J. (2011). Modeling arbitrary layers of continuous-level defenses in facing with strategic attackers. *Risk Analysis, 31*, 533–547.

Golany, B., Kaplan, E. H., Marmur, A., & Rothblum, U. G. (2007). Nature plays with dice–terrorists do not: Allocating resources to counter strategic versus probabilistic risks. *European Journal of Operational Research, 192*, 198–208.

Golmakani, H. R., & Moakedi, H. (2012). Periodic inspection optimization model for a two-component repairable system with failure interaction. *Computers & Industrial Engineering, 63*, 540–545.

Guikema, S. D. (2009). Game theory models of intelligent actors in reliability analysis: An overview of the state of the art. In *Game theoretic risk analysis of security threats* (pp. 1–19). Springer.

Guo, H., & Yang, X. (2008). Automatic creation of Markov models for reliability assessment of safety instrumented systems. *Reliability Engineering & System Safety, 93*, 829–837.

Gupta, H., Zongheng, Z., Das, S. R., & Gu, Q. (2006). Connected sensor cover: self-organization of sensor networks for efficient query execution. *IEEE/ACM Transactions on Networking, 14*, 55–67.

Haphuriwat, N., & Bier, V. M. (2011). Trade-offs between target hardening and overarching protection. *European Journal of Operational Research, 213*, 320–328.

Hausken, K. (2011). Game theoretic analysis of two-period-dependent degraded multistate reliability systems. *International Game Theory Review, 13*, 247–267.

Hausken, K., Bier, V., & Zhuang, J. (2009). Defending against terrorism, natural disaster, and all hazards. In *Game theoretic risk analysis of security threats* (pp. 65–97). New York: Springer.

Hausken, K., & Levitin, G. (2011). Shield versus sword resource distribution in K-round duels. *Central European Journal of Operations Research, 19*, 589–603.

Hausken, K., & Levitin, G. (2012). Review of systems defense and attack models. *International Journal of Performability Engineering, 8*, 355.

Hou, W. R., & Jiang, Z. H. (2012). An optimization opportunistic maintenance policy of multi-unit series production system. *Advanced Materials Research, 421*, 617–624.

Ibarra, N. (2012). Security system fails to detect man on runway at NY's Kennedy Airport. http://articles.cnn.com/2012-08-13/travel/travel_airport-intrusion_1_security-system-perimeter-intrusion-detection-system-raytheon.

IEC 61508. (2010). Functional safety of electric/electronic/programmable electronic safety-related systems, vol. 61508.

Jiang, L., Feng, Q., & Coit, D. W. (2011). Reliability analysis for dependent failure processes and dependent failure threshold. In *International Conference on Quality, Reliability, Risk, Maintenance, and Safety Engineering* (pp. 30–34).

Junbin, L., Sridharan, S., Fookes, C., & Wark, T. (2014). Optimal camera planning under versatile user constraints in multi-camera image processing systems. *IEEE Transactions on Image Processing, 23*, 171–184.

Kanturska, U., Schmocker, J.-D., Fonzone, A., & Bell, M. G. H. (2009). Improving reliability through multi-path routing and link defence: An application of game theory to transport. In *Game Theoretic Risk Analysis of Security Threats* (pp. 1–29). New York: Springer.

Katersky, A. (2010). Newark airport's security cameras were broken. http://abcnews.go.com/Travel/newark-airports-security-cameras-broken-slowed-tsa-security/story?id=9484216#.UG8li5jYHoF.

Kharoufeh, J. P., & Cox, S. M. (2005). Stochastic models for degradation-based reliability. *IIE Transactions, 37*, 533–542.

Kim, I., Choi, H., Yi, K., Choi, J., & Kong, S. (2010). Intelligent visual surveillance—A survey. *International Journal of Control, Automation and Systems, 8*, 926–939 (2010/10/01).

Knight, J. C. (2002). Safety critical systems: Challenges and directions. In *Proceedings of the 24rd International Conference on Software Engineering* (pp. 547–550).

Kobbacy, K. A. H., & Murthy, D. P. (2008). *Complex system maintenance handbook*. New York: Springer.

Koochaki, J., Bokhorst, J. A. C., Wortmann, H., & Klingenberg, W. (2012). Condition based maintenance in the context of opportunistic maintenance. *International Journal of Production Research, 50*, 6918–6929.

Krishnamachari, B., & Iyengar, S. (2004). Distributed Bayesian algorithms for fault-tolerant event region detection in wireless sensor networks. *IEEE Transactions on Computers, 53*, 241–250.

Levitin, G., & Amari, S. V. (2010). Approximation algorithm for evaluating time-to-failure distribution of k-out-of-n system with shared standby elements. *Reliability Engineering & System Safety, 95*, 396–401.

Levitin, G., & Hausken, K. (2009a). False targets efficiency in defense strategy. *European Journal of Operational Research, 194*, 155–162.

Levitin, G., & Hausken, K. (2009b). Redundancy vs. protection vs. false targets for systems under attack. *IEEE Transactions on Reliability, 58*, 58–68.

Levitin, G., & Hausken, K. (2009c). Intelligence and impact contests in systems with redundancy, false targets, and partial protection. *Reliability Engineering & System Safety, 94*, 1927–1941.

Levitin, G., & Hausken, K. (2012). Resource distribution in multiple attacks with imperfect detection of the attack outcome. *Risk Analysis, 32*, 304–318.

Levitin, G., Zhang, T., & Xie, M. (2006). State probability of a series-parallel repairable system with two-types of failure states. *International Journal of Systems Science, 37*, 1011–1020.

Lewis, P. A. W., & Shedler, G. S. (1976). Statistical analysis of non-stationary series of events in a data base system. *IBM Journal of Research and Development, 20*, 465–482.

Li, W. J., & Pham, H. (2005a). An inspection-maintenance model for systems with multiple competing processes. *IEEE Transactions on Reliability, 54*, 318–327.

Li, W., & Pham, H. (2005b). Reliability modeling of multi-state degraded systems with multi-competing failures and random shocks. *IEEE Transactions on Reliability, 54*, 297–303.

Liang, L., Xi, Z., & Huadong, M. (2011). Localization-oriented coverage in wireless camera sensor networks. *IEEE Transactions on Wireless Communications, 10*, 484–494.

Lin, F. Y. S., Tsang, P.-H., & Lin, Y.-L. (2007). Near optimal protection strategies against targeted attacks on the core node of a network. In *The Second International Conference on Availability, Reliability and Security* (pp. 213–222).

Liu, G.-S. (2012). Three m-failure group maintenance models for M/M/N unreliable queuing service systems. *Computers & Industrial Engineering, 62*, 1011–1024.

Liu, Y., & Huang, H. (2010). Optimal replacement policy for multi-state system under imperfect maintenance. *IEEE Transactions on Reliability, 59*, 483–495.

Liu, Y., Huang, H., & Pham, H. (2008). Reliability evaluation of systems with degradation and random shocks. In *Annual Reliability and Maintainability Symposium* (pp. 328–333).

Malik, A., Salcic, Z., Chong, C., & Javed, S. (2013). System-level approach to the design of a smart distributed surveillance system using systemj. *ACM Transactions on Embedded Computing Systems, 11*, 1–24.

Marcenaro, L., Oberti, F., Foresti, G. L., & Regazzoni, C. S. (2001). Distributed architectures and logical-task decomposition in multimedia surveillance systems. *Proceedings of the IEEE, 89*, 1419–1440.

Marchesotti, L., Marcenaro, L., & Regazzoni, C. (2002). A video surveillance architecture for alarm generation and video sequences retrieval. In *Proceedings 2002 International Conference on Image Processing* (vol. 1, pp. I-892–I-895).

Mariano, V. Y., Junghye, M., Jin-Hyeong, P., Kasturi, R., Mihalcik, D., Li, H., Doermann, D., & Drayer, T. (2002). Performance evaluation of object detection algorithms. In *Proceedings. 16th International Conference on Pattern Recognition* (vol. 3, pp. 965–969).

Martorell, S., Sanchez, A., & Serradell, V. (1999). Age-dependent reliability model considering effects of maintenance and working conditions. *Reliability Engineering & System Safety, 64*, 19–31.

Mathur, F. P. (1971). On reliability modeling and analysis of ultrareliable fault-tolerant digital systems. *IEEE Transactions on Computers, C-20*, 1376–1382.

Mathur, F. P., & de Sousa, P. T. (1975). Reliability modeling and analysis of general modular redundant systems. *IEEE Transactions on Reliability, R-24*, 296–299.

Mathur, F. P., & de Sousa, P. T. (1975). Reliability models of NMR systems. *IEEE Transactions on Reliability, R-24*, 108–113.

Mavrinac, A., & Chen, X. (2013). Modeling coverage in camera networks: A survey. *International Journal of Computer Vision, 101*, 205–226. (2013/01/01).

Muralidharan, K. (2008). A review of repairable systems and point process models. In *ProbStat Forum* (pp. 26–49).

Nam, Y., & Hong, S. (2012). Optimal placement of multiple visual sensors considering space coverage and cost constraints. *Multimedia Tools and Applications*, 1–22 (2012/11/01).

Nicolai, R., & Dekker, R. (2008). Optimal maintenance of multi-component systems: A review. In *Complex System Maintenance Handbook* (pp. 263–286). London: Springer.

Nikoofal, M. E., & Zhuang, J. (2012). Robust allocation of a defensive budget considering an attacker's private information. *Risk Analysis, 32*, 930–943.

Nordmann, L., & Pham, H. (1997). Reliability of decision making in human-organizations. *IEEE Transactions on Systems, Man, and Cybernetics Part A: Systems and Humans, 27*, 543–549.

Nordmann, L., & Pham, H. (1999). Weighted voting systems. *IEEE Transactions on Reliability, 48*, 42–49.

Nowakowski, T., & Werbińka, S. (2009). On problems of multicomponent system maintenance modelling. *International Journal of Automation and Computing, 6*, 364–378.

Park, M., & Pham, H. (2012). A generalized block replacement policy for a k-out-of-n system with respect to threshold number of failed components and risk costs. *IEEE Transactions on Systems, Man, and Cybernetics Part A: Systems and Humans, 42*, 453–463.

Paté-Cornell, E., & Guikema, S. (2002). Probabilistic modeling of terrorist threats: a systems analysis approach to setting priorities among countermeasures. *Military Operations Research, 7*, 5–20.

Patriksson, M., Strömberg, A.-B., & Wojciechowski, A. (2012). The stochastic opportunistic replacement problem, part II: a two-stage solution approach. *Annals of Operations Research*, 1–25.

Peng, H., Feng, Q., & Coit, D. W. (2010a). Reliability and maintenance modeling for systems subject to multiple dependent competing failure processes. *IIE Transactions, 43*, 12–22.

Peng, R., Levitin, G., Xie, M., & Ng, S. H. (2010b). Optimal defence of single object with imperfect false targets. *Journal of the Operational Research Society, 62*, 134–141.

Pham, H. (1997). Reliability analysis of digital communication systems with imperfect voters. *Mathematical and Computer Modelling, 26*, 103–112.

Pham, H. (1999). Reliability analysis for dynamic configurations of systems with three failure modes. *Reliability Engineering & System Safety, 63*, 13–23.

Pham, H. (2010). On the estimation of reliability of k-out-of-n systems. *International Journal of Systems Assurance Engineering and Management, 1*, 32–35.

Pham, H., & Xie, M. (2002). A generalized surveillance model with applications to systems safety. *IEEE Transactions on Systems, Man, and Cybernetics Part C: Applications and Reviews, 32*, 485–492.

Qian, C., Nakamura, S., & Nakagawa, T. (2003). Replacement and minimal repair policies for a cumulative damage model with maintenance. *Computers & Mathematics with Applications, 46*, 1111–1118.

Ram, S., Ramakrishnan, K. R., Atrey, P. K., Singh, V. K., & Kankanhalli, M. S. (2006). A design methodology for selection and placement of sensors in multimedia surveillance systems. Presented at the Proceedings of the 4th ACM International Workshop on Video Surveillance and Sensor Networks. California: Santa Barbara.

Rangan, A., & Tansu, A. (2008). A new shock model for system subject to random threshold failure. *Proceedings of World Academy of Science, Engineering and Technology, 30*, 1065–1070.

Rashmi, R., & Latha, B. (2013). Video surveillance system and facility to access Pc from remote areas using smart phone. In *International Conference on Information Communication and Embedded Systems (ICICES)* (pp. 491–495).

Riveiro, M., Falkman, G., & Ziemke, T. (2008). Improving maritime anomaly detection and situation awareness through interactive visualization. In *11th International Conference on Information Fusion* (pp. 1–8).

Saini, M., Natraj, Y., & Kankanhalli, M. (2009). Performance modeling of multimedia surveillance systems. In *11th IEEE International Symposium on Multimedia, 2009. ISM '09* (pp. 179–186).

Sarkara, A., Beherab, D. K., & Kumarc, S. (2012). Maintenance policies of single and multi-unit systems in the past and present. *Asian Review of Mechnical Engineering, 15*.

She, J., & Pecht, M. G. (1992). Reliability of a k-out-of-n warm-standby system. *IEEE Transactions on Reliability, 41*, 72–75.

Sheu, S., & Jhang, J. (1997). A generalized group maintenance policy. *European Journal of Operational Research, 96*, 232–247.

Sinaki, G. (1994). Ultra-reliable fault-tolerant inertial reference unit for spacecraft. *Advances in the Astronautical Sciences, 86*, 239–240.

Singh, V. K., Atrey, P. K., & Kankanhalli, M. S. (2008). Coopetitive multi-camera surveillance using model predictive control. *Machine Vision and Applications, 19*, 375–393.

Singh, V. K., & Kankanhalli, M. S. (2009). Adversary aware surveillance systems. *IEEE Transactions on Information Forensics and Security, 4*, 552–563.

Singh, C., & Mitra, J. (1997). Reliability analysis of emergency and standby power systems. *Industry Applications Magazine, IEEE, 3*, 41–47.

Siqueira, K., & Sandler, T. (2010). Terrorist networks, support, and delegation. *Public Choice, 142*, 237–253.

Stauffer, C., & Grimson, W. E. L. (1999). Adaptive background mixture models for real-time tracking. In *IEEE Computer Society Conference on Computer Vision and Pattern Recognition* (Vol. 2, p. 252).

Sung, B., & Schrage, D. P. (2009). Optimal maintenance of a multi-unit system under dependencies. In *Annual Reliability and Maintainability Symposium* (pp. 118–123).

Szczodrak, M., Dalka, P., & Czyzewski, A. (2010). Performance evaluation of video object tracking algorithm in autonomous surveillance system. In *2nd International Conference on Information Technology (ICIT)* (pp. 31–34).

Torres-Echeverría, A. C., Martorell, S., & Thompson, H. A. (2011). Modeling safety instrumented systems with MooN voting architectures addressing system reconfiguration for testing. *Reliability Engineering & System Safety, 96*, 545–563.

Trivedi, M. M., Gandhi, T. L., & Huang, K. S. (2005). Distributed interactive video arrays for event capture and enhanced situational awareness. *Intelligent Systems, IEEE, 20*, 58–66.

Tsai, Y. T., Wang, K. S., & Tsai, L. C. (2004). A study of availability-centered preventive maintenance for multi-component systems. *Reliability Engineering & System Safety, 84*, 261–270.

Ushakov, I. (1986). Universal generating function. *Soviet Journal of Computer and System Sciences, 24*, 37–49.

Valera, M., & Velastin, S. A. (2005). Intelligent distributed surveillance systems: a review. *IEE Proceedings-Vision Image and Signal Processing, 152*, 192–204.

Vaughan, T. S. (2005). Failure replacement and preventive maintenance spare parts ordering policy. *European Journal of Operational Research, 161*, 183–190.

Vu, H. C., Do Van, P., Barros, A., & Bérenguer, C. (2012). Maintenance activities planning and grouping for complex structure systems. In *Annual Conference of the European Safety and Reliability Association*.

Wallace, E., & Diffley, C. (1988). *CCTV control room ergonomics*. Publication: Published by Police Scientific Development Branch of the Home Office.

Wang, H. (2002). A survey of maintenance policies of deteriorating systems. *European Journal of Operational Research, 139*, 469–489.

Wang, B. (2011). Coverage problems in sensor networks: A survey. *ACM Computing Surveys, 43*, 1–53.

Wang, P., & Coit, D. W. (2004). Reliability prediction based on degradation modeling for systems with multiple degradation measures. In *Annual Symposium on Reliability and Maintainability* (pp. 302–307).

Wang, H., & Pham, H. (2006a). Availability and maintenance of series systems subject to imperfect repair and correlated failure and repair. *European Journal of Operational Research, 174*, 1706–1722.

Wang, H., & Pham, H. (2006). *Reliability and optimal maintenance*. Springer.

Wang, Y., & Pham, H. (2011a). Dependent competing-risk degradation systems. In H. Pham (Ed.), *Safety and Risk Modeling and Its Applications* (pp. 197–218). London: Springer.

Wang, Y., & Pham, H. (2011b). Imperfect preventive maintenance policies for two-process cumulative damage model of degradation and random shocks. *International Journal of System Assurance Engineering and Management, 2*, 66–77.

Wang,Y., & Pham, H. (2011c). A multi-objective optimization of imperfect preventive maintenance policy for dependent competing risk systems with hidden failure. *IEEE Transactions on Reliability, 60*, 770–781.

Wang, Y., & Pham, H. (2012). Modeling the dependent competing risks with multiple degradation processes and random shock using time-varying copulas. *IEEE Transactions on Reliability, 61*, 13–22.

Wang, L., Ren, S., Korel, B., Kwiat,K. A., & Salerno, E. (2013). Improving system reliability against rational attacks under given resources. *IEEE Transactions on Systems, Man, and Cybernetics: Systems*, 1–1.

Wang, X., Xing, G., Zhang, Y., Lu, C., Pless, R., & Gill, C. (2003). Integrated coverage and connectivity configuration in wireless sensor networks. *Presented at the Proceedings of the 1st international conference on Embedded networked sensor systems* Los Angeles.

Wang, G. J., & Zhang, Y. L. (2005). A shock model with two-type failures and optimal replacement policy. *International Journal of Systems Science, 36*, 209–214.

Xu, M., Chen, T., & Yang, X. (2012). Optimal replacement policy for safety-related multi-component multi-state systems. *Reliability Engineering & System Safety, 99*, 87–95.

Yi, Z., & Chakrabarty, K. (2003). Sensor deployment and target localization based on virtual forces. In *INFOCOM 2003. Twenty-Second Annual Joint Conference of the IEEE Computer and Communications. IEEE Societies* (vol. 2, pp. 1293–1303).

Yi, Y., Chung-Hao, C., Abidi, B., Page, D., Koschan, A., & Abidi, M. (2010). Can you see me now? Sensor positioning for automated and persistent surveillance. *IEEE Transactions on Systems, Man, and Cybernetics; Part B: Cybernetics, 40*, 101–115.

Ying-Wen, B., Zong-Han, L., & Zi-Li, X. (2010). Use of multi-frequency ultrasonic sensors with PIR sensors to enhance the sensing probability of an embedded surveillance system. In *International Symposium on Communications and Information Technologies (ISCIT)* (pp. 170–175).

Yuan, X., & Pandey, M. D. (2009). A nonlinear mixed-effects model for degradation data obtained from in-service inspections. *Reliability Engineering & System Safety, 94*, 509–519.

Yun, W. Y., & Cha, J. H. (2010). Optimal design of a general warm standby system. *Reliability Engineering & System Safety, 95*, 880–886.

Zhang, Y., & Pham, H. (2013a). A dual-stochastic process model for surveillance systems with the uncertainty of operating environments subject to the incident arrival and system failure processes. *International Journal of Performability Engineering*.

Zhang, Y., & Pham, H. (2013b). Modeling the effects of two stochastic-process on the reliability of k-out-of-n surveillance systems with two competing failure modes. *IEEE Transactions on Reliability*.

Zhang,Y., & Pham, H. (2015a). A cost model of an opportunistic maintenance policy on k-out-of-n surveillance systems considering two stochastic processes.

Zhang, Y., & Pham, H. (2015b). Reliability analysis of k-out-of-n surveillance systems subject to dual stochastic process and (m, d, t) opportunistic maintenance policy.

Zhang, T., Wang, Y., & Xie, M. (2008). Analysis of the performance of safety-critical systems with diagnosis and periodic inspection. In *Annual Reliability and Maintainability Symposium* (pp. 143–148).

Zhao, J. (2011). Camera planning and fusion in a heterogeneous camera network.

Zhao, J., Cheung, S., & Nguyen, T. (2009). Optimal visual sensor network configuration. *Multi-camera networks: Principles and applications*, 139–162.

Zhou, X., Lu, Z., & Xi, L. (2012). Preventive maintenance optimization for a multi-component system under changing job shop schedule. *Reliability Engineering & System Safety, 101*, 14–20.

Zio, E. (2009). Reliability engineering: Old problems and new challenges. *Reliability Engineering & System Safety, 94*, 125–141.

Zuo, M. J., Jiang, R., & Yam, R. (1999). Approaches for reliability modeling of continuous-state devices. *IEEE Transactions on Reliability, 48*, 9–18.

Part IV
Design, Applications and Practices

Reliability Management

Fred Schenkelberg

1 Overview

Reliability is an aspect of a product that exists whether or not it is actively managed, monitored, or controlled. Product failures do sometimes occur. A product's reliability performance primarily results from decisions made within the entire product development team. The product's reliability performance also impacts every aspect of an organization, including its profitability.

Each organization has a different view of product reliability. Its view varies based on the market, its products, and the organization's internal culture. The focus may be on one or more of the following elements:

- warranty,
- design for reliability,
- reverse logistics,
- failure analysis,
- product testing,
- vendor management,
- marketing, and/or
- manufacturing.

The ability to coordinate the creation and delivery of reliable products that meet market expectations is a function of the organization's management team, the members of which often do not have specific training in reliability engineering. They do know that many considerations go into creating a new product and that reliability is one important concern. By its action or inaction, the management team crafts the results achieved. By focusing on product reliability, it elevates the

F. Schenkelberg (✉)
Reliability Engineering and Management Consultant,
FMS Reliability, California, USA
e-mail: fms@fmsreliability.com

© Springer-Verlag London 2016
H. Pham (ed.), *Quality and Reliability Management and Its Applications*,
Springer Series in Reliability Engineering, DOI 10.1007/978-1-4471-6778-5_11

importance of reliability. By encouraging positive steps toward achieving a reliable product, the management team can create an effective reliability culture within the organization.

Reliability management is the organization's process to achieve the product's reliability outcome. The process may be well crafted, involving most of the organization, or it may be little more than an afterthought. It may involve only select areas of the organization, such as customer service, which reports on trends related to product failures and customer complaints, or it may include isolated teams attempting to improve the product's reliability performance.

Organizations that achieve consistent low field failure rates and do so economically take a proactive approach to the management of product reliability. They do not use different tools or techniques than other organizations; yet, they do consistently understand the value of each tool or technique to making better decisions. There is no single set of tools or techniques that mark a "good" reliability program: The best performing programs may appear to actually do very little to achieve low field failure rates. It is in the application and use of the tools that make all the difference. The best performers use the appropriate tool to make decisions.

In this chapter, we will explore the difference between reactive and proactive reliability programs, describe the roles and duties of those assigned to manage reliability, and provide an introduction to the reliability maturity matrix and how to take specific steps to move your organization to proactively managing reliability.

The difference between reactive and proactive reliability programs is best revealed in how and why decisions are deliberately made related to improving product reliability.

Upstairs, Downstairs

While conducting reliability assessments of divisions within an organization I had the opportunity to assess two similar-sized groups that created very similar products. Two years previously, both organizations lost their reliability professional from their staffs because these individuals left to start a new position in a different organization. Furthermore, both teams were located in one building, one upstairs the other downstairs, which made scheduling the assessment interviews convenient.

Though the course of the interviews I enjoyed the conversations upstairs. These interviews started on time and were not interrupted. The engineers and managers knew how to use a wide range of reliability tools to accomplish their tasks. For example, the electrical design engineer knew about derating and accelerated life testing; she also knew about the goal and how it was apportioned to her elements of the product. Each person I talked to upstairs understood the overall objective and how they provided and received information using a range of reliability tools to make decisions. They enjoyed a very low field failure rate and simply went about the business of creating products.

Downstairs was a different story. The interviews rarely started on time and most were interrupted by an urgent request, usually involving an emerging major field issue or customer complaint. The engineers and managers knew that the former reliability engineer with the team did most of what I was asking about. Most did not know what a highly accelerated life test (HALT) or an accelerated life test

Table 1 Comparison of upstairs and downstairs

	Upstairs	Downstairs
Goals	Clearly stated for each project and apportioned	Make product "as good or better" than last one
Measurements	Elements regularly measured and input to reliability block diagram model	Not sure what or how to measure during development
HALT or ALT	Only used when appropriate, with understand of when to use each tool	Not sure what those terms mean
Vendors	Work closely with them to select appropriate parts	Blamed them for sending bad parts
Atmosphere	Calm	Chaotic
Former reliability professional	Was our mentor and coach	Did all that reliability stuff for us

(ALT) was and did not have time to find out. There was a vague goal and all agreed that because this goal was not measured during product development it was meaningless. The downstairs team had a very high field failure rate and the design team often spent 50 % or more of its time addressing customer complaints.

Table 1 provides a summary of findings from these interviews.

The only salient difference between the teams and their history was the behavior of the former reliability professionals with each team. Upstairs, the reliability professional was well versed in the use of a wide range of reliability tools and processes. She provided direct support along with coaching and mentoring across the organization. She encouraged every member of the team to learn and use the appropriate tools to make decisions. This empowerment enabled them to make decisions that led to products meeting their reliability goals.

Downstairs, the former reliability engineer was a reliability professional who was also well versed in the use of a wide range of reliability tools and processes. He directly supported the team by doing the derating calculations, asking vendors for reliability estimates, designing, and conducting HALTs or ALTs as needed, and performing the myriad other tasks related to creating a reliable product. He provided input and recommendations for design changes that would improve reliability. He was a key member of the team. However, because he did not coach or mentor his team, when he moved to a new role, his knowledge and skills went with him.

The difference between these organizations was in their culture. When the entire team possesses knowledge appropriate for their role on the team, they can apply those tools to assist in making design decisions. Without that knowledge, design teams will use the tools and knowledge they have to make design decisions. Without the consideration of reliability-related information, the design decisions may or may not be beneficial to the product's reliability performance.

Reliability occurs at the point of decision during the design process: when components are selected, when structures are finalized, or when all risks have been addressed. Near the end of any product development process, the team asks whether the product is "good enough" to start production and introduce into the market. Having a clear goal with appropriate measure of the current design's ability to meet that goal provides the reliability aspect that quantifies "good enough."

1.1 Reliability Value

What is reliability management? What is reliability engineering? Would a product design or an organization benefit from focusing on reliability management and engineering? What is the value of a focus on reliability?

Any product development and producing organization has resource limits. It may be talent, capabilities, time, funding, or some combination thereof. Yet, the goal to create a product that meets customer expectations includes the concept of product reliability. The product should provide the expected functions over time without failure. This expected product reliability exists even if the design requirements and advertising do not explicitly mention product reliability.

For example, consider a laptop that needs a new power supply. Your first thought how old is the machine might be? Is it still under warranty? But then you consider the inconvenience of either being without your laptop during the repair period or the time you would need to take to setup a new machine. If the machine is only a few months old, it may be under warranty; yet, your dissatisfaction is higher: It should not have failed so soon.

If the machine is 5 years old, that is a different story. You had many years of use and, if this was the first failure, you have gotten considerable value. Besides, it may well be time to upgrade to a new machine. The inconvenience of a having to make a repair or set up a new machine, although not totally alleviated, is much less.

1.2 Product Reliability

Product reliability's primary value lies in meeting the customer's expectation that the product will work as intended for sufficient time. The market rejects products that fail often whereas it desires products that "just work." Creating a reputation for a reliable product assists in increased sales.

An extension of the value that consumers place on reliability is the willingness to pay a premium for products with high reliability. Automobiles, computers, printers, appliances, and test equipment are all examples of products whose known high reliability can warrant a premium. It is worth it because the cost of downtime during a failure can more than outweigh the additional purchase expense.

A business that creates reliable products creates value in a similar manner. Products that are sought after and command a price premium lead to higher sales and higher profit margins. Additionally, the lower failure rates reduce warranty expenses, which increases future profit margin. Yes; it may cost more in materials to create a durable product, but the business will be rewarded by higher customer satisfaction, market share, and profit margins.

1.3 Reliability Engineering

As defined by the Institute of Electrical and Electronics Engineers, *reliability engineering* is "an engineering field that deals with the study, evaluation, and lifecycle management of reliability: the ability of a system or component to perform its required functions under stated conditions for a specified period of time." (IEEE 1990).

Reliability engineering includes the use of statistics, data analysis, experimental design, customer and environmental surveys, component and product testing, failure analysis, design, manufacturing, procurement, and at times marketing and finance. Reliability engineers must possess a broad set of skills and be able to apply tools and techniques generally to enable an organization to create a reliable product.

The reliability engineer's role can span across tasks and disciplines. Although some reliability engineers will specialize on one area of the field, say accelerated testing, others may find a role that extends over nearly every function of an organization. The ability to influence and create a product that meets the customer's reliability performance expectations is both challenging and rewarding.

1.4 Reliability Management

Management is responsible for the oversight and control of reliability activities. In some organizations, these duties are performed by a dedicated reliability manager or a senior reliability engineer; but in others, reliability management is merely part of the organization's management functions. There is no one right way to organize to accomplish improved product reliability. More important is the focus brought across the organization on the effect of decisions on the resulting product's reliability performance. The management of reliability, like reliability engineering, may entail working closely with staff across the organization.

Reliability engineering and management are very similar. The former involves implementing activities and analysis that enable the creation of a reliable product. The latter accomplishes the same though the allocation of resources to enable the activities.

Organizations that include reliability considerations (i.e., requirements, predictions, risks, evaluations, and analysis) deliberately and use the information gained to guide decisions across the organization will create reliable products. Those that ignore or isolate reliability to a limited role within the organization are less likely to create a reliable product. The actual individual titles are less important than the reliability engineering activities and decisions. Reliability engineering skills are part of any engineering discipline; with some practice and encouragement nearly all engineers have the capability. Management skills are similar to any other product-producing organization set of management skills: The ability to coordinate activities, allocate resources, and focus on reliability is augmented with a solid understanding of reliability engineering tools and techniques, just as with any other management task.

Most organizations will agree that product reliability is important to their customers and to their bottom line. In general, there are two basic approaches to managing product reliability: reactive and proactive. Reactive organizations operate as a fire department. They respond to each field failure, to each product testing failure, and to each vendor component failure, attempting to prevent the failure from spreading. Proactive organizations design and build products with an acceptable reliability and anticipate the type and number of field failures. The former is regularly surprised whereas the latter is rarely surprised.

The reactive approach may exhibit many of the following behaviors:

- Waiting for failure to occur before making improvements.
- Setting reliability goals but not measuring them.
- Discounting accelerated stress testing failures as not like-use conditions.
- Discounting possible failures as conjecture or speculation.
- Awarding those who quickly solve field issues.
- Using field reliability measures that smooth or average results.
- Performing failure analysis partially as the team jumps to a possible solution.
- Working in functional silos and not sharing product failure information.
- Running the same set of "reliability" tests for all products as standard policy.
- Setting up and running tests that exhibit no failures.
- Adding testing routines in response to failures.
- Allocating over 25 % of engineering talent to address product failures.

In contrast, the proactive approach exhibits many of these behaviors:

- Anticipating failures and making improvements.
- Setting and measuring reliability goals.
- Using accelerated stress testing to discover failure mechanisms.
- Rigorously investigating failures to obtain a fundamental understanding.
- Awarding those who produce reliable products the first time.
- Using field reliability measures that provide sufficient details to make decisions.
- Regularly sharing reliability information across elements of an organization
- Only setting up and running "reliability" tests to address specific questions or provide specific information for decisions.
- Setting up and running testing that provides value (i.e., information, failures, and valid life estimates).
- Regularly reviewing test plans and removing unneeded tests while adding valuable tests.
- Allocating less than 10 % of engineering talent to address product failures.

The reactive organization generally has less time to design new products because of the demands of the current product requiring attention to address field failures. The demands of vendor visits, redesign, testing, change orders, and customer audits take time away from the focus on creating a reliable product. Because teams have less time, they may streamline testing by creating a standard set of tests (independent of any changes that may affect the validity of the tests). Engineering talent spends time deflecting the growing list of product faults by discounting stress testing or customer returns as caused by unrealistic overstress conditions. The lack of design time may result in an "as best we could" mentality.

The proactive organization generally has more time to design a new product because of fewer demands to address failures. Failures that occur during development either confirm anticipated weaknesses or provide the opportunity to learn about the design. The proactive team works closely with vendors, suppliers, and manufacturers to understand the possible causes of failures, thus permitting product designs that minimize field failures. Design teams make decisions, balancing the expected reliability performance with other costs, time to market, and functionality requirements. The proactive team celebrates failures when they lead to knowledge and acknowledge the value of failures that confirm design margins and expected reliability performance.

The difference may be quite visible in the number and type of meetings held to address field issues. Another indicator is to what extent the product design team uses testing results to make design improvements. The difference also becomes apparent when tracing the results of a reliability risk analysis, prototype stress testing, or vendor component qualification. If the objective of the tasks is to provide information to make reliability-related decisions, rather than to check off the action item as accomplished, then the organization is extracting value from the task and is behaving proactively.

Personally surveying hundreds of product design engineering managers, I have found that about 25 % of engineering time is spent addressing field failures. This is higher for lower maturity organizations and on average lower for proactive organizations. For reliability maturity stage 4 and 5 (proactive) organizations, the average engineering time spent resolving customer-identified issues is about 10 %.

In general, the product design team focuses on creating the next product. They are also the experts on troubleshooting and resolving issues found by customers. Some organizations create positions to 'protect' the design team for the diversion from new product development. These protection groups, sometimes called sustaining or new product introduction engineering, address manufacturing or customer-identified deficiencies in the design.

Every reliability program involves myriad decisions. Those decision makers who are enabled with meaningful information related to reliability can make decisions that directly impact and improve the fielded product's reliability performance. If the decision makers do not have the necessary information, they will still have to act, and the likelihood of mistakes that reduce the reliability performance of fielded products increases. The value of a task relates to the efficacy of the decisions related to that task to avoid future product failures (Schenkelberg 2012)

The task of creating and maintaining a reliability program may reside in a single individual. This "reliability champion" may or may not have the title of reliability manager and is likely to be someone who has found value in the use of reliability tools. Such individuals are often self-taught in reliability engineering. Of course, the same task of creating a reliability program may fall to a very experienced reliability professional with a Ph.D. in reliability engineering. Regardless of which person fulfills this leadership role, what is important is the knowledge, coaching, and mentoring available to everyone involved in creating a reliable product.

Part of the role of the reliability manager is to provide structure and guidance for the sequence of steps to take that will enable decision making across the organization that includes product reliability information. The role is evolutionary, from creating the basic structure and expectations, to teaching and training each step of the way, up to providing advanced reliability engineering skills as needed to supplement the overall program's progress. The primary role of the reliability champion is to instill upon everyone the capability to fully consider product reliability in nearly every decision during product development and production. Another role is to determine the current state of the organization's reliability capability and to take steps to improve the maturity to an appropriate level. Later in the chapter, we will describe the reliability maturity matrix and the specific steps to be taken to improve maturity. Leading this improvement effort across organizational boundaries has the rewards of improved product reliability but the leader faces plenty of challenges.

2 Typical Reliability Participants Within an Organization

Many employees within an organization participate in the product reliability process. These include members of the design team, design managers, quality and reliability engineers and managers, procurement engineers and managers, warranty managers, failure analysis specialists, members of the marketing and sales staff, members of the finance and manufacturing teams, and field service and call center staffs. These will be addressed in turn.

2.1 Design Team

To a large degree, the final performance of a product relies on the skill of the product design team. The industrial, electrical, mechanical, and other design engineers attempt to create a product that operates as intended, providing the functionality the customer expects. Successful designs do this elegantly, balancing cost, performance, time to market, and reliability along with a long list of other considerations, such as sustainability, recyclability, safety, manufacturability, and maintainability.

The design team creates the solution that attempts to meet all the constraints. Most design engineers intuitively understand that a product that fails to function before the end of the customers' expected operating duration is considered a failure. If a printer is expected to operate for 5 years in a home or office environment and fails to print after 2 years, it has failed to meet the customers' expectation of a 5-year life. Design engineers generally design to avoid failures (Petroski 1994a, b). They use their judgment and experience to identify design weaknesses and to anticipate use conditions and the possible adverse effect on product performance.

The reliability role of design engineers is to make design decisions that provide an acceptable balance among all the constraints and demands on the design along with the product reliability expectations. A key role for the design engineer is to determine and understand the risks to reliability performance. This role may include performing failure mode and effect analysis (FMEA) and HALTs, modeling, simulation, and prototype testing. It is the design engineer who often understands the elements of a design with the most unknowns, the most risk, and the least robustness. It is this knowledge that should prompt product testing, modeling, and simulations to aid in understanding the design decision options and strike the right balance for the final design.

2.2 Design Managers

Product development and design team managers provide the prioritization and resources that enable the design team to create a new product. A primary role related to reliability is the reinforcement of the importance of reliability by ensuring the use of suitable data and information for each decision under full consideration of the influence various options have on the product's reliability performance. For example, when considering two power supply options, the capabilities of the supplies such as weight, power output, and stability, along with cost, may dominate the decision of which option to incorporate into the design. The consideration of reliability here includes whether either or both options meet the reliability goal allocation.

Another consideration is the tradeoff between cost and expected cost of field failures. If one product has a markedly higher expected failure rate than another it may cost more in warranty expenses and customer dissatisfaction than the other option. Many design managers encourage design tradeoff calculations related to performance, cost, and time to market. It is an additional duty to require full consideration of reliability for nearly all decisions made during development.

Consider the following example. In a product development team meeting, a design team manager listened to a report on an early reliability prediction of the design. It was broken down by group within the team: display, motherboard, power supply, etc. The first report indicated that the power supply was the weakest link or the most likely element to fail. So, the design team manager asked the power supply team lead to do something about the low reliability and develop a plan to tackle the issue at the next week's meeting.

Each week the team focused on the element that limited the product reliability as being the weakest link. The team considered tradeoffs between cost and reliability. The team made steady progress and the leads learned to prepare a plan, in case the prediction tagged their area as the weakest link.

The lesson here is that with a little reliability information and the simple question, "What are you going to do to improve your reliability?" the team's focus on product reliability enhanced the design and its reliability performance.

2.3   Quality and Reliability Engineers

The role for quality and reliability (Q & R) engineers is often only limited by the individual's initiative. The focus is on product Q & R and includes influences across the entire organization. The design team requires advice and feedback to design a product that meets customer expectations. The manufacturing teams require insights and measurements that enable stable and capable processes. The procurement teams require knowledge, guidance, and assistance when selecting vendors that will provide or improve supply of valuable materials.

The Q & R engineer's primary role can often be design or manufacturing centric, often being associated with performance of statistical analysis and conduction of environmental stress testing and qualifications. When the focus turns to the product goals for performance in the hands of customers, then the scope of Q & R engineers spans the organization.

A reliability engineer may begin with the creation of reliability predictions for a new product design. This is not the only task expected though. An organization requires a reliability prediction to make decisions, inform customers, and plan production, yet the organization also needs to know what will fail and when it is likely to fail in greater detail to avoid or mitigate those risks.

Q & R engineers are in the business of identifying and resolving product performance risks. This can be accomplished once field failures occur or during the early design process. The most successful engineers tend to be proactive and work to avoid field failures.

Another essential role played by the Q & R engineer involves teaching, coaching, or mentoring members of the rest of the organization to fully consider the impact of their decisions on product Q & R performance. Transferring the unique education and experience of the Q & R engineer to design, manufacturing, procurement, and other disciplines will greatly expand the effectiveness of a single Q & R engineer. The ability of others to identify risk, consider Q & R fully during tradeoff decision making, and become aware of Q & R goals and progress permits the Q & R engineer to effectively influence an entire organization.

The real test of the effectiveness of an individual Q & R engineer takes place when that engineer leaves the organization. If the product's Q & R performance declines, we may conclude that the organization did not adopt and incorporate the breadth of reliability practices. If, however, the Q & R performance remains stable or continues to improve, the organization successfully created sustaining business processes coupled with sufficient reliability engineering knowledge to continue developing reliable products.

2.4   Quality and Reliability Managers

Like the Q & R engineer the Q & R manager plays a role with a broad scope within the organization, providing advice, guidance, and feedback to the rest of the organization on the Q & R objectives. The scope may include developing strategy, facilitating interdepartmental cooperation, and providing oversight of a team of Q & R professionals. Q & R managers often work directly with customers to understand their requirements, needs, and objectives. They also work closely with suppliers to avoid or resolve supply chain impact on Q & R product performance. With customers or suppliers, the work may be proactively minimizing the adverse impact to Q & R or reacting to Q & R issues. Like the Q & R engineer, the Q & R manager's most valuable service to the organization and customer is accomplished proactively.

Setting clear Q & R goals and developing systems to predict and monitor progress toward those goals is essential for a proactive role. Identifying risks, allocating resources, and promoting progress for the organizations approach to Q & R product performance may involve changing the culture of the organization.

Consider the following example. After an expensive field failure episode, the Q & R manager was tasked with avoiding a similar product return situation in the future. He is given no budget and no personal and has a week to come up with a plan. The failure analysis of the current field failure cause revealed that the organization knew about the issue in a previous product development project. In this program, the product design team was unaware of the previous discovery and failed to avoid the problem.

So the Q & R manager created a short list of field failure lessons learned. For example, ceramic capacitors are like glass and their fracture reminded design teams to consider the fragility of these common components. The manager's first list had 16 items highlighting previous major field failure events. He then visited each product design team program manager and asked what procedures needed to be implemented to avoid the 16 classes of failure causes. For each item on the list, the development team included a specific task, study, or test in the product development plan. This process raised awareness of the lessons learned.

The Q & R manager then returned to the product design team at the final design review, just prior to launch, and asked "What did you do to avoid each issue?" and "Did your team actually proactively review and verify the risk?" If the team successfully remembered the potential risk to increased field failures, he would approve the design for production.

In 5 years, the list has grown to 20 items and they have successfully not repeated any of the previous design errors that lead to major field failure events. The process also created a culture in which product reliability was deemed important and worth the investment to achieve the Q & R goals, as it directly impacted the field failure rate and company profitability.

2.5 Procurement Engineers and Managers

Procurement entails working with suppliers to obtain a supply of components or materials that meet the design requirements. The procurement team comprises engineers, managers, and support staff. One of its primary goals is to minimize cost. Often, this means procuring the lowest price for a component that meets the design requirements.

The reliability role for procurement professionals is to obtain the best price for a component that meets or exceeds the reliability requirements. They should know the specific goal, including the probability of success, duration, local environment, and use, along with the design specifications. The conversations with suppliers should include conveying the reliability requirements along with learning about potential failure mechanisms.

Unfortunately, the focus on price often outweighs the perceived value of focusing on component reliability. Rather than building a working relationship with suppliers to quickly solve field issues, the focus should be on preventing design and component selection errors. Adding the cost of failure to the price equation often resets the procurement focus to include finding the most reliable components.

2.6 Warranty Managers

Products with warranties often include a warranty policy (terms and conditions of the offered warranty) plus some means to estimate future warranty costs and monitor warranty returns and payments.

The reliability role may include the following:

- creating and using accurate predictions of future failure rates,
- minimizing the costs of the reverse supply chain, and
- conducting failure analysis of returns and working to eliminate future failures.

A critical role is the feedback function that warranty tracking provides for the organization, as it represents the customer experience of the product's reliability performance. Warranty acts as the scorecard.

Consider a medical device organization in which no one managed warranty. The Chief Financial Officer explained that warranty was not a major issue because it was an insignificant annual expense. However, about half of the products sold were returned within the warranty period. Although the manufacturing cost of each product was thousands of dollars, the cost of warranty was limited to the cost of the spare parts consumed during repair. Thus, senior management was unaware of the very large field failure rate. A warranty manager would have been responsible for providing clear and appropriate measures of field performance. By increasing the warranty expense to include a call center, field service engineers, and the cost of replacement units, warranty for this organization would have accurately reflected the impact of such a high field failure rate.

2.7 Failure Analysis Specialists

Many organizations do little more than isolate the faulty components and return them to supplier for analysis. This process rarely results in accurate or rapid resolution of product failures resulting from design, manufacturing, or supplier issues. The role of a failure analyst is to collect failure information and determine the root cause. Some organizations have large well-staffed failure analysis (FA) labs and provide detailed root cause analysis. Every organization should have basic diagnostic and troubleshooting equipment and staff trained in basic FA techniques.

For issues that go beyond the internal failure analysis capabilities, the role of the FA specialists is to work with outside FA labs (not the supplier lab) to determine the root cause of the product failure.

Failure analysis is not to be confused with Failure Mode and Effect Analysis (FMEA). FMEA is a design tool to identify optional design improvements. While a thorough FMEA may provide someone conducting a failure analysis information on possible causes of an observed failure mode, it is generally not sufficient to identify the root cause.

With or without a FA specialist, the basic process for failure analysis is summarized by the eight disciplines of problem solving, 8D:

1. Prepare for failure analysis project

   a. Collect information on failure circumstance
   b. Emergency response if needed
   c. Form a team

2. Describe the Problem

   a. Why approach
   b. Problem statement
   c. Is/Is not boundary definition

3. Interim Containment Action

   a. Determine scope of failures
   b. Determine severity of consequence of failures
   c. Isolate suspected items

4. Root Cause Analysis

    a. Determine fundamental causes (software code, physics or chemistry level)
    b. Experiment to verify cause theory
    c. Determine options to resolve

5. Permanent Corrective Action

    a. Determine best course of action for resolution
    b. Determine risk of solution (FMEA)
    c. Balance risk, business needs, and customer expectation

6. Implement and Validate

    a. Create plan for implementation
    b. Monitor effectiveness of solution

7. Prevention

    a. Examine other products for similar circumstances leading to failure
    b. Examine and remove management and engineering processes that contributed to cause(s) to failure.
    c. Provide documentation and education on problem and resolution.

8. Closure and Team Celebration

    a. Archive documents
    b. Celebrate accomplishment

The FA specialist brings tools and knowledge especially useful in step 4 above.

2.8 Marketing and Sales Staff

The primary role of the marketing and sales teams is to create demand and book sales, but they also act as both consumers and providers of reliability information. First, these teams should understand and convey accurate information about the product's reliability performance. Second, these teams should understand and convey accurate information about customer expectations and requirements to the rest of the organization.

2.9 Finance Team

Warranty accruals and expenses often reside solely in the domain of finance. An understanding of basic reliability principles and prediction information can dramatically increase the accuracy of the accruals. In one organization, the product development teams created detailed and accurate models of future field failures and sent the finance team Weibull cumulative distribution plots covering the expected product lifetime. Because the finance team did not know how to read these plots, they arbitrarily selected the midpoint from each graph for use in warranty accrual estimates. With just a little training and understanding, the accrual accuracy increased 100 fold, thereby saving the organization from major swings in warranty expense accounting.

2.10   Manufacturing Team

Manufacturing can only make a product's Q & R worse. It is impossible to create a product as good as the design intent owing to material, assembly, and environmental variation. Therefore, the role of the manufacturing team is to minimize variation that adversely impacts field reliability.

Understanding the critical Q & R elements of design enables the manufacturing team to focus on monitoring and controlling elements that have a high impact on Q & R performance. The common focus on production yield is often related to field Q & R performance, when the production testing includes the ability to detect latent defects or significant adverse changes to the expected product durability.

2.11   Field Service and Call Center Staffs

The primary role of field service and call center organizations is to support installation, operation, and restoration of product for customers. Members of these teams must also understand the Q & R risks to better enable rapid troubleshooting and restoration for customers. The Q & R information required should not come from customer complaints; rather, it should come from product development, new product introduction, or manufacturing teams.

Another critical role played by field service and call centers is to provide early detection of field failure issues. To enhance this capability, these teams should understand what is expected to fail or cause complaints, thus permitting the detection of unusual events or trends. Another role is to secure information and returned products that exhibit field failure or unusual behavior. It is often time consuming and frustrating for the consumer to work though a detailed diagnosis and troubleshooting procedure, so replacing the product can restore the customer's use of the product and also permit a detailed failure analysis of the faulty unit.

Setting up an aggressive replacement policy for limited periods of time, along with creating a dedicated failure analysis team to analyze returns and implement improvements, enables reliability improvements. This practice is best done at the beta testing phase or early during the product launch when the number of units in the field is limited.

3   Organizational Structure and Decision Making

Both organizational structure and decision-making policies have an impact on improving product reliability. The former is more quantifiable whereas the latter involves more intangible subtleties. These topics are addressed next.

3.1   Organization

There is no single organizational structure that leads to improved product reliability performance over any other structure. Both centrally and distributed reliability teams have been successful and have failed to create reliable products. Both small cross-functional teams and large functional silo organizations have been successful and failed. Even the presence or absence of reliability professionals on staff is not an indicator of reliability performance.

Top performing organizations use a common language around product reliability and possess a culture that encourages and enables individuals to make informed decisions related to reliability. Individuals across the organization know their role to both use and share information essential to making decisions. There is an overriding context for reliability decisions that balances the needs to meet customer expectations for reliability along with other criteria. Alignment exists among the organization's mission, plans, priorities, and behaviors related to reliability.

Product reliability is not the only element that benefits from a proactive culture. Whether top performing organizations enjoy a proactive culture that naturally includes reliability activities to make decisions or evolved while improving product reliability to become a proactive organization with collateral benefits for other areas of running the business remains unclear. The latter is more likely, since it takes leadership to build and maintain a proactive organization, although some organizations focus on building a proactive reliability program and develop the benefits later in other functions of the business.

Moving the organizational block around the organizational chart may have some value, although it is not directly related to improving product reliability. It entails a more fundamental change than developing the reporting structures to transition from a reactive to proactive reliability program.

3.2   Decision Focus and Value

An essential element of a successful reliability program is the notion that all reliability activity relates to decisions. If you are performing a HALT because it is listed on the product development guidelines, or because it was carried over from the last program's plan, and the HALT results are not part of the design improvement decision-making process, then you probably should not be doing so. If the HALT results yield little or no information (e.g., it is just being checked off the list as accomplished) then the HALT itself provides little or no value.

If performance of a HALT is on the plan because the new product has new materials, vendors, or design elements, then it may reveal weaknesses. If that list of weaknesses is made available to the design team members and they are permitted and encouraged to improve the design based on that input, then the HALT data can provide input to decisions about design improvements. The value that the HALT plays here is related to reduced field failures from design improvement opportunities discovered by such highly accelerated life testing.

If the HALT is done too late to permit any decisions to improve the product, it has no value. If the HALT is done to facilitate decision making concerning design improvements, it may have great value (Hobbs 2000).

Another example of decision making is life testing to estimate the expected durability of a product. At some point, in most product development processes, there is a meeting to decide whether the product is ready for production and shipment. One element of this decision is the ability of the design to meet or exceed product reliability objectives.

As an example, say a motor is the key element that will determine the life of the product and currently there is uncertainty concerning the motor's expected reliability performance. Therefore, the team decides to conduct an accelerated life test. If the test provides a meaningful estimate prior to the decision point on readiness, it adds value. If the ALT provides results a few months after products start shipping, it adds little value for the readiness decision prior to launch.

During product development or maintenance planning two basic questions are often asked:

1. What will fail?
2. When will it fail?

The various reliability tools provide information to address these two basic questions. If that information is meaningful and timely (prior to the decision point), then the reliability tasks have value. Performing reliability tasks simply to be doing reliability activities is unlikely to add value for decision makers. Purposefully using reliability engineering tools to support decisions is very likely to add value.

4 Maturity Matrix

The concept of a maturity model is not new (Crosby 1979a, b; West 2004; IEEE Std. 1624–2008 2008). The maturity matrix in this chapter follows the IEEE Std. 1624–2008 model with additional text and modification based on the experience of the author. The matrix provides a means to identify the current state and illuminate the possible improvements to a reliability program. The matrix serves a guide to assist an organization in improving its program.

The matrix has five stages. In general, the higher stages are most cost effective and efficient at achieving higher rates of product reliability performance. These stages—uncertainty, awaking, enlightenment, wisdom, and certainty—are described in the following and summarized in Table 2.

Stage 1: Uncertainty

"We don't know why we have problems with reliability."

Reliability is rarely discussed or considered during design and production. Product returns resulting from failure are considered a part of doing business. Field failures are rarely investigated, and often blame is assigned to customers. The few people who consider reliability improvements gain little support. Reliability testing is done in an ad hoc fashion and often just to meet customer requirements or basic industry standards.

Stage 2: Awakening

"Is it absolutely necessary to always have problems with reliability?"

Reliability is discussed by managers but not supported by funding or training. Some elements of a reliability program are implemented, yet generally not in a coordinated fashion. Some tools such as FMEA and accelerated and highly accelerated life testing are experimented with, but most effort still focuses on standards-based testing and meeting customer requirements. Some analysis is done to estimate reliability or understand field failure rates, yet limited use is made of these data in making product decisions. There is, however, an increasing emphasis on understanding failures and resolving them. Failure analysis is typically accomplished by component vendors with little result.

Stage 3: Enlightenment[1]

"Through commitment and reliability improvement we are identifying and resolving our problems."

A robust reliability program exists and includes many tools and processes. Generally, significant effort is directed to resolving prototype and field reliability issues. Increasing reliance is placed on root cause analysis to determine appropriate solutions. Some tools are not used to their full potential owing to lack of understanding of reliability and how the various tools apply. Some reliance is placed on

---

[1]Stages 2 and 3 are reactive in nature. When an issue arises it is considered and addressed. Stages 4 and 5 are proactive. As the design evolves the design team exerts concerted effort to discover and resolve issues before they manifest themselves in product failures both in the development prototypes and during customer use. In reactive organizations, management pays attention to prototype or field failures. In proactive organizations, management seeks weaknesses in products and improves product reliability before prototype or field issues arise.

Table 2 Maturity matrix

	Reliability:	Stage 1: uncertainty	Stage 2: awakening	Stage 3: enlightenment	Stage 4: wisdom	Stage 5: certainty
Product Requirements	Requirement & planning	Informal or nonexistent	Basic requirements based on customer requirements or standards. Plans have required activities.	Requirements include environment and use profiles. Some apportionment. Plans have more details with regular reviews.	Plans are tailored for each project and projected risks. Use of distributions for environmental and use conditions.	Contingency planning occurs. Decisions based on business or market considerations. Part of strategic business plan.
	Training & development	Informally available to some, if requested	Select individuals trained in concepts and data analysis. Available training for design engineers	Training for engineering community for key reliability-related processes. Managers training on reliability and lifecycle impact.	Reliability and statistics courses tailored for design and manufacturing engineers. Senior managers trained on reliability impact on business.	New technologies and reliability tools tracked and training adjusted to accommodate. Reliability training actively supported by top management.
Engineering	Analysis	Nonexistent or solely based on manufacturing issues	Point estimates and reliance on handbook parts count methods. Basic identification and listing of failure modes and impact	Formal use of FMEA. Field data analysis of similar products used to adjust predictions. Design changes cause re-evaluation of product reliability	Predictions are expressed as distributions and include confidence limits. Environmental and use conditions used for simulation and testing.	Lifecycle cost considered during design. Stress and damage models created and used. Extensive risk analysis for new technologies.

(continued)

Table 2 (continued)

Reliability:	Stage 1: uncertainty	Stage 2: awakening	Stage 3: enlightenment	Stage 4: wisdom	Stage 5: certainty
Testing	Primarily functional	Generic test plan exist with reliability testing only to meet customer or standards specifications	Detailed reliability test plan with sample size, and confidence limits. Results used for design changes and vendor evaluations.	Accelerated tests and supporting models used. Testing to failure or destruct limits conducted	Test results used to update component stress and damage models. New technologies characterized.
Supply chain management	Selection based on function and price	Approved vendor list maintained. Audits based on issues or with critical parts. Qualification primarily based on vendor datasheets.	Assessments and audit results used to update AVL. Field data and failure analysis related to specific vendors.	Vendor selection includes analysis of vendor's reliability data. Suppliers conduct assessments and audit of their suppliers.	Changes in environment, use profile, or design, trigger vendor reliability assessment. Component parameters and reliability monitored for stability
Feedback Process					
Failure data tracking & analysis	Failures during function testing may be addressed	Pareto analysis of field return and internal testing. Failure analysis relies on vendor support.	Root cause analysis used to update AVL and prediction models. Summary of analysis results disseminated.	Focus is on failure mechanisms. Failure distribution models updated based on failure data	Customer satisfaction relationship to product failures is understood. Use of prognostic methods to forestall failure.

(continued)

Table 2 (continued)

Reliability:	Stage 1: uncertainty	Stage 2: awakening	Stage 3: enlightenment	Stage 4: wisdom	Stage 5: certainty
Validation & verification	Informal and based on individuals rather than process	Basic verification plans are followed. Field failure data regularly reported.	Supplier agreements around reliability monitored. Failure modes regularly monitored.	Internal reviews of reliability processes and tools. Failure mechanisms regularly monitored and used to update models and test methods	Reliability predictions match observed field reliability.
Improvements	Nonexistent or informal	Design and process change process followed. Corrective action process includes internal and vendor engagement.	Effectiveness of corrective actions tracked over time. Identified failure modes addressed in other product. Improvement opportunities identified as environment and use profiles change.	Identified failure mechanisms addressed in all products. Advanced modeling techniques explored and adopted. Formal and effective lessons-learned process exists.	New technologies evaluated and adopted to improve reliability. Design rules updated based on field failure analysis.
Management — Understanding & attitude	No comprehension of reliability as a management tool. Tend to blame engineering for 'reliability problems'	Recognizing that reliability management may be of value but not willing to provide money or time to make it happen.	Still learning more about reliability management. Becoming supportive and helpful.	Participating. Understand absolutes of reliability management. Recognize their personal role in continuing emphasis.	Consider reliability management an essential part of company system.

(continued)

Table 2 (continued)

Reliability:	Stage 1: uncertainty	Stage 2: awakening	Stage 3: enlightenment	Stage 4: wisdom	Stage 5: certainty
Status	Reliability is hidden in manufacturing or engineering departments. Reliability testing probably not done. Emphasis on initial product functionality.	A stronger reliability leader appointed, yet main emphasis is still on an audit of initial product functionality. Reliability testing still not performed.	Reliability manager reports to top management, with role in management of division.	Reliability manager is an officer of company; effective status reporting and preventive action. Involved with consumer affairs.	Reliability manager is on the board of directors. Prevention is the main concern. Reliability is a thought leader.
Measured cost of unreliability	Not done other than anecdotally	Direct warranty expenses only	Warranty, corrective action materials, and engineering costs monitored	Customer and lifecycle unreliability costs determined and tracked	Lifecycle cost reduction done through product reliability improvements
Prevailing sentiment	"We don't know why we have problems with reliability"	"Is it absolutely necessary to always have problems with reliability?"	"Through commitment and reliability improvement we are identifying and resolving our problems."	"Failure prevention is a routine part of our operation."	"We know why we do not have problems with reliability."

establishing standard testing and procedures for all products. Only some use of these testing results is made for estimating product reliability to supplement predictions. Predictions are primarily made to address customer requests and not as feedback to design teams.

Stage 4: Wisdom

"Failure prevention is a routine part of our operation."

Each product program or project has a tailored reliability program that can be adjusted as the understanding of product reliability risks changes. Reliability tools and tasks are selected and implemented because they will provide needed information for decisions. Testing focuses on either discovering failure mechanisms or characterizing failure mechanisms. Testing often proceeds to failure, if possible. Advanced data analysis tools are regularly employed and reports are distributed widely. There is increasing cooperation with key suppliers and vendors to incorporate the appropriate reliability tools upstream.

Stage 5: Certainty

"We know why we do not have problems with reliability."

Product reliability is a strategic business activity across the organization. There is widespread understanding and acceptance of design for reliability and how it fits into the overall business. Product reliability is accurately predicted prior to product launch using a mix of appropriate techniques. New materials, processes, and vendors are carefully considered for their ability to meet internally established reliability requirements. The few failures that do occur are expected and analysis is done to identify early signs of material or process changes. Customers and suppliers are regularly consulted on ways to improve reliability.

Moving toward a Proactive Program

While driving toward the Houston airport and my flight home, I received a call from the Vice President of Quality for a medical device company. I had been working with one division for a few years and did not know the person calling. He had heard of the results the division had achieved in improving their product's reliability and their market share. After identifying himself he asked, "I want you to change our company's culture." After regaining control of my car, I continued to talk to him about the next steps and arranged a meeting in person.

Over the next year, we conducted assessments of each division in the organization, recommended changes to the various programs, and installed supporting elements across the corporation. Each division had different paths toward improvement, yet each made progress. Each division also found support and gained focus from corporate management.

5   Reliability Maturity Matrix Guide

The basic approach—whether for an individual product development team or for a multidivision corporation—is the same. What is the current state? What are the strengths to reinforce? What are the weaknesses to improve? And, overall, what specific actions are needed to move the organization to a higher level of maturity?

In this section, we expand upon each cell of the maturity matrix and suggest a few specific actions that would move the organization to the right on the matrix. As you consider your organization first conduct an assessment. Determine which cells in the matrix best describe your organization. For the highest stage cells, highlight the value that behavior provides to the organization and reinforce that behavior. For the lower stage cells, consider specific steps to improve the organization's performance. The actions may involve improving data collection and analysis, providing additional training, procuring better equipment, or improving risk assessment. The specific tasks will vary depending on the maturity across the matrix along with resources and priorities.

An organizational reliability program assessment is only of value when the resulting action creates a more effective reliability program. Moving to the right (increasing maturity) on the matrix provides value to the organization. Some examples include reduced field failures, reduced cost of product development and testing, increased ability to meet market introduction deadlines, and increased market share.

Each organization's culture, history, capabilities, and priorities will influence its reliability improvement program. Local effective change management and the internal influence of thought leaders and champions will also affect any improvement effort. Therefore, any effort to improve an organization's reliability maturity must account for the local culture and norms; thus each improvement program will be different. Yet, the basic tools, approaches, and processes related to reliability engineering remain largely the same across organizations. The particular product and market may place unique constraints on specific tools, but the basics tend to remain the same.

The Reliability Maturity Matrix will provide the structure for the guidelines we develop (see Appendix A). The IEEE Std. 1624–2008 Standard for Organizational Reliability Capability (IEEE Std. 1624–2008 2008), Crosby's *Quality Is Free* (Crosby 1979a, b), and other works related to reliability engineering provide further guidance. The intention is to provide the recommended tasks to facilitate a transition from one maturity level to the next across each measurement category.

In general, organizations tend to have fairly consistent reliability maturity across categories. There may be some variation, yet commonly only one level higher or lower from the overall average maturity. The maturity matrix consistency reflects the cultural elements and the overall organization's approach or policy toward reliability. The consistency also reflects the interconnectedness between categories.

The assessment is the tool to clearly identify the maturity level of an organization as well as the cultural aspects of why. The recommendations generated by the assessment focus on reinforcing strengths and improving weaknesses. Also, the specific recommendations focus on moving the average maturity to the right or up in maturity. Given the interconnected nature of the categories, it is often difficult to only improve one category to a higher maturity without affecting other related categories. Likewise, categories that are not as mature reveal those weaknesses as they do not support activities in other categories, and categories that are more mature tend to expect mature practices across the other categories.

5.1 Product Requirements

All products have a list of requirements for customer expectations, design parameters, critical functionality, and in some cases reliability objectives. Every product has some form of reliability expectations, whether or not they are articulated. Customers have an expectation for the duration of use and will either complain, return the product, or avoid future purchases when their expectations are not met.

Goals without Apportionment or Measures

A life-support-equipment company manager desires to conduct a reliability program assessment. The company is experiencing about a 50 %/year failure rate and at least the Director of Quality thought it should do better. One of the findings was related to reliability goal setting and how it was used within the organization.

Nearly everyone knew that the product had a 5,000 h Mean Time Before Failure (MTBF) reliability goal, but very few knew what that actually mean. It was how this team used the product goal that was even more surprising. There were five elements to the product with five different teams working to design those elements: a circuit board, a case, and another three elements. Within each team, team members designed and attempted to achieve the reliability goal of the product, the 5,000 h MTBF goal. Upon data analysis of the field failures, they actually did achieve their goal as each element was just a little better than 5,000 h MTBF in performance.

Reliability statistics stipulates that in a series system one has to have higher reliability for each of the elements than for the whole-system goal. For example, if each element achieves 99 % reliability over 1 year, the reliability values of product's five elements, 99^5, would produce a system-level reliability performance of approximately 95 % at 1 year. We call it apportionment when we divvy up the goal to the various subsystems or elements within a product.

This team skipped that step and designed each element to the same goal intended for the system.

Compounding the issue was the simplistic attempts to measure reliability of the various elements and total lack of measurement at the system level. For each component, the team primarily relied on using the weakest component within the subsystem to estimate the subsystem's reliability. For example, the circuit board had about 100 parts, one of which the vendor claimed had about a 5,000 h MTBF. Thus that team surmised that, because it was the weakest element, nothing would fail before 5,000 h and thus this was all the information the team members needed to consider. They did not consider the cumulative effective of all the other components nor the uncertainty of the vendors estimate within their design and use environment.

This logic was repeated for each subsystem.

The result was a product that achieved about the same reliability it achieved in the field. The estimated use of the product was about 750 h/year; thus each element would achieve about 85 % reliability for a year, which seemed to be an adequate reliability goal. However, this is a series system, meaning that a failure in one element would cause the system to fail. The math works out as follows:

$$\text{Reliability}(750 \text{ h}) = \left(e^{-750/5,000}\right)^5 = 0.47, \text{ or } 47 \%.$$

Because the product of the reliabilities of the individual five elements was overlooked, the system reliability turned out to be less than 50 %, not the expected 85 %. The field performance was the result of how the product was designed to meet the reliability goal for each subsystem. The team got what it designed. Its members had forgotten or ignored a basic, yet critical element of reliability engineering knowledge.

5.1.1 Requirements and Planning

Designing and producing a product that meets customer expectations requires some level of understanding of customer expectations for functionality, use and environmental conditions, and durability. These requirements influence every facet of product design and production. The overall plan to achieve the reliability requirements establishes the sequence of reliability activities and decision points over the product lifecycle.

Stage 1

Planning is either informal or nonexistent

Stages 1–2

Publish and highlight customer requirements related to product reliability.

Gather and highlight information about customer use and environmental conditions.

Create a reliability program plan including a list of reliability activities to accomplish.

Stage 2

Basic requirements are based on customer requirements or standards. Plans incorporate the required activities.

Stages 2–3

Create fully stated reliability requirements including those for function, environment, duration, and probability of success.

Gather and publish customer profiles including a range and distribution of environmental and use conditions.

Apportion reliability requirements to product subsystems and major components.

Create a detailed reliability program plan including budgets for resources, personal, and capital equipment.

Evaluate designs and suppliers for new materials or processes that may increase reliability risk.

Stage 3

Requirements include environment and use profiles along with some apportionment. Plans have more details and regular reviews are performed.

Stages 3–4

Express reliability objectives as distribution rather than point estimates, when applicable.

Incorporate reliability plans within product development plans.

Create decision points within the reliability plan to adjust activities based on current information.

Review supplier and vendor reliability programs to identify potential risk areas.

Create an overall reliability program strategy and implementation plan.

Stage 4

Plans are tailored for each project and projected risks. Use is made of distributions for environmental and use conditions.

Stages 4–5

Create reliability plans that include contingency plans for range of design, supply chain, and requirements disruptions.

Create reliability strategic plans that are integrated with overall business strategic plans.

Stage 5

Contingency planning occurs. Decisions are based on business or market considerations. Requirement planning is part of the strategic business plan.

5.1.2 Training and Development

The technical skills and knowledge to design and produce a product span a wide range of reliability engineering activities. Individuals across the organization need to understand the reliability-related goals, plans, tasks, and measures and their importance to effectively create a reliable product. Specific training opportunities abound (e.g., see reliabilitycalendar.org to find a calendar full of events from a wide range of sources covering nearly all topics in reliability engineering).

Only One Person Performs All Reliability Tasks

Having been invited to evaluate reliability practices within a company, I conducted a series of interviews with various staff members. When asked any question on the reliability techniques used, members of the engineering, procurement, operations, and quality departments all responded with nearly the same comment: "Oh, the reliability guy does that." It appeared that the organization had one reliability engineer who did everything related to reliability. His interview was scheduled last that day. I was looking forward to meeting him.

The sole reliability engineer supported three design teams working on similar products. He set the reliability goals, worked out the apportionment, calculated the derating and safety factors on most elements of the design, worked with vendors to secure parts for accelerated life testing, conducted HALTs and a range of environmental testing, established any testing related to reliability for the manufacturing team, and monitored field issues, failure analysis, warranty estimates, and other aspects. He was working long days and was often unable to address all the issues that arose. He knew reliability engineering but did not have time to conduct all the tasks necessary to help produce reliable products.

He and I agreed that this situation was neither sustainable nor beneficial to the rest of the team. Although he felt valuable and sought after by nearly everyone in the organization, the expectation was that he would do everything related to reliability. We agreed that many of the engineers and managers within the organization had the capability to take on most of the reliability work. They just needed to know how. He then sighed, lamenting on the fact that one more time-intensive task had just been added to his day.

We talked about ways to facilitate this transition. I talked to the engineering and operations managers and they agreed they needed to spread out the work and that many of the tasks would be best done by their engineer once they learned a little more about reliability engineering.

About a month later, I heard that the reliability engineer had left the group to take a new position. We talked briefly and he said that the former team meant well but never found time to learn or take on any tasks, citing the need to get the product out. "Could you do it this one time?" had become the common repetitive request. Nothing really changed. When he left he took the entire reliability program with him.

Stage 1

Training is informally available to some, if requested.

Stages 1–2

Create reliability overview seminars for designers and product development teams.

Create a list of reliability training resources related to industry or technology.

Provide training opportunities for reliability practitioners with an emphasis on reliability concepts and statistical methods.

Stage 2

Select individuals trained in concepts and data analysis. Make training available for design engineers

Stages 2–3

Create and provide regular classes for engineers on root cause analysis and corrective action methods.

Create and provide regular seminars for managers on reliability activities and use and value of those activities for improvement of product reliability.

Stage 3
Training is provided for engineering community for key reliability-related processes. Managers are trained on aspects of reliability and lifecycle impact.
Stages 3–4
Create tailored reliability courses for key reliability tasks including when and how to determine the need to accomplish the task.
Create and provide seminars and workshops to senior managers on how reliability impacts the business
Encourage reliability practitioners to learn how to identify failure modes and mechanisms related to the product and industry
Create a reliability training program for engineers and associated managers focused on design for reliability and implementation of critical reliability activities.
Stage 4
Reliability and statistics courses are tailored for design and manufacturing engineers. Senior managers are trained on how reliability affects the business.
Stages 4–5
Create a means to learn about industry trends, new materials and processes, and reliability modeling and analysis tools that may have a meaningful impact on the business.
Create a comprehensive reliability training program for everyone in the organization with visible management support and involvement.
Stage 5
New technologies and reliability tools are tracked and training is adjusted to accommodate changes. Reliability training is actively supported by top management.

5.2 Engineering

Engineers make decisions that directly influence the reliability of the product. Understanding and addressing the risks to reliability in the design and supply chain lie at the heart of any reliability program. The other elements of the matrix provide information and support to the engineering staff.

5.2.1 Reliability Analysis

Assessing reliability risk with a product's design or field performance illuminates failure modes, mechanisms, and effects. The analysis provides information to create reliability estimates and predictions. The ability to understand, characterize, compare, and judge product reliability enables decisions to be made across the product lifecycle.
Stage 1
Reliability analysis is nonexistent or solely based on manufacturing issues.

Stages 1–2

Poll the design team for reliability risks. Determine what potential risks are known. Create a prediction capability, for example by using a part-counting approach or by drawing simple reliability block diagrams and using vendor data.

Illustrate failure mode impact on the customer.

Stage 2

Carry out point estimates and rely on handbook parts counting methods. Perform basic identification and listing of failure modes and impact.

Stages 2–3

Lead FMEA analysis studies with willing teams.

Conduct field data reliability analysis to estimate reliability performance.

Review design changes to ascertain the broader impact on product reliability.

Use worst-case conditions rather than only nominal conditions.

Use failure mechanism models to design and analyze test results.

Stage 3

Formal use of FMEA is made. Field data analysis of similar products is used to adjust predictions. Design changes cause reevaluation of product reliability.

Stages 3–4

Use distributions rather than point estimates for reliability predictions.

Include confidence intervals or bounds on data analysis results.

Use distributions for use and environmental conditions rather than specification values.

Use failure mechanism models to determine cost–benefit decisions for product changes.

Stage 4

Predictions are expressed as distributions and include confidence limits. Environmental and use conditions are used for simulation and testing.

Stages 4–5

Include lifecycle costs in analyzes for use in decision making.

Create stress–life models for new materials, features, and components when existing models become inadequate.

Create complex simulations or Monte Carlo analysis systems to create predictions and estimate the value of proposed changes.

Stage 5

Lifecycle cost is considered during design. Stress and damage models are created and used. Extensive risk analysis studies are conducted for new technologies.

5.3 Reliability Testing

The intent of physically evaluating product prototypes and production units is to identify design and supply chain weaknesses, explore product limits and potential failure modes, and determine the effects of the expected range of use profiles and environments. Physical testing includes demonstrating that the product's durability (expected reliability) meets the requirements.

Stage 1

Testing is primarily functional.

Stages 1–2

Create a minimum reliability test plan to address primary reliability requirements.

Create design verification testing of functional requirements for use on all products shipped.

Conduct discovery testing to determine the design margin (HALT).

Stage 2

A generic test plan exists with reliability testing used only to meet customer or standards specifications.

Stages 2–3

Create a detailed reliability test plan, including stresses for specific failure mechanisms, samples size calculations, and confidence levels.

Determine failure mechanisms evaluated for each test proposed and verify that all potential failure mechanisms are appropriately exercised within the overall test program

Review vendor testing to determine whether it is adequately connected to expected use and environmental conditions and potential failure mechanisms

Stage 3

A detailed reliability test plan, with proper sample sizes and measured confidence limits, is in place.

Results are used for design changes and vendor evaluations.

Stages 3–4

Conduct reliability testing only when needed to resolve a question or provide information for a decision.

Design accelerated testing that is focused on specific failure mechanisms.

Expand discovery testing to include more stresses related to use conditions and to new vendors or materials under consideration.

Stage 4

Accelerated tests and supporting models are used. Testing to failure or destruct limits is conducted.

Stages 4–5

Use failure mechanism models to design reliability testing, and use test results to improve models.

Characterize the reliability of new vendor components or materials prior to use within a product design.

Stage 5

Test results are used to update component stress and damage models. New technologies are characterized.

5.4 Supply Chain Management

Many products consist of a combination of purchased components and materials assembled into a functional item. The reliability performance is significantly influenced by the reliable performance of the selected components and materials. Reliability is only one aspect of supplier selection, and the active involvement of reliability practitioners permits risk assessment, reliability requirements allocation, joint component reliability testing, and key vendor process-control enhancements. Furthermore, monitoring supplier impact of reliability performance, process variation, change notices, and end-of-manufacture notices permits active management of any effects on product reliability.

Stage 1

Selection is based on function and price.

Stages 1–2

Create approved parts and approved vendors lists (AVLs).

Create a vendor reliability assessment process for use with critical component vendors and new suppliers.

Use vendor data to qualify components for use within a product and environment.

Stage 2

The approved vendor list is maintained. Audits occur when there are issues with critical parts. Qualification is primarily based on vendor datasheets.

Stages 2–3

Include reliability requirements in design specifications and requests for quotes from vendors.

Include assessment information in management of AVLs.

Request and review field reliability performance from critical component vendors.

Evaluate vendor end of production or change notices on product reliability.

Stage 3

Assessments and audit results are used to update AVLs. Field data and failure analysis are related to specific vendors.

Stages 3–4

Create critical-to-reliability criteria for supplier process control and/or ongoing reliability evaluations.

Review reliability testing and failure mechanisms for those tests best performed by vendors (upstream or at the point of least value added).

Require vendors to evaluate the reliability programs of their suppliers.

Evaluate technology maturity and stability of vendor processes and components prior to vendor selection.

Stage 4

Vendor selection includes analysis of vendor's reliability data. Suppliers conduct assessments and audit of their suppliers.

Stages 4–5

Monitor for changes in product environment, use conditions, reliability requirements, or regulatory requirements for impact on product reliability.

Monitor critical-to-reliability parameters and process-control points across the supply chain to identify shifts.

Create contingency plans for possible obsolescence or shortage of parts.

Conduct joint studies with vendors to explore how processes, materials, and technology affect product reliability.

Stage 5

Changes in environment, use profile, or design trigger vendor reliability assessment. Component parameters and reliability are monitored for stability.

5.5   Feedback Process

Feedback provides a closed-loop process. It provides information on the effectiveness of design changes. Finding root cause of failures, understanding the performance and failure modes, and experimenting with improvement options all provide valuable information that enables product reliability improvement. Shortening the feedback process helps to limit field failure exposure. Including timely feedback during the development process helps the development team to avoid field failures.

5.5.1   Failure Data Tracking and Analysis

Each product failure highlights an area for product reliability improvement. Systematically recording, tracking, analyzing, and reporting failures from across the product lifecycle and supply chain permits comprehensive and timely information to be acquired. The product design team needs to understand, prioritize, and design products to minimize product failure. The entire business requires timely and accurate failure data so that decisions concerning, e.g., improvement projects, supplier selection, and warranty policies can be made.

Stage 1

Failures during function testing may be addressed.

Stages 1–2

Collect and report regularly factory yield and field failure data.

Use Pareto charts to identify improvement projects.

Conduct failure analysis and take corrective action for major failures.

Stage 2

Conduct a Pareto analysis of product returns and perform internal testing. Failure analysis relies on vendor support.

Stages 2–3

Collect and analyze failure data to guide component selection.

Revise reliability test plans based in part on field failure data by evaluating test coverage and value in preventing field failures.

Confirm the root cause of failures and the adequacy of product improvement to avoid the failure or to mitigate failure effects.

Collect and analyze time-to-failure information rather than failure counts or percentages.

Stage 3

Root cause analysis is used to update AVLs and prediction models. A summary of analysis results is disseminated.

Stages 3–4

Conduct failure analysis to find the root cause and update the design guidelines and reliability testing to prevent future occurrences.

Analyze failure data for systemic decision-making processes that enabled the failure to occur.

Create part batch, lot, or similar tracking systems.

Stage 4

The focus is on failure mechanisms. Failure distribution models are updated based on failure data.

Stages 4–5

Create links between customer satisfaction and product reliability.

Create a model for determining product reliability readiness for release based on development of a failure reporting and corrective action system (FRACAS) and other business requirements.

Create a prognostic data collection and analysis system within products and manufacturing equipment and processes.

Stage 5

The customer satisfaction relationship to product failures is understood. Use is made of prognostic methods to forestall failure.

5.6   Validation and Verification

This check step in most organizations consists of verifying that the reliability objectives have been met and that planned reliability activities have been completed. A cross-check can support individual results with consistent results from other reliability activities. The process is often part of the overall program management process.

Stage 1

Validation and verification are informal and based on individual cases rather than on a process.

Stages 1–2

Create a management review process for reliability plan implementation.

Compare field reliability data to requirements and predictions.

Create a system to validate the effectiveness of corrective actions.

Stage 2

Basic verification plans are followed. Field failure data are regularly reported.

Stages 2–3

Create a process to verify that supplier corrective actions have the expected effects on product reliability.

Compare stress screening and ongoing reliability testing to field failures and adjust these as needed.

Compare field failure modes to expected failure modes, and modify risk assessment practices to minimize the differences.

Stage 3

Supplier agreements concerning reliability are monitored. Failure modes are regularly monitored.

Stages 3–4

Assess reliability activities and their effectiveness to identify process improvements or best practices.

Verify that risk assessments are a closed-loop process and are updated as new information becomes available.

Compare field failure mechanisms with expected failure mechanisms and adjust risk assessment practices and reliability testing procedures to minimize the difference.

Stage 4

Internal reviews of reliability processes and tools are made regularly. Failure mechanisms are regularly monitored and used to update models and test methods.

Stages 4–5

Validate the use of field failure mechanisms data and analysis to update reliability models and design guidelines.

Create a process to verify the effectiveness of reliability strategy and policies.

Stage 5

Reliability predictions match observed field reliability.

5.7 Reliability Improvements

During this process, one aspires to identify and implement product changes that are designed to improve product reliability. The sources for improvement projects may come from reliability testing and analysis; product failures; customer requests; changes in the supply chain, use, or environmental conditions; or changes in technologies or materials. The implementation of corrective actions includes prioritization, validation of effectiveness, and prevention of the occurrence of similar failure modes or mechanisms.

Stage 1

Reliability improvement activities are nonexistent or informal.

Stages 1–2

Document all design and process changes and their anticipated impact on product reliability.

Implement all design and process changes to address customer complaints and field failures based on an eight disciplines (8D) problem-solving process (Rambaud 2006).

Review field failures for vendor connections and implement vendor improvements or exclude poorly performing vendors from the AVL.

Stage 2

Follow the design and process change process. Include internal and vendor engagement in the corrective action process.

Stages 2–3

Implement corrective actions to internally identified reliability testing failures.

Create a means to track and report corrective action effectiveness.

Create a lessons-learned process based on identified failure modes.

Stage 3

Track the effectiveness of corrective actions over time. Address the identified failure modes in other products. Identify improvement opportunities as environment and use profiles change.

Stages 3–4

Create a lessons-learned process based on identified failure mechanisms.

5.8   Explore a Means to Improve Reliability Predictions, Analysis, and Testing with More Effective or Efficient Techniques or a Combination of Techniques

Create a means to document the value of reliability activities and publish value determination guidelines.

Stage 4

Address the identified failure mechanisms in all products. Explore and adopt advanced modeling techniques. Establish a formal and effective lessons-learned process.

Stages 4–5

Evaluate new vendors, processes, and materials with the intent to improve product reliability.

Update design rules and guidelines based on product reliability performance.

Stage 5

Evaluate and adopt new technologies to improve reliability. Update design rules based on field failure analysis.

5.9 Management

The management team sets the tone for all aspects of an organization. The policies, practices, priorities, all convey the management team's placement of reliability's importance relative to the many priorities within the organization. How the management team's acts is more important than the slogans or official statements—where is the attention and follow up, where are the resources being directed, who is rewarded, and what garners personal involvement?

No Feedback from Field or Distributor

Imagine you are requested to assist a design team in determining how to best improve the reliability of a product. You learn that the organization produces a range of point of sale (POS) devices and they have invited you to a meeting with their staff to discuss the product and ways to improve the field reliability.

To help understand the situation, you may have already started to think of a set of questions whose answers would lead to suitable recommendations:

1. What is the current field failure rate?
2. What is the Pareto of field failure mechanisms or modes?
3. What is the desired level of field failures that is acceptable (i.e., the goal)?
4. How is the product designed with respect to reliability (i.e., design for reliability activities)?
5. What is the current estimate for field reliability based on internal measurements and modeling?
6. What happens when the product fails?
7. What do the failure analysis reports say about the possible causes of field failures?
8. Do field failures match internal testing results?

The meeting included directors of engineering, manufacturing, quality and procurement, and a handful of key engineers from those departments. They each provide a brief introduction to their products and reiterated the desire to improve field reliability. You start to ask the above questions in an attempt to understand the situation.

At first there is little provided by way of response from anyone on the team. Did you hit upon some trade secret? Were you showing your own ignorance by asking such questions?

No; they did not know how many or how the product failed in the field. They had made some assumptions about use, environment, and what could, maybe, possibly go wrong. They had little evidence of field problems. They had not even talked to anyone about the nature of the field issues.

The most interesting part of the product's design was the security feature that destroyed the memory and custom IC when the case was opened or sensed tampering. Destroyed was a pretty accurate description, given the physical damage to the components on the circuit board they showed you. Once the product is assembled and the security system activated, it was nearly impossible to disassemble and conduct a circuit analysis. This would make the field failures difficult to analyze.

Compound this "feature" with the relatively low cost of the device. These two factors lead to a replacement rather than repair strategy when addressing field failures. Furthermore, the failed units were destroyed as they were deemed to have no value for further study.

One other piece of information that pertains to this search for reliability improvements is that the organization has only one customer. Every unit they created went to one customer who bundled the POS device with inventory, payroll, building security, cash register, and various other elements that a small business may require to operate efficiently. The POS is only one piece of a larger kit. The service provides a single point of contact for training, installation, maintenance, and service and support.

The design team did component derating and worked closely with procurement, Q & R, and manufacturing to design as robust a product as they could under the cost and other design constraints. They did derating, qualified their vendors, and conducted a wide range of product testing under a wide range of stresses. They actually did a decent job in creating a reasonably reliable product.

It was just that they did not really know whether any of their assumptions and educated guesses were correct. They really did not know the use environment, the range of expected stresses, or even how often the devices were actually used. They did not know how to relate their internal product design and testing to what would occur with actual use.

Since any fielded unit was destroyed before any failure analysis could be conducted, they did not even have a count of how many failed for any reason nor did they have the basic information a Pareto of field failures would provide. They were blind to how the product actually performed. Also, this team had been producing POS devices for over 5 years and in terms of sales the devices were relatively successful.

Without even having a count of failures, how did they know they needed to improve the reliability? Was this a part-per-million improvement or a 20 % field failure rate problem attributable to first-year product introduction? No one really knew. They were told to make the product more reliable, but it was impacting the warranty costs.

Warranty costs are something tangible that you can analyze. How much were they paying in warranty? What was the cost per unit shipped of warranty? Again, no one had answers to these questions.

The Director of Engineering then spoke up and tried to explain the situation. Remember: They have one customer for all their products. Once a year, the Chief Financial Officer and the customer sit down to discuss pricing, warranty, and sales projections. It was the Chief Financial Officer (CFO) who asked for reliability improvements. It was also the CFO who, if he has the warranty and field failure information, was sharing it, as he considered it company-sensitive information. The CFO did not even talk about the magnitude of the field issues with anyone even in his office. He was not providing any information except to insist that they "make it better."

At this point, you likely would be rather frustrated and at a loss for what to recommend. Surely, no organization should be so blind as to how their product was performing.

After some thought and further discussion, you and the directors decide on two courses of action. First, you would go talk to the CFO and attempt to understand the field failure situation by explaining the importance of the information to the rest of the team. Second, the team would conduct a series of HALTs to attempt to understand the design's weaknesses. In parallel with this testing, an attempt would be made to fully characterize the use environment and use profiles by conducting surveys, field observations, and questionnaires. To effectively conduct HALTs, you need to know the types of stresses the product would experience. Any process operates better when there are a clear goal and a measure of performance. The comparison of the goal and measure provides the necessary feedback that enables design or process improvement.

5.9.1 Understanding and Attitude

Understanding and attitude reflect the level of the management team's comprehension of reliability engineering's role within the organization.
Stage 1
There is no comprehension of reliability as a management tool. Management tends to blame engineering for "reliability problems."
Stages 1–2
Create a basic awareness that product failures occur and can be avoided and that field failures cost the company money and cause customer dissatisfaction.
Create a basic report of the number of field failures and warranty expenses.
Provide training, encourage discussion, and promote learning opportunities for the management team related to basic reliability concepts and activities. Convey that all parts of the organization contribute to the actual product reliability.
Stage 2
Although reliability management is recognized as possibly being of value, the organization is not willing to provide money or time to make it happen.
Stages 2–3
Conduct informal training (e.g., perhaps during lunch sessions) on basic reliability topics and invite the management team to participate.
Highlight and train management in its role in vendor selection, design priorities, product testing, and failure analysis with respect to product reliability. Encourage and coach management team members to ask customers about the importance of product reliability.
Provide regular summary reports on product design progress toward reliability goals and field reliability performance.

Stage 3

Management is learning more about reliability management and is becoming supportive and helpful.

Stages 3–4

Provide the management team with talking points for key reliability program initiatives for use with customers and internal teams.

Provide value statements related to achievement in reliability improvements.

Create a significant element of the senior management bonus structure based on product reliability performance.

Discuss options for proactively addressing major reliability issues.

Develop detailed reliability models that provide a means to conduct "what if" experiments for various reliability activities.

Stage 4

Management personnel are actively participating, understanding the absolutes of reliability management, and recognizing their personal role in continuing emphasis.

Stages 4–5

Provide insights and mentoring concerning approaches to systematically preventing product failures.

Provide reliability reports on reliability predictions and the associated business impact on profit.

Discuss investment areas for product reliability improvements that impact product architecture, technology, and portfolio.

Stage 5

Management considers reliability management an essential part of the company's system.

5.9.2 Status

Within an organization, who are the leaders (independent of position)? What combination of voices tend to drive the company? Who is held in high esteem, rewarded, and promoted? The status of the reliability practitioner may range from nonexistent to esteemed. He or she might be perceived as an obstacle, a necessary part of doing business, a valued team member, or a thought leader. Do people want to become a reliability engineer because it is viewed as important and career enhancing? The status of those identified as reliability practitioners is one indicator of the value placed and found related to reliability engineering activities.

Stage 1

Reliability is hidden in manufacturing or engineering departments. Reliability testing is probably not done. The emphasis lies on initial product functionality.

Stages 1–2

Identify one or more reliability practitioners within the organization to assist in product design decision making.

Highlight individuals and the benefits of reliability-related activities.

Promote an individual to create and manage a reliability program.

Recognize the reliability professional's influence on and benefit to product design and manufacturing decisions.

Stage 2

A stronger reliability leader is appointed, yet the main emphasis remains on an audit of initial product functionality. Reliability testing is still not performed.

Stages 2–3

Invite key reliability practitioners to program and division decision meetings.

Promote a reliability practitioner to report directly to division management.

Recognize the reliability professional's influence on and benefit to product platform decisions.

Stage 3

The reliability manager reports to top management and plays a role in management of the division.

Stages 3–4

Invite key reliability practitioners to critical business and customer meetings.

Invite key managers to lead reliability programs and initiatives as part of a steering committee.

Invite reliability practitioners to early product concept development and major vendor-selection discussions.

Recognize and reward reliability improvement activities outside the ranks of identified reliability professionals.

Recognize the reliability professional's contribution to prevention of product failures.

Stage 4

The reliability manager is an officer of the company, responsible for reporting and preventive action and involved with consumer affairs.

Stages 4–5

Invite key reliability practitioners to provide input to business strategic planning.

Recognize the reliability professional's contribution to customer satisfaction and brand loyalty.

Stage 5

The reliability manager serves on the board of directors and is perceived as a thought leader. Prevention is the key aspect of the position.

5.9.3 Measured Cost of Unreliability

The language of business is money. What does the organization track and value and how is it expressed? The actual measures and their accuracy and relevance to decision making express the importance of product reliability within an organization.

Organizations that have competitors which are U.S. publicly traded companies are required to file information periodically with the Federal Trade Commission (FTC). Deciphering these FTC documents can be difficult and time consuming but there are useful resources to help sort through these (see e.g., Warranty Week online site and weekly newsletter).

Stage 1

The cost of unreliability is not measured.

Stages 1–2

Create a means to collect and report basic product reliability field performance.

Estimate the cost of a product return.

Estimate the cost of warranty at the individual product level.

Track and report the value of reliability activities.

Stage 2

Only direct warranty expenses are measured.

Stages 2–3

Create a means to track costs of failure analysis and re-engineering projects.

Estimate costs of repairs, maintenance, and replacement, along with associated costs.

Create a means to improve resolution (e.g., increase operating hours, determine the root cause of failure, evaluate environmental conditions, etc.) of product reliability field performance reports.

Establish consistent cost calculations and reporting mechanisms within the organization.

Stage 3

Warranty, corrective action materials, and engineering costs are monitored.

Stages 3–4

Establish a means to estimate the return on investment of individual reliability tasks.

Create a means to calculate the cost to the customer for each product failure.

Calculate the cost of product ownership over the entire product lifecycle.

Stage 4

Customer and lifecycle unreliability costs are determined and tracked.

Stages 4–5

Calculate the influence of product reliability improvements on increased sales and brand loyalty (customer satisfaction or net promoter indices).

Calculate the brand value related to product reliability perception or performance

Stage 5

Lifecycle cost reduction is accomplished through product reliability improvements.

6 Summary

Every product will cease to operate at some point. The management of when and how these failures occur will either hearten or dash perceptions. Focusing on improving the overall organization's ability to achieve reliability objectives involves identifying and improving the organization's reliability maturity.

Assessing your organization's maturity using the matrix and working to improve the weak areas and enhancing the strong areas permits your organization to reliably produce reliable products. Creating products that meet both your customer's expectations and do so in a cost-effective manner is a worthy business goal.

Reliability management may have an individual tasked with the role to oversee product reliability. A critical element to understand is that every part of the organization plays a role in reliability management. An individual may help to influence the reliability program, yet it takes the entire organization to perform with enlightenment.

References

Crosby, P. B. (1979). *Quality is free: the art of making quality certain*. New York: Signet.

Hobbs, G. K. (2000). *Accelerated reliability engineering: HALT and HASS*. Chichester: Wiley.

Institute of Electrical and Electronics Engineers. (1990). *IEEE standard computer dictionary: a compilation of IEEE standard computer glossaries*. New York: IEEE.

IEEE Std. 1624–2008. (2008). IEEE *Standard for Organizational Reliability Capability*. New York: IEEE.

Petroski, H. (1994a). *Design paradigms: Case histories of error and judgment in engineering*. Cambridge: Cambridge University Press.

Rambaud, L. (2006). *8D Structured problem solving: A guide to creating high quality 8D reports*. PHRED Solutions.

Schenkelberg, F. (2012). Investment in reliability program versus return—How to decide. In *Reliability and Maintainability Symposium (RAMS), 2012 Proceedings* (pp. 1–5).

West, M. (2004). *Real Process Improvement Using the CMMI*. Auerbach Publications.

Further Reading

O'Connor, P. D. T., & Kleyner, A. (2012). *Practical reliability engineering*. Chicester: Wiley.

Tobias, P. A., & Trindade, D. C. (2012). *Applied reliability*. Boca Raton: CRC/Taylor & Francis.

Meeker, W. Q., & Escobar, L. A. (1998). *Statistical methods for reliability data*. New York: Wiley.

Lalli, V. R. (2000). *Reliability and maintainability (RAM) training*, National Aeronautics and Space Administration, Glenn Research Center.

Crosby, P. B. (1979). *Quality is free: The art of making quality certain*. New York: Signet.

Petroski, H. (1994b). *Design paradigms: Case historyes of error and judgement in engineering*. Cambridge: Cambridge University Press.

Kapur, K. C., IEEE, & E, EI. (2002). The future of reliability engineering as a profession. In Annual Reliability and Maintainability Symposium, 2002 Proceedings (pp. 434–435).

Design for Reliability and Its Application in Automotive Industry

Guangbin Yang

1 Introduction

The global competitive business environment has placed great pressure on manufacturers to deliver products with more features and higher reliability in a shorter time and at a lower cost. Reliability, time to market, and cost are three critical factors that determine if a product is successful in the marketplace. Customers have demanding expectations for reliability because it affects the safety, availability, and ownership cost of the product. It is not surprising that reliability is an important factor that drives customer's decision to purchase a product, especially a large-capital product such as the automobile. As such, manufacturers have been spending every effort to improve reliability.

The conventional approach to reliability improvement relies heavily on test to precipitate failure modes, followed by identification and elimination of the root causes. To grow reliability to an acceptable level, the test-fix-test process often requires multiple repetitions. Obviously, this reactive approach is inefficient in terms of time and cost, and thus can no longer find a place in the current global competition. In fact, it is being replaced by the design for reliability (DFR) process. By this process, a number of reliability tasks are orchestrated to proactively build reliability into products in the product planning, design and development, and validation phases. The focus of DFR should be placed in the design and development of the product life cycle, where the inherent reliability is established. Once a design is released for manufacturing, production for reliability (PFR) process is applied to minimize production process variation to assure that the process does not appreciably degrade the inherent reliability. After the product is deployed to the field, maintenance for reliability (MFR) process is initiated to alleviate performance degradation and prolong product life. The DFR, PFR, and MFR should constitute

G. Yang (✉)
Chrysler Group, Auburn Hills, Michigan, USA
e-mail: hoang84pham@gmail.com

© Springer-Verlag London 2016   353
H. Pham (ed.), *Quality and Reliability Management and Its Applications*,
Springer Series in Reliability Engineering, DOI 10.1007/978-1-4471-6778-5_12

an effective and efficient reliability program, which is implemented throughout the product life cycle.

DFR has generated extensive interest in research and application due to its effectiveness at improving reliability. Several books contain chapters that present DFR process and techniques. For example, Yang (2007) describes development and implementation of effective reliability programs that include DFR as an important constituent. O'Connor and Kleyner (2012) include a chapter on DFR, which describes the authors' DFR process. Crowe and Feinberg (2001) present some reliability techniques commonly used for DFR. Bauer (2010) delineates reliability and robust design, error detection, and DFR case study for information and computer-based systems. In the literature, there have been numerous papers that deal with application of DFR techniques to design components and systems. Examples include De Souza et al. (2007), Lu et al. (2009), and Popovic et al. (2012).

This chapter describes effective DFR process and techniques, and integration of DFR into the product life cycle. The application of DFR to automotive design is discussed in detail. A practical example is presented to illustrate how DFR improves reliability and robustness.

2 DFR in the Product Life Cycle

DFR applies a number of reliability tasks, which are optimally sequenced and integrated with engineering activities of the product life cycle. This section describes phases of the product life cycle, and implementation of DFR in the product design and development phase.

2.1 Phases of the Product Life Cycle

Product life cycle refers to the sequential phases including product planning, design and development, design verification and process validation, production, field deployment, and disposal. The tasks in each phase are described concisely as follows (Yang 2007).

2.1.1 Product Planning Phase

Product planning is the first phase of the product life cycle. The task in this phase is to identify customer expectations, analyze business trends and market competition, and develop product proposals. In the beginning of this phase, a cross-functional team should be established representing different functions within the organization including marketing, financing, research, design, testing, manufacturing, service, and other roles. In this phase, the team is chartered to conduct a number of tasks including business trends analysis, understanding customer expectations, competitive analysis, and market projection. If the initial planning justifies further development of the product, the team shall outline the benefits of the product to customers, determine product features, establish product performances, develop product proposals, and set the time to market and the timelines for completion of tasks such as design, validation, and production.

2.1.2 Design and Development Phase

In this phase, we define the detailed product specifications concerning reliability, features, functionalities, economics, ergonomics, and legality. The specifications shall meet the requirements developed in the product planning phase, ensure the product to satisfy customer expectations, comply with governmental regulations, and embed a strong competitiveness in the marketplace. Once specifications are developed, the next step is to perform the concept design. In this step, we first design a functional structure, which describes the flow of energy and information, and physical interactions. The requirements regarding these functions are cascaded from the product specifications. Functional block diagrams are always useful in this step. Once the architecture design is completed, physical conception begins to fulfill functions of each subsystem. This step benefits from the use of advanced design techniques such as TRIZ and axiom design (Yang and El-Haik 2008), and may result in innovation in technology. Concept design is a fundamental stage in which reliability, robustness, feature, cost, weight, and other competitive potentials are largely determined.

When concept design is completed, we begin to develop detailed design specifications that ensure the subsystem requirements are satisfied. Then the physical details are devised to fulfill the functions of each subsystem within the product. The details may include physical linkage, electrical connection, as well as nominal values and tolerances of functional parameters. In this step, materials and components are also selected.

2.1.3 Design Verification and Process Validation Phase

This phase includes design verification (DV) and process validation (PV). Once a design is successfully completed, a small number of prototypes are built for DV testing to prove that the design achieves the functional, environmental, reliability, regulatory, and other requirements concerning the product. The DV test plan must be carefully developed, which specifies the test conditions, sample sizes, acceptance criteria, test operation procedures, and others. The test conditions should reflect the real-world usage that the product will encounter when used in the field. A large sample size for DV is often unaffordable; however, it should be large enough for test results to be statistically valid. An effective tool for developing DV test plans is the robustness checklist described in Sect. 4.2.5. If functional nonconformance or failure occurs, the root causes must be identified for potential design changes. The redesign must undergo DV testing until all acceptance criteria are completely met.

Once a design passes the DV testing, a pilot production may begin. Then a number of samples are subjected to PV testing. Its purpose is to validate the capability of the production process. The process shall not degrade the inherent reliability to an unacceptable level, and be capable of manufacturing the products that meet all specifications with a minimum variation. By this step, the process has been set up and is intended for volume production. Thus the test units represent the products that customers will see in the marketplace. In other words, the samples and the final products are not differentiable, because both use the same materials, components, production processes, and process monitoring and measuring techniques. The PV test conditions and acceptance criteria are the same as those for DV testing.

2.1.4 Production Phase

A full capacity production may commence after the PV testing is successfully completed. This phase includes a series of interrelated activities such as materials handling, production of parts, assembly, and quality control and management. The end products are subject to final test and then shipped to customers.

2.1.5 Field Deployment Phase

In this phase, products are sold to customers and realize their values. This phase involves marketing advertisement, sales service, technical support, warranty service, field performance monitor, and continuous improvement.

2.1.6 Disposal

This is the terminal phase of a product in the life cycle. A product is discarded, scraped, or recycled when it is unable to or not cost-effective to continue service. A non-repairable product is discarded once it fails. A repairable product may be discarded because it is not worth of repair. The service of some repairable products is discontinued because the performance does not meet the customer demands. The manufacturer shall provide technical supports to dispose of, dismantle, and recycle the product in order to minimize the associated costs and the adverse impacts on the environment.

2.2 Integration of DFR into the Product Life Cycle

The product life cycle can be loosely divided into three stages: design, production, and deployment. As stated earlier, DFR is applied in the design stage, while PFR and MFR are implemented in the production stage and the field deployment stage, respectively, as shown in Fig. 1. The reliability tasks of DFR should be integrated with the engineering activities in the design phase of the product life cycle.

In the product planning phase, a cross-functional reliability team is organized to understand customer expectations, translate the expectations to engineering requirements, set a competitive and feasible reliability target, and conceive and evaluate product proposals from the reliability point of view. The reliability decisions made in this phase have tremendous implications on cost, time to market, and competitiveness. For example, setting an overambitious reliability target would incur unaffordable design and development costs, and thus jeopardizes the competitive advantages. Conversely, a pessimistic reliability target certainly undermines the competitiveness by simply losing customers.

DFR plays a critical role in the design and development phase. The reliability tasks add more values to the product in this phase than in any other phase.

Fig. 1 Integration of DFR, PFR, and MFR into the product life cycle

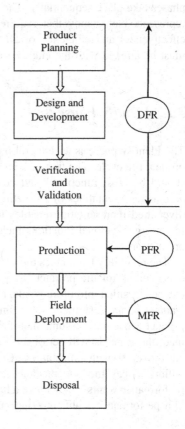

The objective of reliability tasks in this phase is to design-in reliability, while designing-out potential failure modes. The proactive reliability tasks are aimed at designing the things right at the first time. Doing so would cut off the loop of test-fix-test process, which was a typical design model in the old days and unfortunately still finds applications nowadays.

In the design verification and process validation phase, DFR is intended to demonstrate reliability in an economic manner, and to detect potential failure modes. In particular, in the DV stage, reliability verification test is performed to demonstrate that the design meets the reliability requirements. In the PV stage, the test is intended to prove the capability of the production process. The process must be capable of manufacturing the final products that meet the specified reliability target.

3 DFR Process and Techniques

DFR is essentially a process that drives the achievement of the specified reliability target in an economic manner. The DFR process can be divided into five phases: (1) Identify, (2) Characterize, (3) Design, (4) Analyze, and (5) Validate. The five phases take place sequentially. The phases may be iterative; however, any iteration represents inefficiency that requires additional time and cost. To maximize the effectiveness and efficiency of DFSS process, the reliability tasks in each phase must be aligned with the engineering activities.

3.1 Identify Phase

The Identify phase is an integral part of the product planning in the context of the product life cycle. The first task in this phase is to develop a reliability team, which should be cross-functional and comprised of reliability and quality engineers as well as representatives of marketing, design, testing, production, and service. This diversified representation enables the DFR initiatives to be supported across different organizational functions including engineering design, analysis, testing, and production.

The reliability team takes the responsibility of collecting and analyzing customer expectations for the product being planned. Customer expectations for a product can be classified into basic wants, performance wants, and excitement wants. Basic wants describe customers' most fundamental expectations for the functionality of a product. Customers assume that the needs will be automatically satisfied. Failure to meet these needs will seriously dissatisfy the customers. Performance wants are customers' spoken expectations. Customers usually speak out for the needs and are willing to pay more to meet the expectations. A product that better satisfies the performance wants will achieve a higher degree of customer satisfaction. Reliability is a performance want; meeting customer expectation for reliability significantly increases customer satisfaction. Excitement wants represent potential needs whose satisfaction will surprise and delight customers. The customer expectations should be translated into engineering design requirements that drive the design, test, production, and service in the product life cycle. The translation can be accomplished using the quality deployment function (QFD) technique. The QFD analysis also reveals the relationship between reliability and design variables, and thus provides useful information for effectively designing reliability into products.

In the Identify phase, we need to understand and model how customers will operate the products being planned. The customer operation can be characterized by the real-world usage profile, which defines the operational frequency, load, and environment. The profile is also known as the stress distribution, shown in Fig. 2, where $f(S)$ is the probability density function of stress S. The product should be designed to endure a high-percentile stress S_0 to satisfy a large percentage of customers. For example, the 95th percentile is often chosen in the automotive industry. Then this stress level should be used to specify reliability.

An important task in the Identify phase is to specify a competitive and feasible reliability target. In general, there are three methods commonly used to set a reliability target: meeting customer satisfaction, achieving warranty cost objective, and minimizing total life cycle cost.

In the competitive business environment, satisfying customers means sustaining and gaining market share. Customers are usually dissatisfied when product performance degrades appreciably. The degree of dissatisfaction increases with degradation. For products whose performance degrades over time, a failure is said to have occurred if the performance characteristic (y) exceeds a threshold (G). For a the-larger-the-better characteristic, the degree of customer dissatisfaction at time t is the probability of $y \leq G$, shown in Fig. 3. The probability at time t (say, design life) is often specified from customer surveys, benchmarking, or other methods. Then the reliability $R(t)$ of the product is

$$R(t) = 1 - \Pr[y(t) \leq G]. \tag{1}$$

In the product planning stage, some manufacturers set a warranty cost objective. This objective can be translated into a reliability target. Suppose that the planned sales volume is n, the warranty time is t_0, the average cost per repair is c_0, and the maximum allowable warranty cost is C_0. If the product is subject to a minimal

Fig. 2 Design stress specified at S_0 of the real-world usage profile

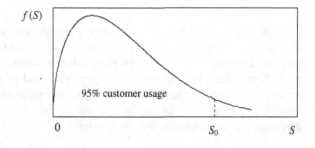

Fig. 3 Customer
dissatisfaction representing
failure probability

$$R(t) = 1 - \Pr[y(t) \leq G].$$

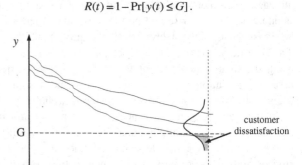

repair, meaning the failure rate after repair is equal to that right before repair, the expected warranty cost C_w of the n units is

$$C_w = c_0 n \ln\left[\frac{1}{R(t_0)}\right].$$ (2)

Then the reliability target at the warranty time is

$$R(t_0) = \exp\left(-\frac{C_w}{c_0 n}\right).$$ (3)

In some applications, the reliability target may be derived from the total life cycle cost, which includes all costs incurred in product planning, design, development, test, production, maintenance, and disposal (Asiedu and Gu 1998). These cost elements are closely related to reliability (Seger 1983). Minimizing the total life cost results in an optimal reliability target. In some situations, the cost elements are difficult to model, and a quantitative reliability target cannot be attained. Nevertheless, the principle of minimizing total life cycle cost is applicable and useful in justifying a reliability target.

3.2 Characterize Phase

The Characterize phase takes place during the product design and development. The purpose of this phase is to characterize the product from the reliability perspective. In particular, we describe the product function, the interaction within the product, and the interaction between the product and other systems and the environment. As a result, the factors that impact the reliability of the product will be identified. The purpose of this phase can be achieved by developing boundary diagram, interface analysis, and P-diagram.

3.2.1 Boundary Diagram

A product may consist of several subsystems, and interacts with other systems and the environment. A boundary diagram is used to illustrate the boundary of the product or system. In particular, it shows the flow of energy, material, and information between the subsystems within the product; and between the product and the environment, and the components and subsystems of other systems. Figure 4 shows a generic boundary diagram, where the one-directional arrow represents a one-way interaction, and the two-directional arrow means a two-way interaction. Defining the boundary of a product is the process of identifying the signals to the product, the outputs from the product, and the noise sources disturbing the function of the product. Therefore, a boundary diagram provides useful information to the subsequent creation of P-diagram and interface analysis. In addition, a boundary diagram is a valuable input to the failure mode and effects analysis (FMEA).

3.2.2 Interface Analysis

After a boundary diagram is developed, an interface analysis should be performed. The purpose of this analysis is to describe how the subsystems within a product interface with each other, and how the product interfaces with the environment and other systems. There are four types of interface: (1) physical contact, (2) energy transfer, (3) information transfer, and (4) material transfer. An interface between two units may be necessary for the product to perform its intended function. Such an interface is desired, and should be made reliable. For example, electrical energy is a required input from power supply (other system) to electronic module (product). On the other hand, an interface between two units may do harm to the product. Examples include noise and vibration generated by a car running on the

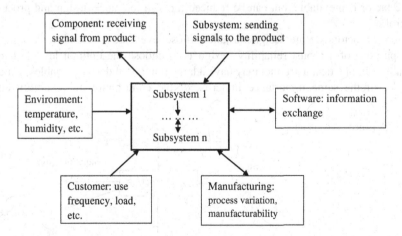

Fig. 4 A boundary diagram of a typical product

road. Such an interface should be avoided, minimized, or managed in the product design process. An example of interface analysis is described in Sect. 4.2.3.

3.2.3 P-Diagram

A P-diagram is usually created at the end of the Characterize phase. The diagram graphically illustrates the inputs (signals), outputs (intended function or responses), control factors, noise factors, and failure modes of the product. Figure 5 shows a generic P-diagram, where the noise factors, signals, and functional responses may be carried over from a boundary diagram. A P-diagram contains the necessary information for subsequent robust reliability design. An example of P-diagram is given in Sect. 4.2.4. The major elements of a P-diagram are described below.

Signals are the inputs from the customers or other systems, subsystems, or components to the product. The product then transforms the signals into the functional responses, and of course, the failure modes. Signals are essential for fulfilling the function of a product. For example, steering angle is a signal to the steering system of an automobile. The steering system then transfers the angle into a turn of the vehicle.

Noise Factors are the variables that have adverse effects on robustness and are impossible or impractical to control. Generally, there are three types of noise factors: (1) internal noise, (2) external noise, and (3) unit-to-unit noise. Internal noise is the performance degradation or deterioration as a result of product aging. For example, abrasion is an internal noise of the automobile steering system. External noise is the operating conditions that disturb the functions of a product. It includes the environmental stresses such as the temperature, humidity and vibration, as well as the operating load. For instance, road condition is an external noise factor disturbing the steering system. Unit-to-unit noise is the variation in performance, dimension, and geometry due to an imperfect material or production process. This noise factor is inevitable, but can be reduced through tolerance design and process control.

Control factors are the design variables whose levels are specified by designers. The purpose of a robust reliability design is to choose the optimal levels of the variables. In practice, a product may have a large number of design variables, which are not of the same importance in terms of the contribution to reliability and

Fig. 5 A generic P-diagram

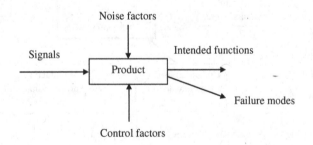

robustness. Often only the key ones are included in the P-diagram. These factors are identified by using engineering judgment, analytical study, preliminary test, or historical data analysis.

Intended functions are the functionalities that a product is intended to perform. The functions are impacted by the signals, noise factors, and control factors. The noise factors and control factors influence both the average value and variability of the functional responses, whereas the signals determine the average value and not the variability.

3.3 Design Phase

The Design phase occurs in the product design and development stage and before prototypes are created. In this phase, reliability tasks are integrated with the engineering activities to build reliability into products, and to prevent potential failure modes from occurrence. For example, reliability prediction is conducted to support selection of design options; robust reliability design is performed to select the optimal levels of product design variables. The reliability techniques commonly applied in this phase are described below.

3.3.1 Robust Reliability Design

A failure is attributed to either the lack of robustness or the presence of mistakes induced in design or production. The purpose of robust reliability design is to build reliability and robustness into products in the design stage through the parameter design. This technique uses the P-diagram created earlier.

Parameter design aims at minimizing the performance sensitivity of a product to noise factors by setting its design variables at the optimal levels. In this step, designed experiments, which often involve accelerated testing, are usually conducted to investigate the relationships between the design variables and quality characteristic of the product. The quality characteristic is the performance characteristic if the product fails gradually. The quality characteristic is life or reliability if the failure mode is catastrophic. In any situation, once attaining the relationships, we can determine the optimal setting of the design parameters. Yang (2007) discusses this technique in detail. An application example is presented in Sect. 5.

3.3.2 Reliability Modeling

This technique is to develop reliability model according to the architecture of the product. The architecture lays out the logic connections of components, which may be in series, parallel, or more complex configurations. The product reliability is expressed as a function of component reliabilities. The relation is useful in reliability allocation, prediction, and analysis.

3.3.3 Reliability Allocation

The reliability target established in the product planning stage should be appropriately apportioned to subsystems, modules, and components of the product. The allocated reliability to a lower level becomes the reliability target of that level. The commonly used reliability allocation methods include the equal allocation technique, ARINC method, AGREE technique, and optimal allocation methodology.

3.3.4 Reliability Prediction

In the early design stage, it is frequently desirable to predict reliability for comparing design alternatives and components, identifying potential design issues, determining if a design meets the allocated reliability target, and predicting reliability performance in the field. Several methods are often employed for prediction in this stage. Part count and part stress analysis for electronic equipment well documented in MIL-HDBK-217 was a prevailing approach until the mid-1990s. The approach assumes that components are exponentially distributed (with a constant failure rate) and a system is in logic series of the components. In addition to the assumptions, the part stress analysis overemphases temperature effects and overlooks other stresses such as thermal cycling and transient conditions, which are the primary failure causes of many systems. A more recent prediction methodology, known as the 217Plus, includes component-level reliability prediction models and a process for assessment of system reliability due to non-component variables such as software and process.

Another approach to reliability prediction in the early design stage is modeling system reliability as a function of component reliabilities based on the product configuration, which was described earlier. This approach requires component reliability data, which may be determined from historical test data, warranty data, or other sources.

3.3.5 Stress Derating

This task is to enhance reliability by reducing stresses, which may be applied to a component, to levels below the specified limits. When implemented in an electronic design as it often is, derating technique lowers electrical stress and temperature versus the rated maximum ones. This alleviates parameter variation and degradation, and increases the long-term reliability.

3.4 Analyze Phase

As described earlier, a failure is attributed to either a lack of robustness or the presence of mistakes induced in design or production. The objective of Analyze phase is to identify and eliminate the mistakes that might have been embedded into the design. This phase takes place before the prototypes are built. The most commonly used techniques include the failure mode and effects analysis (FMEA), fault tree analysis (FTA), and design analysis.

3.4.1 FMEA

FMEA is a proactive tool for discovering and correcting design mistakes through the analysis of potential failure modes, effects and mechanisms, followed by the recommendation of corrective actions. It may be described as a systemized group of activities intended to recognize and evaluate the potential failure of a product and its effects, identify actions which could eliminate or reduce the likelihood of the potential failure occurrence, and document the process (SAE 2009). Essentially, FMEA is a bottom-up process consisting of a series of steps. It begins with identifying the failure mode at the lowest level (e.g., component), and works its way up to determine the effects at the highest level (e.g., end customer). The process involves the inductive approach to consider how a low-level failure can lead to one or more effects at high level.

FMEA has been extensively used in various private industries. The overwhelming implementation lies in the automotive sector, where the FMEA has been standardized as SAE J1739. Although originated in the automotive industry, the standard is prevalent in other sectors of industry. Another influential standard is the IEC 60812, which finds its application mainly in electrical engineering. In the defense industry, failure mode, effects, and criticality analysis (FMECA) is more common and performed by following MIL-STD-1629A. FMECA is similar to FMEA except that each potential failure effect is classified according to its severity.

Performing FMEA often begins with the lowest level component within the product. For each component, its functions and respective failure criteria must be completely defined. The next step is to identify the failure modes of the component. This is followed by revealing the effects of the failure mode and evaluating the severity of each associated effect. For the failure mode under consideration, the responsible failure mechanisms and their occurrences are determined. The subsequent step is to develop the control plans that help obviate or detect the failure mechanisms, modes, or effects. The effectiveness of each plan is evaluated by detection ranking. Then the next step is to assess the overall risk of a failure mode. The overall risk is measured by the Risk Priority Number (RPN), which is the product of severity, occurrence, and detection. A high value of RPN indicates a high risk of the failure mode, and requires corrective actions. In many applications, a high severity value warrants a review and possible design change regardless of RPN value.

3.4.2 FTA

FMEA identifies the failure modes that have high level of severity. For each of these failure modes, it is necessary to understand the failure paths (e.g., single-point failure) and the root causes, and then to determine if corrective actions are required. FTA is an effective technique for the analysis. A fault tree is a graphical representation of logical relationships between failure events, where the top event is logically branched into contributing events through cause-and-effect analysis. Obviously, FTA is a top-down process. The steps of performing FTA are summarized below.

- Define the boundary of the product, assumptions, and failure criteria. The interactions between the product and neighbors including the human interface should be fully analyzed in order to take account of all potential failure causes in the FTA. For this purpose the boundary diagram is helpful.
- Understand the hierarchical structure of the product and functional relationships between subsystems and components. A block diagram representing the product function is often instrumental for this purpose.
- Identify and prioritize the top-level fault events of the product. When FTA is performed in conjunction with FMEA, the top events should include the failure modes that have high severity values. A separate fault tree is needed for a selected top event.
- Construct a fault tree for a selected top event using gate symbols and logics. Identify all possible causes leading to the occurrence of the top event. These causes can be considered as the intermediate effects.
- List all possible causes that can result in the intermediate effects and expand the fault tree accordingly. Continue the identification of all possible causes at lower level until all possible root causes are determined.
- Once the fault tree is completed, perform analysis of it in order to understand the cause-and-effect logic and interrelationships among the fault paths.
- Determine whether corrective actions are required. If necessary, develop measures to eradicate fault paths or minimize the probability of fault occurrence.
- Document the analysis and then follow up to ensure that the proposed corrective actions have been implemented. Update the analysis whenever a design change takes place.

3.4.3 Design Analysis

When a design is completed and before prototyped, the design should be analyzed to identify potential design deficiencies. The common analyses include mechanical stress analysis, thermal analysis, vibration analysis, electromagnetic compatibility analysis, geometric dimensioning and tolerancing (GD&T) analysis, and others. The analyses employ analytical approaches and often require complicated mathematical models. As a result, computer simulation is usually used. Readers are referred to, for example, Yang (2007), Sergent and Krum (1998), Petyt (2010), for details.

3.5 Validate Phase

In the product design and development phase, various reliability techniques are applied to proactively build reliability into product. The next phase is to validate that the product has achieved the specified reliability. The reliability validation can be divided into two stages, which are aligned with the design verification (DV) and process validation (PV). In the DV stage, reliability validation is to prove that the design has attained the reliability required through testing prototypes. In the PV stage, reliability validation is to demonstrate that the production process is capable of manufacturing products that meet the reliability required. In the two stages, reliability validation test methods are the same; the commonly used methods are bogey test, test to failure, and degradation test.

Bogey testing is to test a sample of predetermined size for a certain period of time. Required reliability is verified if no failures occur in the testing. This type of test is simple to implement; it does not require failure monitoring and performance measurement during testing. Thus it is widely applied by the commercial industry including the automotive sector. However, this test method is inefficient because it requires a large sample size and a long test time. For example, to demonstrate at 95 % confidence level that a product has 95 % reliability at 1 million cycles, we need to test 59 samples, each run to 1 million cycles. In addition, this test method does not yield data for reliability estimation. Yang (2009a, 2010) improves the efficiency of bogey test using degradation measurements when the product life is Weibull or lognormal.

Test-to-failure test method is to test most samples to failure. The test may take a longer time; however, it requires fewer samples and generates considerably more information. The test is usually conducted at higher stress levels. Using the failure data and an appropriate acceleration model, we can estimate the reliability at the design stress level. If the lower bound of one-sided confidence interval for reliability is greater than the reliability required, then we conclude that the product meets the reliability requirement.

For some products whose performance characteristic degrades over time, a failure is said to have occurred if the performance characteristic reaches a specified threshold. For such products, we can measure the performance characteristic during testing. The measurement data can be used to compute the reliability at a given time and its confidence interval. The test can be terminated even if zero failures occur. Apparently, this test method is efficient.

4 Automotive Design for Reliability

Automobiles are expensive and represent a major expenditure for most customers. It is not surprising that customers are cautious and picky in their decision-making; they always choose the products that work best for them and cost least. The primary factors that usually drive customers' decision include reliability, price, feature, warranty, and fuel economy. Manufactures must compete on these factors. On the other hand, automobiles are complicated systems, which affect human safety and the environment. As such, the governmental agencies have imposed a number of stringent regulations on safety, emission, and fuel economy, which the manufacturers must comply with. It is worth noting that reliability is not only a sales point, but it affects other factors such as warranty, safety, emission, and price. To survive and grow in such a competitive environment, automotive manufactures have spent every effort to improve reliability.

4.1 Automotive DFR Process

The traditional reliability process of test-fix-test is not effective in the automotive industry. In recent years, the design for reliability process thrives, which has proven to yield a shorter time to market, lower development cost, higher reliability, and increased customer satisfaction. A typical DFR process in the automotive industry is illustrated in Fig. 6, where CAE standards for computer-aided engineering and is

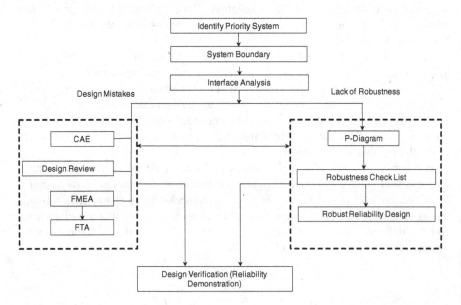

Fig. 6 Automotive DFR process

a tool for design analysis. Many techniques shown in Fig. 6 have been described earlier. This process is based on the fact that a failure is due to either a mistake or lack of robustness. In Fig. 6, the branch on the left-hand side is intended to eliminate the design mistakes, while the right-hand side is aimed at building reliability and robustness into the product in the design and development phase. The tasks in the two branches are not independent; for example, P-diagram is a useful input to FMEA. In addition, the tasks in the process can iterate. However, any repetition represents inefficiency. Doing right things right at the first place is the essence of DFR. Once the two branches are successfully executed, the design is subject to verification, including the reliability demonstration.

4.2 DFR Implementation

This subsection explains application of the primary techniques shown in Fig. 6.

4.2.1 Identification of Priority System

An automobile is a complicated product, which consists of body system, powertrain system, electrical system, and chassis system. Each of these systems can be further broken into subsystems, and then to modules. Practically, the DFR process cannot be implemented on all systems, subsystems, and modules (hereafter, products). As shown in Fig. 6, the first step in the DFR process is to identify a product that requires DFR. The selected product should be the one that contains new technologies, is an old design to be used in a new environment, or has a high failure probability in the past.

4.2.2 Product Boundary Diagram

The product identified in the previous step may have multiple components, and interfaces with neighboring products, the environment, and driver. Therefore, a boundary diagram is created to define the boundary of product under study. Figure 7 shows the boundary diagram of an automotive catalytic converter assembly (Ford Motor Company 2004), which converts toxic chemicals in the exhaust of an internal combustion engine into less toxic substances. The diagram illustrates the components within the converter, and the systems out of the study scope but interact with the converter. This diagram provides useful inputs to subsequent tasks such as interface analysis, P-diagram development, and FMEA.

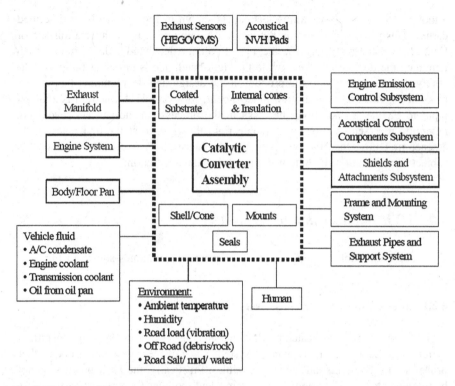

Fig. 7 Boundary diagram of catalytic converter

4.2.3 Interface Matrix

Interface analysis yields an interface matrix. Figure 8 shows an interface matrix for the catalytic converter (Ford Motor Company 2004). In this example, interfaces are divided into five categories. An interface necessary for intended functions is assigned +2, while an interface that must be prevented or avoided to achieve the functions is given −2. If an interface is beneficial but not absolutely necessary for functionality, it receives +1. An interface is assigned −1 if it causes negative effects but does not prevent functionality. If an interface does not affect functionality, it is given 0. The design should make positive interface reliable and minimize the effects of negative interfaces.

4.2.4 P-Diagram of Catalytic Converter

P-diagram is described in Sect. 3.2.3. As an example, the P-diagram of the catalytic converter is shown in Fig. 9 (Ford Motor Company 2004). In developing automotive P-diagrams, the noise factors are often divided into five categories: (1) piece-to-piece variation, (2) change in dimension, performance, or strength over

	Shell/Cone - Catalytic Converter	Seals - Catalytic Converter	Coated Substrate - Catalytic Converter	Mounts - Catalytic Converter	Internal Cones & Insulation - Catalytic Converter	Environment	Exhaust Manifold	Engine Emission Control Subsystem	Acoustical NVH Pads
Shell/Cone - Catalytic Converter		2		-1 -1	2		-1 2		-2 -2
Seals - Catalytic Converter	2		2 -1	2 -1	2		-1	-1	-1 -1
Coated Substrate - Catalytic Converter		2 -1		2 -1			-1	2 2 2	
Mounts - Catalytic Converter	-1 -1	2 -1	2 -1				-1		-2 -2
Internal Cones & Insulation - Catalytic Converter	2	2					-1		-1 -1
Environment	-1	-1	-1	-1	-1				
Exhaust Manifold	2		-1	2					
Engine Emission Control Subsystem			2 2						
Acoustical NVH Pads	-2 -2	-1 -1		-2 -2	-1 -1				

P E P: Physically touching   E: Energy transfer
I M I: Information exchange   M: Material exchange

Fig. 8 Interface matrix of catalytic converter

time and mileage, (3) customer usage and duty cycle, (4) external environment, including climate and road conditions, and (5) system interactions. Although the last three types belong to the external noise described in Sect. 3.2.3, the itemization is helpful in identifying all relevant noise factors.

4.2.5 Robustness Checklist

From a P-diagram, a robustness checklist can be created, which lists the noise factors, error states (failure modes), intended functions (response). More important, it identifies what noise factors cause each failure mode and disturb each intended function.

The robustness checklist also identifies an appropriate method for management of each critical noise factor. The selected method should be effective and least expensive. In general, there are six methods for noise effect management.

- Change technology and system architecture. It is effective in alleviating the effects of internal and external noise factors.
- Apply parameter design. This method is inexpensive and effective. Section 5 presents an example.
- Upgrade design specifications. It is expensive and should be avoided if possible.

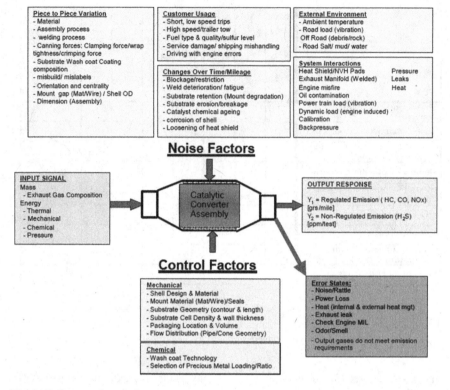

Fig. 9 P-diagram of catalytic converter

- Remove and reduce noise. This technique needs special design aimed at particular noise factors.
- Add compensation device. This method is passive, but useful in many applications. The effectiveness of the method depends on the reliability of the compensation device. Once the device fails, the noise effects will be active.
- Disguise or divert noise. This technique bypasses the noise to unimportant systems or environment.

An important function of the robustness checklist is to identify all appropriate DV and PV tests. Each test should include at least one of the noise factors as test stress, and produce at least one relevant failure mode (error state). Otherwise, the test should be eliminated. On the other hand, each and every failure mode should be covered by at least one test.

As an example, Fig. 10 shows a snapshot of the robustness checklist of the catalytic converter (Ford Motor Company 2004).

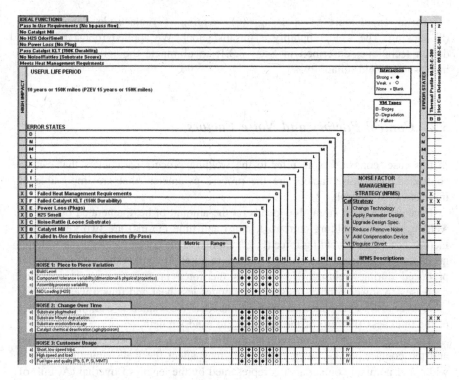

Fig. 10 Robustness checklist of catalytic converter

5 Robust Reliability of Weld

This section presents an example of design for reliability and robustness of weld (Yang 2009b).

5.1 Problem Statement

In many applications, welds are created by resistance welding. This technique bonds two pieces of metal materials together by melting the materials in the contact area. An elevated temperature is generated by contact resistance when electrical current flows through the contact area of the two materials (Zhang and Senkara 2005). Generally, a current is applied through two close electrodes. In application, however, it is necessary to set the two electrodes far apart because the contact area is physically unreachable. The need resulted in a new technique called the remote resistance welding. The far separation of the electrodes may generate a number of failure modes including: (1) weld overheat, (2) weak weld, (3) discoloration,

(4) metal expulsion, (5) sparking, (6) warping, (7) electrode damage, and (8) sticking. The weld reliability is unsatisfactory. We applied robust reliability design technique, described in Sect. 3.3.1, to eliminate the failure modes and increase the weld reliability.

5.2 Control Factors and Noise Factors

In this case, control factors refer to the welding process parameters that can be controlled, and noise factors are the process parameters that are impossible or impractical to control.

Because the weld strength depends on the amount of heat generated during the welding process, welding current (I, amp) and time (T, ms) are two key control factors. In addition, welding force (F, gf) applied to the contact area during welding needs to be optimized. A sufficient force is required to contain the molten material produced during welding. However, as the force increases, contact resistance decreases. A lower contact resistance generates less heat and thus a weaker weld.

A layer of Sn60 is plated on the base material. We wanted to examine the effect of the thickness (H, mil) of the layer on the weld strength.

The magnitude of contact area and the distance (D, inch) between the electrode and the contact point, shown in Fig. 11, may affect the weld strength. Given a width, the nominal contact area is represented by the length of overlap (A, mil) of the Sn60 layers. D and A are likely to vary in manufacturing; they are considered as noise factors.

The levels of the control factors and noise factors are shown in Table 1.

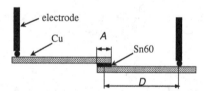

Fig. 11 Remote resistance welding

Table 1 Levels of control and noise factors

Control factor				Noise factor	
I	T	H	F	A	D
270	3	4	100	40	0.14
300	4	6	500	60	0.54
330	5	8			

5.3 Experimental Design and Analysis

Orthogonal arrays are used to design the experiments. It is expected that no interactions exist between the control factors. An $L_9(3^4)$ array is selected as an inner array to accommodate the control factors, and an $L_4(2^2)$ array is used as an outer array for the noise factors.

The experiments were conducted on workpieces of 2 mil thick, 30 mil wide, and 1.65 in. long, which represented the production intent. The experiments of each group were carried out according to the experimental layout. Each experiment was replicated four times. The welds were sheared, and the strength of each weld was measured. For each run, there were 16 measurements.

The robustness of the weld strength against the production variation can be measured by the signal-to-noise (SN) ratio. Therefore, it is used as an experimental response to analyze the factor effects. Because the strength is a the-larger-the-better characteristic, the SN ratio is

$$SN = -10 \log \left[\frac{1}{16} \sum_{i=1}^{16} \frac{1}{y_i^2} \right], \tag{4}$$

where y_i is the shear strength of sample i.

The SN ratios are calculated for each run using (4), and are summarized in Table 2. The analysis of variance (ANOVA) technique is used to identify significant factors. The analysis indicates that the welding current, time, and force have significant effects on robustness, while the thickness has almost no contributions. The optimal levels of welding current, time, and force are 330 apms, 5 ms, 100 gf, respectively. Because thickness is an insignificant factor, it is set to 4 mil to reduce the material cost.

5.4 Weld Reliability

To estimate the reliability of the weld, 20 samples welded at the optimal levels of the process parameters were tested under a thermal cycle profile. The low and high temperatures of the profile are −25 and 110 °C, at which the dwell times are 1.5 and 2.5 h, respectively. The test was terminated after 58 cycles, which is equivalent to the design life at the use stress level.

After the test, each weld was sheared. The maximum and minimum shear strengths are 827 and 338 gf, respectively. As shown in Fig. 12, the shear strength

Table 2 SN value for each run

Run #	1	2	3	4	5	6	7	8	9
SN Value	48.3	48.5	50.4	50.1	51.2	50.5	51.8	53.7	56.5

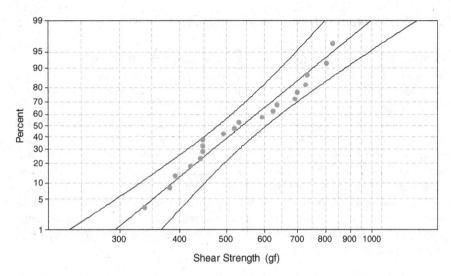

Fig. 12 Lognormal fit to the shear strength and the 95 % confidence intervals for the shear strength

can be adequately modeled with a lognormal distribution with scale parameter 6.293 and shape parameter 0.269. The 95 % confidence intervals for the scale parameter and the shape parameters are (6.178, 6.408) and (0.192, 0.357), respectively. When the shear strength of a weld is less than 250 gf, the weld is said to have failed. The reliability R_0 of the weld at the design life is

$$\hat{R}_0 = 1 - \Phi\left[\frac{\ln(250) - 6.293}{0.269}\right] = 0.9984,$$

where $\Phi(\cdot)$ is the cumulative distribution function of the standard normal distribution. The one-sided 95 % lower confidence bound for the reliability is 0.9819. That is, we have 95 % confidence to expect that the failure probability at the design life will not exceed 1.8 %. Therefore, the weld has a high reliability.

References

Asiedu, Y., & Gu, P. (1998). Product life cycle cost analysis: State of the art review. *International Journal of Production Research, 36*(4), 883–908.
Bauer, E. (2010). *Design for reliability: Information and computer-based systems*. Hoboken: Wiley-IEEE Press.
Crowe, D., & Feinberg, A. (Eds.). (2001). *Design for reliability*. NY: CRC.
De Souza, M. M., Fioravanti, P., Cao, G., & Hinchley, D. (2007). Design for reliability: The RF power LDMOSFET. *IEEE Transactions on Devices and Material Reliability, 7*(1), 162–174.

Ford Motor Company (2004). *Failure Mode and Effects Analysis: FMEA Handbook* (*with Robustness Linkages*), Dearborn, MI. http://cfile205.uf.daum.net/attach/170321414F3C726E2EFA29.

Lu, H., Bailey, C., & Yin, C. (2009). Design for reliability of power electronics modules. *Microelectronics Reliability, 49*(9–11), 1250–1255.

O'Connor, P. O., & Kleyner, A. (2012). *Practical reliability engineering* (5th ed.). Hoboken: Wiley.

Petyt, M. (2010). *Introduction to finite element vibration analysis* (2nd ed.). Cambridge: Cambridge University Press.

Popovic, P., Ivanovic, G., Mitrovic, R., & Subic, A. (2012). Design for reliability of a vehicle transmission system. *Proceedings of the Institution of Mechanical Engineers, Part D: Journal of Automobile Engineering, 226*(2), 194–209.

SAE (2009). *Potential Failure Mode and Effects Analysis in Design (Design FMEA), Potential Failure Mode and Effects Analysis in Manufacturing and Assembly Processes (Process FMEA), and Potential Failure Mode and Effects Analysis for Machinery (Machinery FMEA)*, SAE J1739, Society of Automotive Engineers, Warrendale. www.sae.org.

Seger, J. K. (1983). Reliability investment and life-cycle cost. *IEEE Transactions on Reliability, R-32*(3), 259–263.

Sergent, J. E., & Krum, A. (1998). *Thermal management handbook: For electronic assemblies.* NY: McGraw-Hill.

Yang, G. (2007). *Life cycle reliability engineering.* Hoboken: Wiley.

Yang, G. (2009a). Reliability demonstration through degradation bogey testing. *IEEE Transactions on Reliability, 58*(4), 604–610.

Yang, G. (2009b). Design for reliability and robustness: A case study. *International Journal of Reliability, Quality and Safety Engineering, 16*(5), 403–411.

Yang, G. (2010). Test time reduction through optimal degradation testing. *International Journal of Reliability, Quality and Safety Engineering, 17*(5), 495–503.

Yang, K., & El-Haik, B. (2008). *Design for six sigma: A roadmap for product development* (2nd ed.). NY: McGraw-Hill Professional.

Zhang, H., & Senkara, J. (2005). *Resistance welding: fundamentals and applications.* FL: CRC.

Product Durability/Reliability Design and Validation Based on Test Data Analysis

Zhigang Wei, Limin Luo, Fulun Yang, Burt Lin and Dmitri Konson

1 Introduction

Better quality leads to less waste, improved competitiveness, higher customer satisfaction, higher sales and revenues, and eventually higher profitability. Meeting the quality and performance goals requires that decisions be based on reliable tests and quantitative test data analysis. Statistical process control (SPC) is a fundamental quantitative approach to quality control and improvement. In 1920s and 1930s pioneered the use of statistical methods as a tool to manage and the control production. Walter Shewhart, William Edwards Deming was also a strong advocate of SPC but could not convince US companies until around 1980, and then the original equipment manufactures (OEMs) in the US began to adopt his approaches requiring their suppliers to show statistical evidence of the quality of their products. Six sigma, originally initiated by Motorola in 1980s, is also the widely used as a quantitative and statistical tool for quality control. The term "statistical" simply means organizing, describing, and drawing conclusions from data with statistics methods. A major difference between the around "old-style" approach and the statistics-based quality control and management is that, in the past, quality control was product driven Montgomery 2009. Inspectors would measure critical dimensions carefully, and either scrap or rework the parts that did not conform. Although this practice resulted in good quality of the final product, it was wasteful and did not lead to improvements in overall quality, cost reduction, and productivity Montgomery 2009. Statistical methods provide engineers and managers with the tools needed to quantify variation, and identify causes, and find solutions to reduce or remove unwanted variation (Pham 2006).

Durability and reliability performance is one of the most important concerns of almost all engineering systems (Lee et al. 2005; Yang 2007). Modern durability and

Z. Wei (✉) · L. Luo · F. Yang · B. Lin · D. Konson
Tenneco Inc., Grass Lake, MI, USA
e-mail: ZWei@Tenneco.com

© Springer-Verlag London 2016   379
H. Pham (ed.), *Quality and Reliability Management and Its Applications*,
Springer Series in Reliability Engineering, DOI 10.1007/978-1-4471-6778-5_13

reliability analysis of products is essentially based on SPC approach and the accurate interpretation of the test data. To ensure the accuracy in durability/reliability assessment, sophisticated and efficient testing and analysis methods are required to obtain statistically sound results and conclusions from test data. Testing and statistical analysis can be complicated by many factors such as heteroscedasticity (unequal variance), unknown distribution, multiple failure modes, censored data, nonlinearity, etc. Additionally, the time and efforts involved in product validation, can be very expensive and usually only a small number of samples are available for testing. Therefore, it is highly desirable to use methods which can be applied to data from a small number of samples while maintaining adequate accuracy.

In this chapter, the most recent practices in product durability/reliability design and validation is reviewed. Several new concepts, approaches, and procedures recently developed, mainly by the authors, are also introduced. The focus of this Chapter is on the following five related aspects, which are essentially the backbones of all a durability/reliability analysis and design methods: (1) failure mechanisms and modes, (2) linear data analysis, (3) design curve construction, (4) Bayesian statistics for sample size reduction, and (5) accelerated testing. These approaches can serve as a practical guide for product design and validation engineers in their test planning and data analysis.

2 Failure Mechanisms and Modes

Failure mode and effect analysis (FMEA) is often the first step and a core task in durability/reliability engineering, safety engineering, and quality engineering. It involves reviewing parts to identify failure modes, their causes and effects. It is widely used in development and manufacturing industries in various phases of a product life cycle. Failure probability can only be estimated or reduced by first understanding the failure mechanisms and failure modes.

Failure mechanisms of components/systems depend on materials, loading condition, and operating environment. For automotive exhaust systems, the most common failure mechanisms are fatigue and corrosion (SAE 1997). Other mechanisms, such as creep, oxidation, erosion, wear, or some combination, are responsible for the remaining failures. Fatigue is essentially a cycle-dependent failure mechanism caused by engine vibration, road condition, thermal cycling, etc. Corrosion, creep, and oxidation are basically time-dependent failure phenomena. Creep and oxidation of metals are usually the issues for components such as auto manifolds operated at high temperature. Corrosion in auto exhaust systems can be caused by salt, condensate, urea, and other corrosive agents. Figure 1a, b shows the examples of fatigue and thermal fatigue cracks observed in tests for exhaust systems development. Figure 2a–c shows, respectively, the corrosion features in a muffler, a cross-sectional view of a corrosion pit, and a 3-D profile of a corrosion pit.

Increasingly stringent government emission regulations and the need for fuel economy drive vehicle exhaust systems toward increased engine efficiency, reduced

(a)   (b)

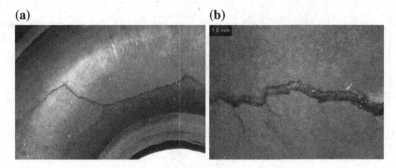

Fig. 1 Fatigue crack **a** in a component subjected to step stresses at intermediate temperature; **b** in a component under high-temperature thermal cycling

(a)   (b)   (c)

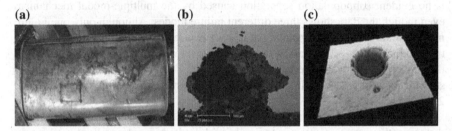

Fig. 2 **a** Pitting corrosion in a muffler, **b** profile of a corrosion pit, **c** 3-D profile of a corrosion pit

weight, and advanced aftertreatment strategies. This requires materials to perform under higher temperature (up to 1000 °C or more), more severe mechanical loading, and potentially, in a more aggressive environment.

Fatigue failure is a probabilistic process and the cycles to failure usually show a big scatter band. The uncertainty of the cycles to failure of vehicle exhaust components and systems comes from many sources, including material uncertainty, loading uncertainty, and the uncertainty of the initial damage distribution. Additionally, the failure mechanisms of a component may be caused by a single failure mechanism or simultaneously by multiple mechanisms, such as fatigue, creep, and oxidation mechanisms. For a single failure mechanism, several failure modes (failure locations) can occur in a component. Years of experience have shown that the prevailing failure modes in exhaust systems are cracked welds at joints between pipes and muffler/resonator/converter (360° welds), cracked hanger-to-pipe welds (line welds), and broken hanger rods (Lin 2011). In testing, several failure modes can be activated or suppressed, depending on the specific geometry and loading condition.

Figure 3a–c shows two-stress level fatigue S-N test data with a single failure mode, two failure modes (Failure mode-A and Failure mode-B), and three failure modes (Failure mode-A, Failure mode-B, and Failure mode-C), respectively (Wei et al. 2012c). It should be noted that for the data shown in Fig. 3b, failure mode-A

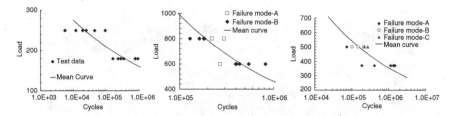

Fig. 3 Fatigue *S-N* test data with **a** a single failure mode, **b** two failure modes, and **c** three failure modes

appears in both stress levels: on left side at the lower stress level and on the right side at the higher stress level. For test data shown in Fig. 3c, the Failure mode-A, mode-B, and mode-C appear on the higher stress level of the data. However, there is no evident subpopulation separation caused by the multiple-modal mechanism even though the data shows three different failure modes. Although only one failure mode (failure mode-A) is operating at lower stress level, further analysis shows that the goodness-of-fit of the data at lower stress level is worse than that of the data at the higher stress level in terms of the Anderson-Darling (AD) statistic. The AD statistic measures how well the data follow a particular distribution (Wei et al. 2012c).

Testings of auto exhaust components and systems are usually expensive and, therefore, the test sample size is typically very limited in order to keep a relatively low budget. For tests with small sample size, a multimodal distribution cannot be clearly distinguished as shown in Fig. 3, and therefore, any multiple-modal behavior is often ignored. However, with more test data, multiple failure modes can be revealed for some materials. Examples with bi-modal failure modes are given in Fig. 4a, b, in which two groups of fatigue data can be clearly separated, indicating two populations with two distinct failure modes (surface and subsurface) (Cashman 2007). The possible physics behind the separation of mechanisms has been discussed and it is suggested that the separation behaviors may be related to the development of different heterogeneity levels in materials (Jha et al. 2009).

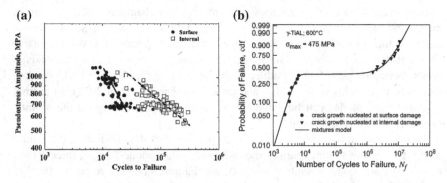

Fig. 4 **a** Fatigue *S-N* curve of René 95 with bi-modal failure modes (Cashman 2007); **b** bi-modal distribution of fatigue life at a stress level for γ—TiAl alloy (Harlow 2011)

3 Test Data Analysis

Data analysis capability is vital to successful durability/reliability engineering designs. Life data, such as fatigue S-N data, can be divided into several types such as complete data, censored data, and multiple censored data. Data analysis of complete life data is the main focus of this chapter. Oftentimes, with proper data transformation and linearization, test data can be curve/surface fitted using a probabilistic distribution function to gain physical understanding and quantitative description. The basic characteristics of life data and the associated probabilistic distribution include mean, scatter, homoscedasticity and heteroscedasticity, skewness (symmetry), kurtosis, entropy (uncertainty), etc. (Neter et al. 1990). The following four fundamental aspects of data analysis will be discussed here: probabilistic distribution function, the equilibrium linear method for curve/surface fitting, the design curve construction method, and the Bayesian statistics interpretation.

3.1 Probabilistic Density Function

One of the most important steps in statistic analysis of test data is to use a probabilistic density function (PDF) to fit and to interpret the test data. Test data can be empirically fitted using several PDFs, which theoretically, may be suitable to specific failure mechanisms. The commonly used continuous PDFs for durability and reliability data analysis are normal, lognormal (two or three parameters), Weibull (two or three parameters), and extreme value distribution functions. The preference of one PDF over another has to be determined by test data correlation. The goodness-of-fit with different distributions can be evaluated and compared using methods such as Kolmogorov–Smirnov (KS) and Anderson–Darling (AD) (Neter et al. 1990). The AD statistic measures how well the data follow a particular distribution, especially in the tails of the distribution. The better the distribution fits the data, the smaller this AD statistic will be.

3.1.1 Commonly Used Probabilistic Density Function for Fatigue
   S-N Data

The lognormal and Weibull PDFs will be briefly described below because they are two of the most commonly used PDFs for fatigue reliability analysis. The three-parameter lognormal PDF is shown in Eq. (1):

$$f(x) = \frac{1}{\sigma(x-\delta)\sqrt{2\pi}} \exp\left[-\frac{1}{2}\left(\frac{\log(x-\delta)-\mu}{\sigma}\right)^2\right]; \quad \sigma > 0, -\infty < x < \infty, x > \delta$$

(1)

μ is the mean, σ is the standard deviation, and δ is threshold or shift parameter. When $\delta = 0$, Eq. (1) is the two-parameter lognormal PDF. If $\log(x)$ is further replaced with x, then we have a normal distribution: $f(x) = \frac{1}{\sigma\sqrt{2\pi}} \exp\left[-\frac{1}{2}\left(\frac{x-\mu}{\sigma}\right)^2\right]$.

The three-parameter Weibull PDF is shown in Eq. (2)

$$f(x) = \frac{\beta}{\eta}\left(\frac{x-\gamma}{\eta}\right)^{\beta-1} \exp\left[-\left(\frac{x-\gamma}{\eta}\right)^{\beta}\right]; \quad x \geq 0, \eta > 0, \beta > 0, x > \gamma \quad (2)$$

η, β, and γ are scale, shape, and location or shift or threshold parameters, respectively. When $\gamma = 0$, the three-parameter Weibull functions are the two-parameter Weibull distributions. The threshold parameters δ and γ give the lower bounds of the PDFs, explicitly indicating the existence of a physical threshold value. As the name implies, the threshold parameters locate the PDF along the abscissa (cycles to failure for the durability data). Changing the values of δ and γ has the effect of "sliding" the PDF to the right because values of δ and γ must be positive. The Weibull distribution function has several different physical implications depending on the value of β: early mortality rate ($\beta < 1$), constant mortality rate ($\beta = 1$), and decreasing mortality rate ($\beta > 1$). Furthermore, the Weibull distribution can be reduced to an exponential distribution function when $\beta = 1$ and to the Raleigh distribution function when $\beta = 2$. For $\beta = 3.2$ the Weibull distribution is very similar to the normal distribution.

The parameters of a distribution function can be estimated using several methods, among which the least square method (LS) and the maximum-likelihood method (ML) are the two most commonly used (Neter et al. 1990). The basic idea of the least square method is to find the parameters, e.g., θ_j for the expected best fit curve by minimizing the sum of the squares of residuals: $R^2(\theta_j) = \sum \left[y_i - f(x_i, \theta_j)\right]^2$ with $\partial(R^2)/\partial\theta_i = 0$. By contrast, the maximum-likelihood method finds the parameters that maximize the likelihood function $L = \prod_{j=1}^{N} f(\theta_j)$, e.g., by setting the partial derivative of the likelihood function to zero: $\partial\text{Log}(L)/\partial\theta_j = 0$. Two examples are provided below.

The likelihood function of the normal distribution is shown in Eq. (3) below

$$L(x_1, \ldots x_n, \mu, \sigma) = \prod_{i=1}^{n} \frac{1}{\sigma\sqrt{2\pi}} \exp\left[-\frac{1}{2}\left(\frac{x_i - \mu}{\sigma}\right)^2\right]$$

(3)

By taking logarithms of Eq. (3), then differentiating it with respect to μ and σ, and equating it to zero: $\partial \ln L / \partial \mu = 0$ and $\partial \ln L / \partial \sigma = 0$, we have the following solutions shown in Eq. (4) below

$$
\begin{cases}
\mu = \frac{1}{n} \sum_{i=1}^{n} x_i \\
\sigma = \left[\frac{1}{n} \sum_{i=1}^{n} (x_i - \mu)^2 \right]^{1/2}
\end{cases}
\tag{4}
$$

where n is the sample number and x_i are the values of data point i. It should be noted that Eq. (4) can be also obtained using the least square method.

The Weibull parameters can also be derived using the maximum-likelihood method for complete data. The likelihood function of the two-parameter Weibull distribution function is shown in Eq. (5) below

$$
L(x_1, \ldots x_n, \eta, \beta) = \prod_{i=1}^{n} \frac{\beta}{\eta} \left(\frac{x_i}{\eta} \right)^{\beta-1} \exp \left[-\left(\frac{x_i}{\eta} \right)^{\beta} \right]
\tag{5}
$$

After taking logarithms of Eq. (5), differentiating it with respect to η and β, we can get the following estimating equations shown in Eq. (6) below

$$
\begin{cases}
\frac{\sum_{i=1}^{n} x_i^{\beta} \ln x_i}{\sum_{i=1}^{n} x_i^{\beta}} - \frac{1}{\beta} - \frac{1}{n} \sum_{i=1}^{n} \ln x_i = 0 \\
\eta = \left(\frac{1}{n} \sum_{i=1}^{n} x_i^{\beta} \right)^{1/\beta}
\end{cases}
\tag{6}
$$

The first formula of Eq. (6) can be estimated using the standard iterative procedures such as the Newton–Raphson method and the bisection method. Once β is found, η can be easily calculated. The formula of η in Eq. (6) is the generalized mean, also known as the power mean or HÖlder mean, which is an abstraction of the quadratic ($\beta = 2$), arithmetic ($\beta = 1$), geometric ($\beta \to 0$), and harmonic means ($\beta = -1$).

3.1.2 Uncertainty and Confidence Interval of the Estimated Parameters

The formulae of confidence intervals for the estimated mean and standard deviation of normal (lognormal) PDF, and the scale and shape parameters of Weibull distributions for a given confidence level and sample size are already available.
Confidence intervals of estimated normal (lognormal)
If the standard deviation σ is known in advance, the confidence intervals of the mean can be calculated as

$$\left(\hat{\mu} - z\frac{\sigma}{\sqrt{n}} < \mu < \hat{\mu} + z\frac{\sigma}{\sqrt{n}} \right) \qquad (7)$$

where $z = F^{-1}(F(z))$ is the inverse cumulative normal function and can be obtained for a given failure probability of F. $\hat{\mu}$ is the estimated mean from test data.

If the standard deviation σ is unknown, the confidence intervals of the mean can be estimated using Student's t-distribution with the estimated sample standard deviation $\hat{\sigma}$.

$$\left(\hat{\mu} - c\frac{\hat{\sigma}}{\sqrt{n}} < \mu < \hat{\mu} + c\frac{\hat{\sigma}}{\sqrt{n}} \right) \qquad (8)$$

c can be calculated from $\Pr(-c \leq T \leq c)$ with the Student's t-distribution.

The confidence interval of the standard deviation σ can be estimated with the following formula based on the obtained $\hat{\sigma}$:

$$\left(\sqrt{\frac{(n-1)\hat{\sigma}^2}{\chi^2_{\alpha/2,n-1}}} < \sigma < \sqrt{\frac{(n-1)\hat{\sigma}^2}{\chi^2_{1-\alpha/2,n-1}}} \right) \qquad (9)$$

where $\chi^2_{,}$ is the Chi-squared distribution.

Confidence intervals of estimated Weibull

The likelihood ratio (LR) method is based on the Chi-squared distribution assumption and it is generally suitable for small sample size. Likelihood ratio bounds are calculated using the likelihood function as follows (Meeker and Escobar 1998):

$$-2\ln\left[\frac{L(\theta_1, \theta_2)}{L\left(\hat{\theta}_1, \hat{\theta}_2 \right)} \right] \geq \chi^2_{\alpha;k} \qquad (10)$$

$L()$ is the likelihood function; $\alpha = 1 - CL$ with CL as the confidence level; $\chi^2_{\alpha;k}$ is the Chi-squared with k degrees of freedom. The task now is to find the values of the parameters so that Eq. (10) is satisfied. For the two-parameter Weibull distribution, the first step is to calculate the likelihood function for the parameters estimates.

$$L\left(\hat{\theta}_1, \hat{\theta}_2 \right) = \prod_{i=1}^{N} f\left(x_i, \hat{\beta}, \hat{\eta} \right) = \prod_{i=1}^{n} \frac{\hat{\beta}}{\hat{\eta}} \left(\frac{x_i}{\hat{\eta}} \right)^{\hat{\beta}-1} \exp\left[-\left(\frac{x_i}{\hat{\eta}} \right)^2 \right] \qquad (11)$$

Equation (10) can be rearranged and an equivalent form is shown in Eq. (12)

$$L(\beta, \eta) \geq L\left(\overset{\wedge}{\beta}, \overset{\wedge}{\eta}\right) \cdot \exp\left[-\frac{\chi^2_{\alpha;1}}{2}\right] \qquad (12)$$

The Chi-squared statistics can be calculated at any confidence levels. For example, the value of the Chi-squared statistics is $\chi^2_{\alpha;1} = 2.705543$ for one-sided confidence level of 90 %. The next step is to find the set of values of β and η that satisfy Eq. (12). The solution is an iterative process that requires setting the value of β and finding the appropriate values of η, and vice versa and these values must be estimated numerically.

Example 1: Confidence intervals of estimated normal (lognormal) parameters from test data

A set of fatigue test data shown in Table 1 is analyzed. The test consists of six data points. Assuming that the fatigue data follows the lognormal distribution, estimate the mean $\overset{\wedge}{\mu}$, standard deviation $\overset{\wedge}{\sigma}$ and the corresponding confidence limits.

Solution:

The confidence intervals at 90 % can be calculated using Eqs. (7) and (8) and are listed in Table 2. In calculating the confidence limits of the mean, the critical value $z^* = 1.645$ is used. Clearly, for an estimated parameter, the true value of the parameter can be located in a wide range, which is dependent on sample size and confidence level.

Example 2: Confidence intervals of estimated Weibull parameters from test data

The same dataset listed in Table 1 is analyzed using the LR plots method to find the bound on the parameters.

Solution:

The confidence limits, calculated from Eq. (12), for each parameter are listed in Table 3. The probability plot and the Weibull parameter contour plots for this dataset are shown in Fig. 5a, b. Since each axis represents the possible values of a given parameter, the boundaries of the contour shown in Fig. 5b represent the

Table 1 Fatigue test data in terms of cycles to failure

	Data-1	Data-2	Data-3	Data-4	Data-5	Data-6
Test	186,555	28,769	25,646	19,650	1608	35,955

Table 2 The estimated lognormal parameters for fatigue test data shown in Table 1

	$\overset{\wedge}{\mu}$ (log)	90 % confidence limits of μ		$\overset{\wedge}{\sigma}$ (log)	90 % confidence limits of σ	
		Lower limit	Upper limit		Lower limit	Upper limit
Test	4.3657	3.9187	4.8127	0.6656	0.4473	1.3907

Table 3 The estimated Weibull parameters for fatigue test data shown in Table 1

	$\hat{\beta}$	90 % confidence limits of β		$\hat{\eta}$	90 % confidence limits of η	
		Lower limit	Upper limit		Lower limit	Upper limit
Test	0.8356	0.39	1.26	44,873	13,000	142,000

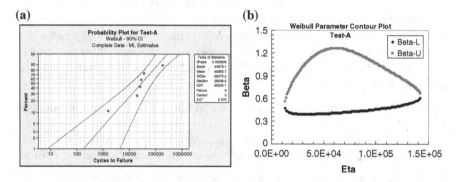

(a)   **(b)**

Fig. 5 Probability plot and the confidence intervals obtained for the scale parameter and the shape parameter of the Weibull distribution

extreme values of the parameters that satisfy Eq. (12). It should be noted that the maximum and minimum points of β do not necessarily correspond with the maximum and minimum η points. The lowest (lower limit) and highest (upper limit) calculated values of β can be determined by finding the maximum peak and the minimum valley.

Example 3: Probabilistic distribution of thermal fatigue test data

The V-shape specimen setup has been developed to simulate manifolds fixed to the engine. The specimen is fixed on both sides creating an initially stress-free condition and then cyclically heated using resistance heating with the maximum temperature zone located at the center of the specimen, Fig. 6a. The specimens are defined as "failed" when the specimen separated into two pieces after a certain cycle N_f. Figure 6b shows the tested specimens, in which total separation occurred. Similar to other life test data, V-shape specimen test data always contain an inherent scatter. Therefore, probabilistic approaches have to be used to interpret the test data in order to implement the observations into new product designs. Two sets of thermal fatigue test data, stainless steel (SS) 409 (ferritic) with 25 data points, and SS309 (austenitic) with 23 data points from V-shape testing, are used for probabilistic analysis. The goal is try to find a proper distribution to fit the test data.

Solution:

Two- and three-parameter lognormal and Weibull PDFs are used and compared (Wei et al. 2013b). Figure 7a, b shows the probability plots for these two datasets using MiniTab (Ryan et al. 1985) for two-parameter lognormal and two-parameter Weibull cumulative distribution functions (CDFs), respectively. The values of fit

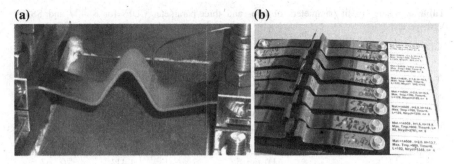

Fig. 6 V-shape specimen for thermal fatigue resistance testing, **a** V-shape test configuration and heating; **b** tested specimens

parameters for these two sets of test data are listed in Table 4. For all of these two steels, the AD value of the lognormal CDF is smaller than that of Weibull CDF, indicating that the lognormal CDF gives a better data correlation because it produces a lower AD value. From all of the data plots shown in Fig. 7a, b, it can be clearly seen that the data does not fall on a straight line. Therefore, a more sophisticated CDF which can capture the major control parameters is needed to give a better data correlation. Further examination reveals that the major feature of these

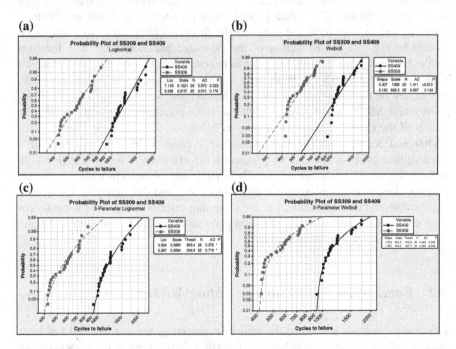

Fig. 7 Probability plots of **a** two-parameter lognormal CDF, **b** two-parameter Weibull CDF, **c** three-parameter lognormal CDF, **d** three-parameter Weibull CDF for the test data

Table 4 Values of fit parameters of two- and three-parameter CDFs for SS409 and SS309 V-shape test

CDFs	Parameters	Materials	
		SS409	SS309
Two-parameter Lognormal	Mean (μ)	7.105	6.338
	Standard deviation (σ)	0.182	0.213
	AD statistic	0.870	0.510
Two-parameter Weibull	Shape (β)	5.307	5.120
	Scale (η)	1339.0	628.3
	AD statistic	1.411	0.567
Three-parameter Lognormal	Mean (μ)	5.824	5.267
	Standard deviation (σ)	0.589	0.5984
	Threshold (δ)	838.4	349.9
	AD statistic	0.372	0.719
Three-parameter Weibull	Shape (β)	1.378	1.302
	Scale (η)	352.3	184.0
	Threshold (γ)	916.6	407.7
	AD statistic	0.449	0.646

curves is the concave downward trend, and theoretically, subtraction of a positive threshold δ or γ in a three-parameter CDF function can improve the data fit. Based on the characteristics of the data pattern, three-parameter lognormal and Weibull distributions are used to fit the data and the fit curves obtained from MinTab are plotted in Fig. 7c, d. The values of the estimated parameters are also listed in Table 4. Clearly, the overall fits are much better than that of the two-parameter CDFs, especially for SS409. With the help of the three-parameter lognormal CDF, the values of the threshold parameters obtained are $\delta = 838.4$ and $\delta = 349$ for SS409 and SS309, respectively. With the three-parameter lognormal CDF, the values of the threshold parameters are $\gamma = 916.4$ and $\gamma = 407.7$, respectively, for SS409 and SS309 with the three-parameter Weibull CDF. The existence of a positive threshold parameter δ or γ indicates the existence of a physical threshold value below which no failure occurs. These parameters δ and γ provide an estimate of the earliest time-to-failure of the units under test, and they must be less than or equal to the first time-to-failure, i.e., the minimum extreme value. This is consistent with the observations that crack initiation plays an important role in thermal fatigue life of the V-shape specimen testing.

3.2 Equilibrium Curve/Surface Fitting Method

Linear data analysis is commonly used in engineering applications and has been standardized, e.g., ASTM standard (ASTM 1962, 2010). Data fit parameters can be obtained using the least squares (LS) method or the maximum-likelihood estimations (MLE). Some conventions are made in the ASTM standard for metallic fatigue (ASTM 2010). The stress range is defined as the independent variable and the cycle to failure N_f is considered as the dependent variable. It is also recommended in the ASTM standard that the cycle to failure N_f be plotted on abscissa, while the stress range S is plotted on ordinate. Therefore, the horizontal offset method is eventually used in the ASTM standard to evaluate the variation of cycles to failure N_f. Vertical and perpendicular offsets methods are also commonly used in engineering application. For a given set of test data with small scatter band, the difference of the results from the three methods may be negligible. However, for data with a big scatter band, the difference of the results from these three methods may be significant. Therefore, a guideline must be given for properly selecting a method. For this purpose, a new equilibrium, based data analysis method which is analogous to the equilibrium of force and angular moment in the classic mechanics has been recently developed (Wei et al. 2013d). The formulae derived from the equilibrium method are exactly the same as that obtained from the LS method. However, based on the equilibrium concept, the identification of data pattern is added as an indispensable pre-processing procedure before any data analysis can be conducted as well as a post-processing tool to examine the goodness-of-fit.

3.2.1 Equilibrium Curve Fitting Method

With the equilibrium method, three "standard patterns" have been identified for curve fitting: (1) vertical pattern, Fig. 8a, (2) horizontal pattern, Fig. 8b, and (3) perpendicular pattern, Fig. 8c, have been identified for curve fitting. Similarly, two "standard patterns," i.e., vertical pattern, Fig. 9a, perpendicular pattern, Fig. 9b, have been identified for surface fitting. For datasets with these patterns, based on the equilibrium mechanism, the best fit curve/surface must be the middle curve (line)/middle surface (plane), so that the data can be symmetrically distributed around the expected curve/surface and balanced to make sure that the net 'force' and 'moment' are zero as required from the equilibrium principle. Therefore, the best mean curve fit can be guaranteed for these ideal data patterns.

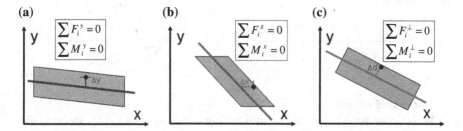

Fig. 8 Standard data patterns **a** vertical, **b** horizontal and, **c** perpendicular for curve fitting

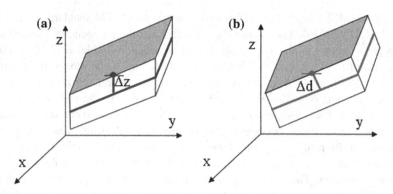

Fig. 9 Vertical and perpendicular offsets directions for surface fitting

Vertical offsets

As shown in Fig. 8a, F_i^y $(i = 1, \ldots n)$ is written in terms of 'force' from a point (x_i, y_i) to the expected curve $y = a + bx$ along the vertical direction. For a point i, F_i^y is derived as

$$F_i^y = y_i - (a + bx_i) \tag{13}$$

The total force equilibrium is then shown in Eq. (14).

$$\sum_{i=1}^{n} F_i^y = \sum_{i=1}^{n} [y_i - (a + bx_i)] = 0 \tag{14}$$

The total moment balance, with arms along x direction, of all these data around a point (x_c, y_c), which could be the centroid of the points system, is shown in Eq. (15)

$$\sum_{i=1}^{n} M_i^y = \sum_{i=1}^{n} \{[y_i - (a + bx_i)](x_i - x_c)\} = \sum_{i=1}^{n} \{[y_i - (a + bx_i)]x_i\} = 0 \tag{15}$$

In deriving Eq. (15), the terms with x_c is canceled out because of Eq. (14). The final solution is

$$a = \frac{\sum_{i=1}^{n} y_i \sum_{i=1}^{n} x_i^2 - \sum_{i=1}^{n} x_i \sum_{i=1}^{n} x_i y_i}{n \sum_{i=1}^{n} x_i^2 - \left(\sum_{i=1}^{n} x_i\right)^2} \quad b = \frac{n \sum_{i=1}^{n} x_i y_i - \sum_{i=1}^{n} x_i \sum_{i=1}^{n} y_i}{n \sum_{i=1}^{n} x_i^2 - \left(\sum_{i=1}^{n} x_i\right)^2}$$

$$\tag{16}$$

Horizontal offsets

In the same way, the equations of fitted curve can be obtained by applying the equilibrium concept along the x-direction, Fig. 8b.

$$a = \frac{\sum_{i=1}^{n} y_i \sum_{i=1}^{n} x_i y_i - \sum_{i=1}^{n} x_i \sum_{i=1}^{n} y_i^2}{n \sum_{i=1}^{n} x_i y_i - \sum_{i=1}^{n} x_i \sum_{i=1}^{n} y_i} \quad b = \frac{n \sum_{i=1}^{n} y_i^2 - \left(\sum_{i=1}^{n} y_i\right)^2}{n \sum_{i=1}^{n} x_i y_i - \sum_{i=1}^{n} x_i \sum_{i=1}^{n} y_i}$$

$$(17)$$

Perpendicular offsets

Similarly, the equation of a fitted curve based on perpendicular offsets is shown in Eq. (18) below

$$a = \frac{1}{n}\left[\sum_{i=1}^{n} y_i - b \sum_{i=1}^{n} x_i\right] \quad b = \frac{-P \pm \sqrt{P^2 + 4}}{2} \tag{18}$$

where

$$P = \frac{Q - R}{S}, \quad S = \sum_{i=1}^{n} x_i y_i - \frac{1}{n}\sum_{i=1}^{n} x_i \sum_{i=1}^{n} y_i, \quad Q = \sum_{i=1}^{n} x_i^2 - \frac{1}{n}\left(\sum_{i=1}^{n} x_i\right)^2,$$

$$R = \sum_{i=1}^{n} y_i^2 - \frac{1}{n}\left(\sum_{i=1}^{n} y_i\right)^2$$

3.2.2 Vertical Surface Fitting

For the linear function with two independent variables, Eq. (19), with the vertical offsets method, Fig. 9a, the parameters a, b, and c can be uniquely solved with Eq. (20) in matrix form (Wei 2013d):

$$z = f(x, y) = a + bx + cy \tag{19}$$

$$MX = K \text{ or } X = M^{-1}K \tag{20}$$

where

$$M = \begin{vmatrix} n & \sum_{i=1}^{n} x_i & \sum_{i=1}^{n} y_i \\ \sum_{i=1}^{n} x_i & \sum_{i=1}^{n} x_i^2 & \sum_{i=1}^{n} x_i y_i \\ \sum_{i=1}^{n} y_i & \sum_{i=1}^{n} x_i y_i & \sum_{i=1}^{n} y_i^2 \end{vmatrix} \quad X = \begin{vmatrix} a \\ b \\ c \end{vmatrix} \quad K = \begin{vmatrix} \sum_{i=1}^{n} z_i \\ \sum_{i=1}^{n} x_i z_i \\ \sum_{i=1}^{n} y_i z_i \end{vmatrix}$$

Perpendicular surface fitting

For the linear function, Eq. (19), with the perpendicular offsets method, Fig. 9b, the parameters a, b, and c can be obtained by solving Eq. (21) (Wei 2013d). Iterative procedures have to be used in solving Eq. (21).

$$a = \frac{1}{n}\left(\sum_{i=1}^{n} z_i - b \sum_{i=1}^{n} x_i - c \sum_{i=1}^{n} y_i\right)$$

$$b = \frac{1}{\left(\sum_{i=1}^{n} z_i^2 - \sum_{i=1}^{n} x_i^2\right)}\left[\begin{array}{l} -\sum_{i=1}^{n} x_i z_i + a \sum_{i=1}^{n} x_i + c \sum_{i=1}^{n} x_i y_i + b^2 \sum_{i=1}^{n} x_i z_i - ab^2 \sum_{i=1}^{n} x_i - b^2 c \sum_{i=1}^{n} x_i y_i \\ -c^2 \sum_{i=1}^{n} x_i z_i - bc^2\left(\sum_{i=1}^{n} y_i^2 - \sum_{i=1}^{n} x_i^2\right) + c^3 \sum_{i=1}^{n} x_i y_i + 2ab \sum_{i=1}^{n} z_i + 2bc \sum_{i=1}^{n} y_i z_i \\ + ac^2 \sum_{i=1}^{n} x_i - a^2 bn - 2abc \sum_{i=1}^{n} y_i \end{array}\right]$$

$$c = \frac{1}{\sum_{i=1}^{n} x_i z_i}\left[ac \sum_{i=1}^{n} x_i + bc\left(\sum_{i=1}^{n} x_i^2 - \sum_{i=1}^{n} y_i^2\right) + c^2 \sum_{i=1}^{n} x_i y_i + b \sum_{i=1}^{n} y_i z_i - ab \sum_{i=1}^{n} y_i - b^2 \sum_{i=1}^{n} x_i y_i\right]$$

$$(21)$$

Example 4: Why data pattern is important in curve/surface fitting?

Solution: It can be demonstrated that any deviation from a standard pattern will result in a deviated fit curve/surface, which is not accurate and is therefore undesired. An example is given in Fig. 10. If two triangle data blocks are symmetrically added to the lower and upper bounds of the existing dataset with the standard vertical pattern, Fig. 8a or Fig. 10a, the fit curve will go down to meet the new 'force' equilibrium because of the added two blocks of data. The net 'angular moment' can be canceled out in the case shown in Fig. 10b. If the two blocks added to the existing standard pattern are anti-symmetrical, Fig. 8c, then the fit curve will rotate around the centroid to a certain degree to establish a new equilibrium because of the added net 'angular moment.' In the case shown in Fig. 10c, the net 'force' contributed from the two blocks are canceled out. For cases of more general added data blocks, such as the case with only one triangle block, both 'force' and 'moment' will cause the fit curve to make both translation and rotation movements, which will result in inaccurate fit curves. Therefore, with a certain equilibrium direction, any deviation in data pattern from a standard pattern will lead to an inaccurate fit curve.

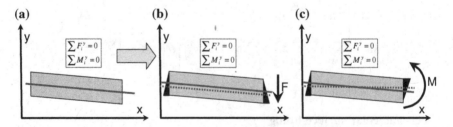

Fig. 10 Equilibrium establishment of **a** vertical pattern; **b** vertical patterned data with added symmetrical data blocks; **c** vertical patterned data with added anti-symmetrical data blocks

Example 5: Horizontal offset (standard) method

Figure 11 shows a set of fatigue data of welded automotive exhaust components made of steel materials. Tests were conducted by controlling the applied force at two force levels with six data points at each force level. Wide scatter bands can be observed for both force levels due to many factors involved in the failure of the exhaust components. Since the data pattern in Fig. 11 is similar to that shown in Fig. 8b, the horizontal offsets method, which is the ASTM standard recommended method (ASTM 2010), should provide a reasonable fit curve. The fit curves with the three fitting methods are plotted in Fig. 11 and the fit parameters are listed in Table 5. It is clear that the results of horizontal offsets method are very different from other two methods, while the vertical and the perpendicular are almost identical.

Example 6: Perpendicular offset method

It is clear from Fig. 12 that the data pattern of the fatigue test data of welded structures does not belong to the standard vertical pattern or the horizontal pattern. Additionally, it is difficult to trim the data into standard vertical and horizontal patterns because of limited test range. Therefore, the perpendicular pattern can be used to obtain a fit curve. The mean curves obtained from both perpendicular (solid line) and standard horizontal (dash-dot line) methods are also plotted in Fig. 12. The values of fit parameters are listed in Table 6. It is clear that the difference between these two methods is significant for the welds data, and the fit curve obtained with the horizontal method (standard) is not accurate and

Fig. 11 Vertical, horizontal, and perpendicular offsets methods for fatigue data of an automotive exhaust component

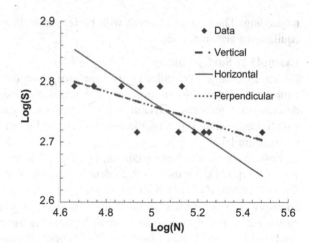

Table 5 Calculated fit parameters with $y = a + bx$, where $a = \log(C)$ and $b = -1/m$ for the power law $S = CN^{-1/m}$

	Vertical offsets	Horizontal offsets	Perpendicular offsets
a	3.345	4.037	3.355
b	−0.117	−0.254	−0.119

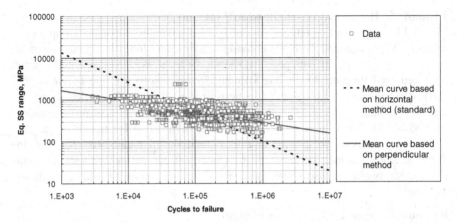

Fig. 12 Fatigue *S-N* data of 360° welded structures and fit curves

Table 6 The values of fit parameters of welds data with three curve fitting methods $y = a + bx$, where $a = \log(C)$ and $b = -1/m$ for the power law $S = CN^{-1/m}$

Tenneco 360° welds data	Vertical	Horizontal	Perpendicular
a	3.8604	6.2573	3.9931
b	−0.2292	−0.7094	−0.2558

misleading. The rotated fit curve with horizontal method can be described by the equilibrium argument, see Fig. 10.

Example 7: Surface fitting

Figure 13 plots a collection of average creep rate data r of a steel at various temperature T and stress σ levels and it is found that linear function can be used to describe the creep rate. Overall, the data pattern of each dataset belongs to the pattern identified in Fig. 8a for individual data, or Fig. 9a for overall dataset at each temperature level.

Both the vertical offsets methods, i.e., Eq. (20), and the perpendicular offsets method, Eq. (21), are used here to demonstrate the surface fitting capability. The linear function Eq. (19) with $z = \log(r)$, $x = T$, $y = \log(\sigma)$ is used here. The corresponding power law form of the function is $r = 10^a 10^{bT} \sigma^c$. The predicted curves for each temperature level are also plotted in Fig. 13, and the fit parameters are listed in Table 7. It is found that the fit curves obtained with the perpendicular method rotate counter-clockwise with respect to the respective curves obtained with vertical offsets methods, with degrees depending on the orientations of the given data. Furthermore, the individual curves obtained for each method are parallel to each other because of the use of the unified linear function $z = a + bx + cy$, which is linear in log–log plot.

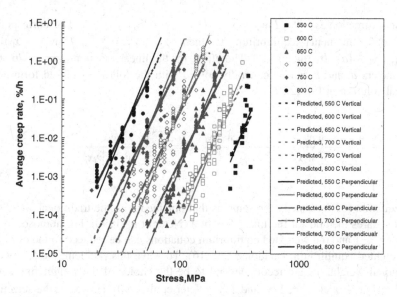

Fig. 13 Results of vertical and perpendicular offsets method with the unified linear surface $z = a + bx + cy$ where $z = \log(r), x = T, y = \log(\sigma)$

Table 7 Fit parameters for creep test data for all testing temperatures (°C) with $z = a + bx + cy$, where $z = \log(r)$, $x = T, y = \log(\sigma)$

Vertical offsets			Perpendicular offsets		
a	b	c	a	b	c
-31.66	0.025	6.27	-38.43	0.031	7.74

Example 8: How to handle heteroscedastic data with the equilibrium method?

Large amounts of fatigue and creep test data, are generally heteroscedastic (unequal variance and a "funnel shape" visually in a plot). With conventional methods originally developed for homoscedastic data, larger deviations tend to influence (weight) the regression line more than smaller deviations, and thus the accuracy in the lower variance end of the range is impaired. To deal with heteroscedastic data, several concepts have been proposed and the simplest and the most effective way is the weighted least squares linear regression (Neter et al. 1990). With a weight function, the data could be transformed to homoscedastic data and the weight function w_i is usually supposed to be the function of x, y, say $w_i = w(x_i)$ or $w_i = w(y_i)$. The concept of "weight" can be simply implemented into the "equilibrium" data analysis method (Wei et al. 2012d). Take the vertical offsets direction as an example and follow the procedure for the equilibrium method: the signed weighted distance or 'force' F_i^y, from a point x_i to the expected curve along the vertical direction is $\sum_{i=1}^{n} F_i w_i = \sum_{i=1}^{n} \{[y_i - (a + bx_i)]w_i\} = 0$. The corresponding total 'moment' equilibrium is then $\sum_{i=1}^{n} M_i^y = \sum_{i=1}^{n} [F_i^y w_i (x_i - x_0)] = \sum_{i=1}^{n} \{[y_i - (a + bx_i)]w_i x_i\} = 0$. Therefore, for the linear function, $y = a + bx$, the parameters a, and b can be generally solved with the following closed-form analytical solution in Eq. (22).

$$a = \frac{\sum w_i x_i^2 \sum w_i y_i - \sum w_i x_i \sum w_i x_i y_i}{\sum w_i \sum w_i x_i^2 - (\sum w_i x_i)^2} \quad b = \frac{\sum w_i \sum w_i x_i y_i - \sum w_i x_i \sum w_i y_i}{\sum w_i \sum w_i x_i^2 - (\sum w_i x_i)^2}$$

$$(22)$$

Equation (22) is exactly the same as that obtained with the traditional weighted least squares method for the linear function (Neter et al. 1990). For homoscedastic data, $w_i = 1$ and the weighted equilibrium equations can be reduced to the ones for unweighted equilibrium equations, Eq. (16). To solve this problem, the following empirical weights w_i are recommended to be the choice of the weight functions: $1/x^{1/2}$, $1/x$, $1/x^2$, $1/y^{1/2}$, $1/y$, and $1/y^2$ (Neter et al. 1990). However, the selection of the weighted functions is arbitrary and the number of the potential weighted functions is infinite.

The equilibrium method has been extended to heteroscedastic data with the introduction of a linear weight function $w(x_i) = c + dx_i$, in which the unknown c and d as well as the standard deviation are determined with two more equilibrium equations (Wei et al. 2012d). Figure 14 shows the schematic procedure.

A set of two-stress level fatigue S-N data with six data points at each stress level and a set of creep rate data with multiple stress levels are plotted in Fig. 15a, b, respectively, with the estimated mean and the lower/upper bounds $\pm 2\sigma$. The scatter plot of the data strongly suggests the heteroscedastic nature of the data. It should be noted that the predicted curves should be valid within the ranges studied and should not be extrapolated beyond the ranges studied.

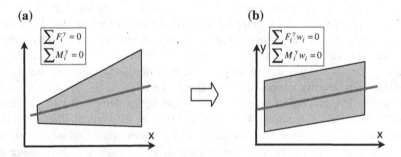

Fig. 14 The schematic of the equilibrium establishment of **a** the original data with; **b** transformed data with a weight function

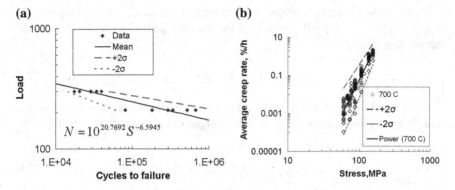

Fig. 15 The mean and the design curves obtained with the equilibrium mechanism for **a** two-stress level heteroscedastic fatigue data, and **b** multiple stress levels creep data

3.3 Design Curve Construction

In addition to the mean curve and standard deviation, design curves, such as fatigue design S-N curves, are also important in product design and subsequent durability/reliability validation. A design curve is constructed to ensure that a majority of the fatigue data or other failure data falls above the lower bound value with a certain failure probability and confidence (Owen 1968; Shen et al. 1996). Therefore, the common practice to define a design curve, such as a S-N fatigue design curve, is to shift the mean curve by a distance to the left based on the magnitude of the scatter. Several methods are already available to construct a design curve and some of these methods, with varying degrees of conservativeness, accuracy, and simplicity, have been adopted by engineering standards, codes, and guidelines, such as the ASME code (ASME 1969). The traditional ASME code recommends to shift the mean curve leftward by 2X in stress or by 20X in cycles to failure, whichever is more conservative in design. Another commonly used method is to construct a design curve by shifting the mean curve by certain times of the standard deviation, e.g., -2σ or -3σ (BS 1993). However, these two methods ignore the uncertainty or confidence introduced by the sample size. Modern design curve construction is based on tolerance limit concept. Different from the confidence interval, a tolerance limit is a statistical limit below or above which, with some confidence level, a specific proportion of sampled population falls. The determination of reliability (R) and confidence (C) in design curve construction is generally dependent upon material property, safety policy, and industry standards. In the automotive industry, R90C90, R95C95, etc. are often used for safety-related component designs (Lee et al. 2005). For example, the value of R90C90 ensures that there is a 90 % possibility of survival (reliability) with a 90 % confidence level at the specific data point.

Based on the tolerance limit concept and the homoscedastic data assumption, the design limit Y_0 is generally expressed in Eq. (23) with Y as a function of the independent variable X:

$$Y_0 = \overline{Y} - Ks \qquad (23)$$

$\overline{Y}$ is the mean value or curve, and can be easily calculated from a given set of test data. The sample standard deviation s for fatigue data is calculated with Eq. (24):

$$s = \sqrt{\frac{1}{f}\sum_{i=1}^{n}\left[\log(N_i) - \overline{\log(N_i)}\right]^2} \qquad (24)$$

where n is the total sample size and $f = n - 2$ is chosen for problems with two undetermined parameters (two or more stress levels), and $f = n - 1$ should be used for problems with only one undetermined parameter (one stress level); K is a factor generally related to confidence interval, sample size, and failure probability, and is given in Eq. (25) for a specific proportion $(100(1 - p)\%)$, i.e., $R(1 - p)$, of the population be above or below a prescribed confidence level $(100(1 - \gamma)\%)$, i.e., $C(1 - \gamma)$, or overall, R90C90 (Link 1985; Wei et al. 2013a, b, c).

$$K = \frac{z_p(1 - \phi) + \left\{z_p^2(1 - \phi)^2 - \left[(1 - \phi)^2 - z_r^2/(2f)\right]\left(z_p^2 - az_\gamma^2\right)\right\}^{1/2}}{(1 - \phi)^2 - z_\gamma^2/(2f)} \qquad (25)$$

where $\phi = 1/(4f)$. Equation (25) is suitable to data with small sample size. The expressions provided in (Natrella 1966) can be recovered by ignoring the factor ϕ. However, for small n, leaving out this factor underestimates K. The calculation of the standard normal distribution-related z scores z_p and z_r can be easily calculated (Neter et al. 1990). For one stress level problem, the Owen tolerance limit is reduced to the one-dimensional tolerance limit with $a = 1/n$. For regression cases with the Owen tolerance limit, the calculation procedure of a at a given stress level x_0 is suggested as (Shen et al. 1996; Wei 2013d)

$$a(x_0, X) = \{x_0\}^T \left(X^T X\right)^{-1}\{x_0\} \qquad (26)$$

where X is a vector of the values of stress defining the test plan

$$X = \begin{bmatrix} 1 & x_1 & \cdots & x_1^d \\ 1 & x_2 & \cdots & x_2^d \\ \cdots & \cdots & \cdots & \cdots \\ 1 & x_n & \cdots & x_n^d \end{bmatrix} \qquad (27)$$

Table 8 Values of K for one-dimensional tolerance limit

n	R90C75	R90C90	R90C95	R95C75	R95C90	R95C95
3	2.5396	5.7866	24.8836	3.1857	7.2760	31.7304
4	2.1381	3.5324	5.5945	2.6793	4.4043	7.0061
5	1.9604	2.8937	3.9269	2.4577	3.5987	4.8869
6	1.8567	2.5801	3.2792	2.3295	3.2066	4.0703
7	1.7874	2.3894	2.9261	2.2444	2.9699	3.6284
8	1.7372	2.2592	2.7003	2.1831	2.8092	3.3474
9	1.6988	2.1637	2.5417	2.1364	2.6919	3.1509
10	1.6683	2.0901	2.4231	2.0995	2.6019	3.0047
11	1.6433	2.0313	2.3305	2.0693	2.5302	2.8910
12	1.6223	1.9829	2.2559	2.0441	2.4715	2.7997
13	1.6045	1.9424	2.1941	2.0227	2.4223	2.7244
14	1.5891	1.9077	2.1421	2.0042	2.3805	2.6610
15	1.5756	1.8778	2.0975	1.9881	2.3443	2.6068
16	1.5636	1.8515	2.0587	1.9738	2.3127	2.5599
17	1.5530	1.8282	2.0247	1.9611	2.2848	2.5187
18	1.5434	1.8074	1.9945	1.9497	2.2598	2.4823
19	1.5347	1.7887	1.9675	1.9394	2.2375	2.4497
20	1.5268	1.7718	1.9431	1.9300	2.2172	2.4205

and

$$\{x_0\}^T = \begin{bmatrix} 1 & x_0 & \cdots & x_0^d \end{bmatrix} \tag{28}$$

and d is the degree of polynomial chosen $d = f - n + 1$.

The values calculated from Eq. (25) are listed in Table 8 for various RC values and sample size n.

Recent studies (Makam et al. 2013) have shown that the analytical solutions based on Eq. (25) may not be accurate for very small sample size because of the assumptions and approximations introduced to the analytical approach. Monte Carlo simulation methods have been found to be more accurate as they eliminate these assumptions and approximations inherent in the analytical method. The observation has been essentially confirmed (Wei et al. 2013c). The K values calculated from a Monte Carlo simulation are listed in Table 9 (Makam et al. 2013). Cleary, as compared to the analytical solution shown in Table 8, the values of Monte Carlo simulation are generally smaller for small sample size.

Example 9 Design curve construction based on two-stress level test data

Two sets of fatigue test data, shown in Table 10, are analyzed. Test-1 consists of six data points at each of the two stress levels and Test-2 consists of three data points at each stress level. Assuming that the fatigue data follow the lognormal distribution, the estimated mean curves and the corresponding design S-N curves with various

Table 9 Values of K based on Monte Carlo simulation (Makam et al. 2013)

n	R90C75	R90C90	R90C95	R95C75	R95C90	R95C95
3	1.9554	3.3866	4.8181	2.4363	4.1341	6.1806
4	1.7355	2.5405	3.2949	2.1462	3.2021	4.0864
5	1.6136	2.2575	2.7701	2.0299	2.8649	3.4228
6	1.5881	2.1138	2.4866	1.9490	2.6162	3.1086
7	1.5473	1.9952	2.3444	1.9327	2.5031	2.8951
8	1.5001	1.9366	2.2393	1.8915	2.3897	2.7631
9	1.4821	1.8600	2.1500	1.8860	2.3309	2.6462
10	1.4748	1.8317	2.0704	1.8740	2.2702	2.5827
11	1.4592	1.7933	2.0101	1.8428	2.2078	2.4938
12	1.4531	1.7749	1.9799	1.8358	2.1876	2.4666
13	1.4403	1.7550	1.9367	1.8298	2.1486	2.4222
14	1.4438	1.7235	1.8993	1.8093	2.1493	2.3734
15	1.4308	1.6933	1.8880	1.8132	2.1265	2.3344
16	1.4158	1.6841	1.8630	1.8076	2.0959	2.3189
17	1.4180	1.6658	1.8359	1.8028	2.0693	2.2679
18	1.4176	1.6577	1.8213	1.7897	2.0683	2.2616
19	1.4158	1.6427	1.8296	1.7874	2.0621	2.2359
20	1.4130	1.6276	1.7674	1.7952	2.0513	2.2174

Table 10 Cycles to failure as functions of applied constant amplitude cyclic load (lb)

Test			Data-1	Data-2	Data-3	Data-4	Data-5	Data-6
Test-1	Lower load	175	101,654	109,673	132,759	49,834	73,456	193,749
	Higher load	220	46,133	75,692	36,772	40,306	61,020	97,068
Test-2	Lower load	567	149,694	118,743	173,088	–	–	–
	Higher load	750	22,012	29,261	44,729	–	–	–

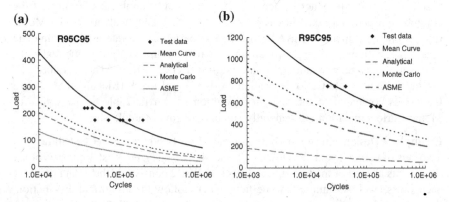

Fig. 16 Fatigue design *S-N* curves constructed for test data with **a** six points and **b** three points at each stress level for R95C95

methods can be constructed and the results of R95C95 are plotted in Fig. 16a, b, respectively, for these two datasets. The design curves constructed using analytical solutions are located below that constructed using the corresponding Monte Carlo method. Additionally, the ASME method (ASME 1969) gives the most conservative design curve for Test-1, whereas the analytical method gives the most conservative design curve for Test-2. The very small sample size makes the analytical solution over-conservative for Test-2.

3.4 Bayesian Statistics

Most of the current life assessment methods utilize only the data observed during the current test and ignore any prior knowledge about the products or its predecessors. Historical failure data can shed light on the current and future designs since they would share some common features when the design changes are not drastic. To effectively utilize the historical information, two parts are necessary: (1) Bayesian statistics, which can provide a rigorous mathematical tool for extracting useful information from the historical data, and (2) historical data. The combination of the historical data and the Bayesian statistics makes the sample size reduction and the accurate life assessment improvement possible.

3.4.1 Bayesian Theory

Bayes's rule (Bayes 1763) was published more 250 years ago, and the general formula can be expressed as

$$p(\theta|x) = \frac{l(\theta; x)p(\theta)}{\int_0^1 l(\theta; x)p(\theta)d\theta} \qquad (29)$$

where $p(\theta|x)$ is posterior PDF for the parameter θ given the data x, $p(\theta)$ is prior PDF for the parameter θ. $l(\theta; x)$ is the likelihood function, which is defined as $l(\theta; x) = \prod_{k=1}^n f(\theta; x_k)$, where x_k is k-th experimental observation and $f(\theta; x_k)$ is the PDF for the experimental data. The denominator in Eq. (29) is simply a normalizing factor which ensures that, over the support of θ, the posterior PDF integrates to one. Analytically, the reference process is to update the prior PDF $p(\theta)$ to the posterior PDF $p(\theta|x)$ through the medium of the likelihood function $l(\theta; x)$. The potential to express information about model parameters as direct probabilistic statements renders the Bayesian approach particularly attractive. The integration operation of Bayesian theorem plays a fundamental role in Bayesian statistics. Except in simple cases, however explicit evaluation of such integrals will rarely be possible, and the use of sophisticated numerical integration or analytical approximation techniques is required.

3.4.2 Numerical Implementation

One such numerical method is the sampling–importance resampling algorithm (SIR) (Smith and Gelfand 1992), in which the important function comprises sampled values from the prior PDF $p(\theta)$ weighted by their relative likelihoods. These sampled values are then resampled to produce a sample from the posterior PDF $p(\theta|x)$. This sampling–resampling approach provides essentially calculus-free use of Bayes' Theorem, and has been advocated as allowing practitioners to perform 'Bayesian statistics without tears' (Smith and Gelfand 1992). The following three algorithms have been provided for Bayesian calculations.

Algorithm-1 (Smith and Gelfand 1992)
In the case where there exists an identifiable constant $M > 0$ such that $l(\theta; x) \leq M$ for all θ, the algorithm is as follows:

(a) generate θ from $p(\theta)$.
(b) generate u from uniform $(0, 1)$.
(c) if $u \leq l(\theta; x)/M$, accept θ; otherwise, reject it and repeat steps (a)–(c) for $i = 1, \ldots n$.

The likelihood therefore acts as a resampling probability, and those θ in the prior sample having high likelihood are more likely to be retained in the posterior sample.

Algorithm-2 (Smith and Gelfand 1992)
In cases where the bound M required in the rejection method is not readily available, Smith and Gelfand (1992) show that Bayes theorem can be implemented as a weighted bootstrap, and approximate resample from $p(\theta|x)$ can be obtained. Actually, since $p(\theta|x) \propto L(\theta; x_i)p(\theta)$, we can also straightforwardly resample using the weighted bootstrap with the following algorithm: Given $\theta_i, i = 1, \ldots n$, a sample from $p(\theta)$, calculate normalized weight for each sample $q_i = l(\theta; x) \Big/ \sum_{j=1}^{n} l(\theta_j; x)$.

Now draw θ^* from the discrete distribution over $\{\theta_1, \ldots, \theta_n\}$, which places probability mass q_i on θ_i. Then θ^* is approximately distributed according to $p(\theta|x)$ with the approximation improving as n increases (ideally $n \rightarrow \infty$). Note that the sample size under such resampling can be as large as desired. The rule of thumb is that the less the $p(\theta|x)$ resembles $p(\theta)$, the larger the sample size n will need to be in order that the distribution of θ^* to sufficiently approximate $p(\theta|x)$.

Algorithm-3 (Gorden et al. 1993)

(a) generate θ from $p(\theta)$.
(b) generate u from uniform $(0, 1)$.
(c) if $\sum_{j=0}^{m-1} q_j < u_i \leq \sum_{j=0}^{m} q_j$, accept θ_m as a sample for the posterior; otherwise, reject it and repeat steps (a)–(c) for $i = 1, \ldots N$.

Table 11 Test data used in Bayesian analysis

	The value of data points (cycles to failure)					
One-data	55,550	–	–	–	–	–
Two-data	55,550	33,665	–	–	–	–
Three-data	55,550	33,665	141,861	–	–	–
Four-data	55,550	33,665	141,861	174,331	–	–
Five-data	55,550	33,665	141,861	174,331	53,047	–
Six-data	55,550	33,665	141,861	174,331	53,047	86,968

Algorithm-3, similar to Algorithm-1, is found to be very effective in Monte Carlos simulation by accepting or rejecting a generated resample random point, and will be used in the following demonstration.

3.4.3 Application of Bayesian Theory in Design Curve Construction

Great efforts have been made to collect and analyze the historical fatigue data of the welded structures. Part of the collected historical data is shown in Fig. 12. However, the *S-N* data and the fitted curves cannot be directly used in engineering design because of its big scatter band, e.g., the data span several orders of magnitude in terms of cycles to failure for a single stress level. The design curves constructed based on the test data are too conservative to use. The database shown in Fig. 12 consists of a large amount of test sets with six data points at each stress level. For each dataset, the scatter band is usually narrow, e.g., within 2X. However, different sets may be located in different locations in Fig. 12. To extract more accurate, representative, and useful information, the distributions of mean and standard deviation of the test data have been analyzed. With the perpendicular offset methods described in Sect. 3, the mean of the test data was found to follow two-parameter lognormal distribution and the standard deviation follows the three-parameter Weibull distribution. The obtained prior information of the sample mean and sample standard deviation can then be used to generate a posterior sample by resampling the prior sample with the help of Bayesian statistics using the Algorithm-3 as described in Sect. 3.4.2.

As an example, the values of the test data with different sample sizes are listed in Table 11.

The scatter plot of prior distribution of mean (abscissa) and standard deviation (ordinate) is shown in Fig. 17a. The estimated results of the posterior sample from Bayesian statistics and the test data points are shown in Fig. 17b–e. The general features of the posterior can easily be identified from the plots, for example, the marginal locations and redistribution of the parameters (mean and standard deviation). As the sample size increases, the updated estimated parameters are clustered into smaller and smaller areas, indicating the increased confidence and more

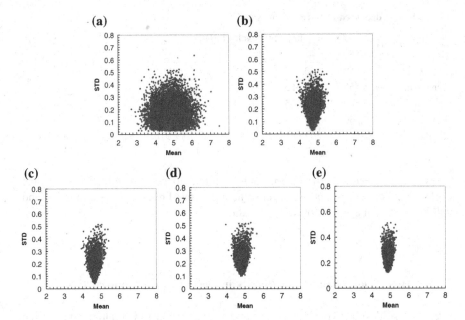

Fig. 17 Sample (10,000 data points) from **a** normal prior distribution, **b** one-data, **c** two-data, **d** three-data, and **e** six-data

Table 12 Test data with two stress levels and six data points at each stress level

Stress level (MPa)	The cycles to failure					
554.0	55,550	33,665	141,861	174,331	53,047	86,968
739.3	23,241	32,986	38,042	31,700	26,976	54,036

accurate estimations. With the generated sample data, posterior distribution parameters such as mean, standard deviation, and design limit/curve can be extracted by analyzing the data points resampled with the help of statistical analysis methods.

Two-stress test data shown in Table 12 are used to illustrate the advantage of the Bayesian approach over the tradition frequentist approach in design curve construction. In this example, the test data with a certain test data number from two stress levels of Table 12 are taken sequentially and the two-stress level design curve construction method (Wei et al. 2013a) is used to construct design curves. The mean curve, design curves (R90C90 and R95C95) for one data point at each stress level, two data points, three data points, and six data points are plotted in Fig. 18a–d, respectively, for both the frequentist and the Bayesian approaches. The frequentist means are calculated directly from the test data with standard procedure. The frequentist design curves are obtained using Eq. (4). The Bayesian means are obtained by averaging the simulated sample values, and the design curves can be

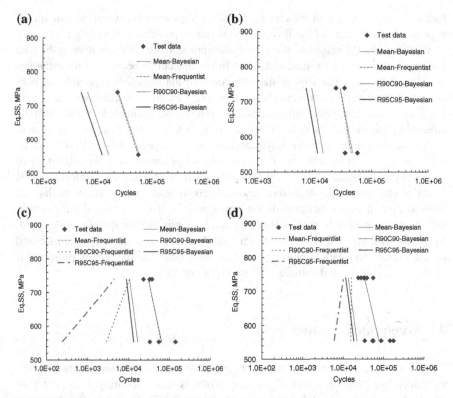

Fig. 18 Comparison between the estimated parameters from the frequentist and Bayesian methods for a set of fatigue data, **a** one data point at each stress level, **b** two data points at each stress level, **c** three data points at each stress level, and **d** six data points at each stress level

constructed using statistical approaches provided in Sect. 3.3. It noted that the design curves for test sample size of 1 and 2 are not provided in either analytical form (Wei et al. 2013a) or Monte Carlo numerical simulation (Makam et al. 2013) because of the limitation of the theory and relatively poor numerical approximation for very small sample size. Additionally, the standard deviation for test data with a sample size of 1 cannot be calculated it can, however, be calculated for sample size of a two, but usually inaccurate statistically. By contrast, the Bayesian approach has no such limitations and Fig. 18a, b demonstrates its capability.

For sample size of three or above, the frequentist approach can provide statistical information about the mean, standard deviation, and design curves; however, it may not provide enough confidence as compared to the Bayesian approach. This can be clearly demonstrated in Fig. 18c, in which the design curves from the lower stress level are located far left as compared to that of the Bayesian approach. It should be noted that the design curves of the frequentist approach shown in Fig. 18c are constructed with K value from the Monte Carlo simulation. If the value of analytical K from Eq. (25) is used, the design curves at both stress levels will be moved further to the left, which makes the curves design over-conservative and meaningless. The advantage of the Bayesian approach in possible cost saving is obvious.

Furthermore, the results of the frequentist approach are very sensitive to the data patterns of the current test data and, often times, inaccurate because of uncertainties introduced by the small sample size. By contrast, the Bayesian approach not only considers the contribution from the current test data but also consider the weight from the historical prior information. Therefore, the estimated results will be affected by the current test data but not as strong as that for the frequentist approach.

It should be noted that the Bayesian approach is especially useful for test data analysis with small sample size. For a test with large sample size, the advantage of it will decrease because the weight contributed from the historical information is heavy. Even worse, the Bayesian approach could lead to inaccurate results and misleading conclusion because of the heavy weight of the historical information, which covers many information that may not be applied to the specific current test. In these cases, the frequentist approach could provide more accurate results based on the current test data. Definitely, the test with large sample size is preferred if testing cost, timing, and sample size are not a concern.

4 Accelerated Testing

Product designers and manufacturers are under continuous and increasing pressure to reduce the "time-to- market" of new products while assuring high levels of durability and reliability of these products (Yang 2007). However, the durability life and warranty time of vehicle components and systems are usually required to last relatively long time, e.g., 10–15 years, which is unbearable for laboratory testing under normal operating conditions. Therefore, the time and efforts involved in testing, especially in early development stage, can be very expensive. To reduce the time-to-market and the cost in product design and validation, accelerated testing is often adopted (Nelson 2004). Accelerated tests are conducted at stress levels higher than that experienced in service condition, and the life of the components and systems under service loading conditions can be estimated using appropriate extrapolation approaches.

Sample size is an important issue in accelerated testing. Usually, many repeats of tests are often required to capture the uncertainty, scatter characteristics, and failure probability of the test data. Additionally, tests with several stress levels are usually conducted to cover a wide range of stress levels, which further lead to more test samples. Therefore, testing plan, sample size and allocation, and testing sequence must be carefully determined in order to find a reasonable compromise between the test accuracy and the cost of the tests. Several sample size determination formulae have been given in Nelson's book (2004). The minimum number of specimens required in fatigue S-N tests depends on the types of test program conducted. The

Table 13 Sample size recommended by the ASTM standard (ASTM 2010)	Preliminary and exploratory (exploratory research and development tests)	6–12
	Research and development testing of components and specimens	6–12
	Design allowable data	12–24
	Reliability data	12–24

sample size for general tests have been documented in the ASTM standard (ASTM 2010), Table 13, for test data generally showing a linear trend against stress.

Several test plans, including the traditional plan, the optimum plan, and the compromised plan, for accelerated testing have been well developed for accelerated testing (Nelson 2004). The traditional test plan has equally spaced test stress levels and equal numbers of specimens at each level. However, traditional plans generally require 25–50 % more specimens for the same accuracy as optimum plans. The optimum test plan is obtained by minimizing the variance of the predicted mean values of the dependent variable and it yields the most accurate estimates of life at the design stress. Yang presented the best compromise three-level (Yang and Jin 1994) and four-level (Yang 1994) constant stress accelerated life test plans for Weibull distributions with different censoring times. The general conclusion of the optimum test plan is that more test samples should be placed on the lower stress level. Two extremes are very informative (Nelson 2004): (1) if the lowest test stress equals the design stress, all specimens should be run at the lowest test stress; and (2) if the service stress is much below the lower stress, the specimens should be equally located at the lower and higher stress levels, which is the case for the traditional test plan.

However, the confidence interval at a certain reliability level obtained using all of the existing methods for, say, fatigue S-N data is generally in concave shape, which is more severe for test data with small sample size, and thus inconvenient and inaccurate for accelerated testing. Based on the tolerance limit concept (Owen 1968), the following two-stress level accelerated testing procedure has been proposed for linear homoscedastic data with arbitrary data points at each stress level (Wei et al. 2012a), Fig. 19.

The procedure:

1. Conduct accelerated test by following the traditional test plans (ASTM 1975);
2. Obtain the mean curve $\overline{Y}$ from test data with the least squares method;
3. Calculate the standard deviation s_H and s_L. The subscripts H and L represent higher and lower stress levels, respectively;
4. Calculate the low bound Y_0 with a certain criterion, e.g., R90C90, for both higher and lower stress levels;
5. Use the Y_0, which has longer distance to the mean curve, to construct a design curve that parallels to the mean curve;

Fig. 19 Accelerated testing
procedure for linear data with
homoscedastic characteristics

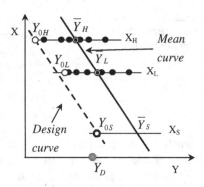

6. Extrapolate the design curve to the service stress X_S to obtain the predicted life Y_{0S} and compare Y_{0S} with the designed life Y_D;
7. If the predicted life Y_{0S} is higher than the design life Y_D, the design is approved; otherwise, a new iteration of design is required.

Many products exhibit multiple failure mechanisms and modes. Accelerated testing methods considering different failure mechanisms and modes and their interaction are important but difficult problems. Nelson devoted a chapter to the issue of competing failure modes in his book. "Interaction" here means the combined effect of all individual failure modes acting either in a "sequential" or "simultaneous" manner. Under some simple conditions, the accelerated testing procedure with multiple failure modes can be simplified. However, for a failure process with two or more failure modes, the inclusion–exclusion principle requires that the failure modes interaction and bivariate or multiple-variate distribution must be provided for a thorough investigation. The Bonferroni inequality can provide a simple and a conservative solution by evaluating the upper bound or the lower bound. An accelerated testing procedure for fatigue–creep testing has been developed (Wei et al. 2012a).

Example 10: Accelerated pure fatigue testing and data analysis

For the pure fatigue analysis of an auto exhaust component or system, with the homoscedastic data assumption, the question is at stress range of 30, can cycle to failure of 10,000,000 be achievable with R90C90? If not, a new round of design iteration would be required. The 10,000,000 cycles would take 11.6 days for only one test to finish at the service stress range of 30 with the loading frequency of 10 Hz, so an accelerated testing procedure was suggested. To do so, a set of two-stress level fatigue S-N test data of auto exhaust components was planned and tested, and the results, including the values of the stress levels and the cycles to failure in non-dimensional form, are listed in Table 14. Due to the cost concern, only six test parts were tested at each stress range level, i.e., 175 and 220. With the accelerated testing plan, the longest time spent for a specimen to fail is about 5.38 h, which is much shorter than the expected time for a test at service stress level.

Table 14 Two-stress level fatigue test data

Stress level	Cycles to failure					
175	49,834	73,456	10,1654	10,9673	13,2759	19,3749
220	36,772	40,306	46,133	61,020	75,692	97,068

First, the parameters needed are calculated as follows: with the two mean points connection method or least squares method (Wei et al. 2012a), the mean curve was obtained as $\log 10(N) = 10.776 - 2.573 \log 10(S)$. The standard deviation calculated using Eq. (24) is 0.186; the factor for R90C90 with sample size of 6 is 2.580 from Table 8. Then, the design curve for R90C90 was obtained as $\log 10(N) = 10.296 - 2.573 \log 10(S)$. Finally, for the stress range of 30, the cycle to failure predicted is 3,126,079, which is less than the target of 10,000,000 cycles, so another iteration of design is needed.

5 Summary

The durability and reliability design, performance assessment, and validation of vehicle exhaust products are essentially based on the observation of test data and advanced statistical and probabilistic analyses. The failure mechanisms and modes as observed in vehicle exhaust components and systems are extremely complex because of the complicated operating conditions such as high temperature, severe load, and corrosive environments. The operating conditions and failure mechanisms result in a variety of data patterns which are difficult to analyze with conventional methods. Therefore, more advanced analysis tools are required.

In this chapter, the durability/reliability performance design and validation of vehicle exhaust products were reviewed followed by a discussion of the challenges of data analysis. The most common failure mechanisms, such as fatigue, creep, and corrosion, as observed in exhaust products were provided. A equilibrium based curve fitting method and the associated data pattern identification methods where introduced, and their importance in accurate data analysis were emphasized. A method for design curve construction was discussed, and several examples were given to demonstrate the design curve construction procedure. Bayesian statistics and associated historical test data were introduced for possible reduction in sample size and testing cost, and accuracy improvement. Finally, an accelerated testing procedure based on linear test data was outlined and a case study was provided to demonstrate the effectiveness of the procedure. In summary, this chapter offers a practical guide on the state-of-the-art methodologies for test data analysis for engineers and managers in their product design and validation.

Acknowledgment The authors would like to thank Prof. Kamran Nikbin, Prof. D. Gary Harlow, Mr. Kay Ellinghaus, Mr. Markus Pieszkalla, Mr. Marek Rybarz, Dr. Pierre Olivier Santacreu, Mr. Maleki Shervin, Mr. Herry Cheng, Mr. Tim Gardner, Mr. Joesph Berkemeier, and Mr. Richard Voltenburg for their helpful comments and contributions to works summarized in this chapter.

References

ASME. (1969). *Criteria of the ASME boiler and pressure vessel code for design by analysis in Sections III and VIII, Division 2.* New York: The American Society of Mechanical Engineers.

ASTM. (1962). *ASTM manual on fitting straight lines, STP 313.* ASTM International.

ASTM. (1975). *Chapter 3-planning S-N and response tests, in manual on statistical planning and analysis for fatigue experiments, STP 588.* ASTM International.

ASTM. (2010). *Standard practice for statistical analysis of linear or linearized stress-life ($S - N$) and strain-life ($\varepsilon - N$) fatigue data.* ASTM Designation: E739-10.

Bayes, T. (1763). An essay towards solving a problem in the doctrine of chances. *Philosophical Transactions of the Royal Society of London, 53,* 370–418.

BS 7608. (1993). Code for practice for fatigue design and assessment of steel structures.

Cashman, G. (2007). A statistical methodology for the preparation of a competing modes fatigue design curve. *Journal of Engineering Materials and Technology, 129,* 159–168.

Gordon, N. J., Salmond, D. J., & Smith, A. F. M. (1993). Novel approach to nonlinear/non-Gaussian Bayesian state estimation. In *Proceedings of the International Conference on Electronics and Electrical Engineering* (vol. 140, pp. 107–113).

Harlow, D. G. (2011). Statistical characterization of bimodal behavior. *Acta Materialia, 59,* 5048–5053.

Jha, S. K., Larsen, J. M. M, & Rosenberger, A. H. (2009). Towards a physics-based description of fatigue variability behavior in probabilistic life-prediction. *Engineering Fracture Mechanics, 76,* 681–694

Lee, Y. L., Pan, J., Hathaway, R., & Barkey, M. (2005). *Fatigue testing and analysis: Theory and practice.* Oxford: Elsevier, Butterworth-Heinemann. ISBN 978-0-12-385204-5.

Lin, S. (2011). Exhaust system reliability evaluation. *International Journal of Reliability, Quality and Safety Engineering, 18,* 327–340.

Link, C. (1985). *An equation for one-sided tolerance limits for normal distributions, research paper FPL 458* (pp. 1–4). Madison, WI: U.S. Department of Agriculture, Forest Service, Forest Products Laboratory.

Makam, S., Lee, Y. L., & Attibele, P. (2013). Estimation of one-sided lower tolerance limits for a Weibull distribution using the Monte Carlo pivotal simulation technique. *SAE International Journal of Materials and Manufacturing, 6*(3). doi:10.4271/2013-01-0329.

Meeker, W. Q., & Escobar, L. A. (1998). *Statistical methods for reliability data.* New York: Wiley Series in Probability and Statistics.

Natrella, M. (1966). *Experimental statistics, handbook 91.* National Bureau of Standards.

Nelson, W. (2004). *Accelerated testing: Statistical models, test plans, and data analysis.* New York: Wiley.

Neter, J., Wasserman, W., & Kutner, M. (1990). *Applied linear statistical models.* Homewood, IL: Richards D. Irwin Inc.

Owen, D. (1968). A survey of properties and applications of the non-central t-distribution. *Technometrics, 10,* 445–472.

Pham, H. (Ed.). (2006). *Springer handbook of engineering statistics.* London: Springer.

Rice, R. C. (Ed.). (1997). *SAE Fatigue design handbook* (3rd ed., AE-22). Warrendale, PA: Society of Automotive Engineers, Inc.

Ryan, B. F., Joiner, B. L., & Ryan, T. A. (1985). *Minitab handbook* (2nd ed.). Boston: Duxbury Press.

Shen, C. L., Wirsching, P. H., & Cashman, G. T. (1996). Design curve to characterize fatigue strength. *Journal of Engineering Materials and Technology, 118,* 535–541.

Smith, A. F. M., & Gelfand, A. E. (1992). Bayesian statistics without tears: A sampling-resampling perspective. *The American Statistician, 46,* 84–88.

Wei, Z., Lin, B., Luo, L., Yang, F., & Dmitri, K. (2012a). Accelerated durability testing and data analysis for products with multiple failure mechanisms. *International Journal of Reliability, Quality and Safety Engineering, 19,* 1240003.

Wei, Z., Yang, F., Luo, L., Avery, K., & Dong, P. (2012b). Fatigue life assessment of welded structures with the linear traction stress analysis approach. *SAE International Journal of Materials and Manufacturing, 5*, 183–194.

Wei, Z., Yang, F., Lin, B., & Harlow, D. G. (2012c). Failure modes analysis of fatigue S-N test data with small sample size. In *Proceedings of the 18th ISSAT International Conference on Reliability and Quality in Design*, 26–28 July 2012, Boston, Massachusetts, USA.

Wei, Z., Yang, F., Maleki, S., & Nikbin, K. (2012d). Equilibrium based curve fitting method for test data with nonuniform variance. In *Proceedings of the ASME 2013 Pressure Vessels & Piping Division Conference*, PVP2012-78234, 15–19 July 2012, Toronto, Canada.

Wei, Z., Dogan, B., Luo, L., Lin, B., & Dmitri, K. (2013a). Design curve construction based on tolerance limit concept. *Journal of Engineering Materials and Technology, 135*, 014501.

Wei, Z., Luo, L., Ellinghaus, K., Pieszkalla, M., Harlow, D.G., & Nikbin, K. (2013b). Statistical and probabilistic analysis of thermal-fatigue test data generated using V-shape specimen testing method. In *Proceedings of the ASME 2013 Pressure Vessels & Piping Division Conference*, PVP2013-97628, 14–18 July 2013, Paris, France.

Wei, Z., Luo, L., Lin, B., Konson, D., & Nikbin, K. (2013c). Design curve construction based on Monte Carlo simulation. In *Proceedings of the ASME 2013 Pressure Vessels & Piping Division Conference*, PVP2013-97631, 14–18 July 2013, Paris, France.

Wei, Z., Yang, F., Cheng, H., Maleki, S., & Nikbin, K. (2013d) Engineering failure data analysis: revisiting the standard linear approach. *Engineering Failure Analysis, 30*, 27–42.

Yang, G. (1994). Optimum constant-stress accelerated life-test plans. *IEEE Transactions on Reliability, 43*, 575–581.

Yang, G. (2007). *Life cycle reliability engineering*. New Jersey: Wiley.

Yang, G., & Jin, L. (1994). Best compromise test plans for Weibull distributions with different censoring times. *Quality and Reliability Engineering International, 10*, 411–415.

Turbine Fatigue Reliability and Life Assessment Using Ultrasonic Inspection: Data Acquisition, Interpretation, and Probabilistic Modeling

Xuefei Guan, El Mahjoub Rasselkorde, Waheed A. Abbasi
and S. Kevin Zhou

1 Introduction

Turbine is one of the most critical components in fossil-fuel power plants. A recent report shows that fossil-fuel power plants provided 82 % of the total energy used in 2011 and will remain a main source of energy generation through 2040, accounting for 78 % of the total energy generation (U.S. Energy Information Administration 2013). Due to the importance and high cost of turbines, it is critical to ensure the operation safety and extend the useful life of existing turbines to reduce the total life-cycle cost. Recent development of turbine engineering set a new record for net efficiency of 60.75 % by a new Siemens gas turbine operated in a combined cycle with a steam turbine. Temperatures within the combustion chamber can be as high as 1500 °C, and the turbine blade tips can rotate at over 1700 km/h, which is much faster than the speed of sound. In such a server working environment, defects in turbine materials such as cracks may initiate, and propagate in a startup–hold–shutdown cycle due to fatigue and creep mechanisms. Ultrasonic inspection as one of the major nondestructive evaluations (NDE) approaches has increasingly been used in several industry sectors (Deng et al. 2004; Drinkwater and Wilcox 2006; Geng 2006; Sposito et al. 2010). In power generation industry, it is widely used due

X. Guan (✉) · S.K. Zhou
Siemens Corporation, Corporate Technology, 755 College Road East,
Princeton, NJ 08540, USA
e-mail: xuefei.guan@siemens.com

S.K. Zhou
e-mail: shaohua.zhou@siemens.com

E.M. Rasselkorde · W.A. Abbasi
Siemens Energy, Inc, 841 Old Frankstown Road, Pittsburgh, PA 15239-2246, USA
e-mail: elmahjoub.rasselkorde@siemens.com

W.A. Abbasi
e-mail: waheed.abbasi@siemens.com

© Springer-Verlag London 2016   415
H. Pham (ed.), *Quality and Reliability Management and Its Applications*,
Springer Series in Reliability Engineering, DOI 10.1007/978-1-4471-6778-5_14

to its capability of detecting flaws deeply embedded in components. Ultrasonic NDE data are analyzed to determine the existence, size, and shape of a flaw. The resulting information of a flaw is used for fatigue life and structural reliability assessments.

The difficulty in NDE data acquisition and interpretation lies in several aspects: Data acquisition requires highly specialized knowledge; acquired data is tremendously intensive for manual interpretations; noise and irrelevant signals can introduce uncertainties in flaw quantification, and spatial information is difficult to analyze since ultrasonic inspection data usually cannot encode detailed geometry information about the scanning path. It is therefore a significant interest to automate the overall process for reliable, efficient, and accurate flaw quantification and fatigue reliability assessment. The objective of this study is to develop a general method and procedure for fatigue reliability assessment integrating automatic ultrasonic nondestructive inspections. In addition, the life prediction and reliability assessment based on the NDE data must scientifically include uncertainties from several major sources such as measurements, sizing, model parameters, and so on. It is therefore highly necessary to develop a systematic methodology for reliable life prediction and reliability assessment using ultrasonic inspection.

The study is organized as follows: First, an automatic ultrasonic non-destructive inspection systems is briefly introduced. NDE data reconstruction and flaw quantification methods are developed. Next, uncertainty quantification models for probability of detection (POD) is established based on a classical log-linear model, which couples the ultrasonic inspection reported flaw size and the actual flaw size. Probability distributions of the actual flaw size are derived. After that, the overall procedure of fatigue reliability assessment incorporating the ultrasonic inspection data is suggested. A realistic steam turbine rotor example with actual ultrasonic inspection data is presented to demonstrate the overall method and procedure.

2 Data Acquisition and Interpretation of Ultrasonic Inspection

Flaw quantification involves flaw identification, grouping, and sizing. Data acquisition is the first step to obtain information for flaw quantification. Computerized automatic acquisition systems are devised and used to improve the efficiency and reduce operation uncertainty. Figure 1a, b illustrate diagrams of ultrasonic NDE data acquisition systems for bored rotors and solid rotors, respectively. The basic system consists of the ultrasonic probe, scan path controller, control terminal, and the inspection target. The probe is attached to the scan path controller allowing for accurate positioning according to programmed scanning paths. Scanning paths can be versatile. For example, in the solid rotor inspection setting the scanning path is around the outside surface of the rotor, starting at one axial position. After finishing one axial position, the probe moves to the next axial position and scans around the

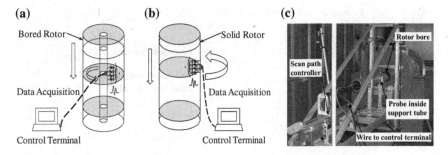

Fig. 1 Automatic ultrasonic inspection systems. **a** Rotor bore inspection diagram, **b** solid rotor inspection diagram, and **c** an automated rotor bore inspection system in action (Abbasi and Metala 2008)

surface again until the axial range of interest is covered. Notice that in the illustration diagrams the scanning planes are parallel to both ends of the rotors. It is also common to screw the probe 90° to set the scanning planes orthogonal to both ends of the rotors. One example system for rotor bore ultrasonic inspections is shown in Fig. 1c. The system consists of a scan path controller, a supporting tube, and a terminal. During data acquisition the probe is fixed on the supporting tube, and is attached to the bore surface. The scan path controller is used to move the probe back and forth supporting tube so different axial position can be inspected. The terminal is responsible for setting up parameters and executing the data acquisition. The rotor is rotated in a certain speed where a position encoder is used to record the probe movement relative to the rotation of the rotor.

2.1 Data Post-processing

Data reconstruction involves loading raw ultrasonic inspection data into a computer, mapping data points to correct physical positions, and producing visualizations with correct scales. The output of the reconstruction is stored as a spatial data grid, which can readily be rendered as a volumetric image. Each cell of the grid is characterized with physical position parameters such as (x, y, z) and the ultrasonic echo intensity. It is possible that multiple data points are mapped to one physical position (i.e., a cell in the data grid). For example, actual paths of ultrasound beams of an phased-array probe can occupy the same cell in the data grid, which is illustrated in Fig. 2. Different data fusion schemes can be used to characterize the cell where multiple data points are mapped to. For instance, using the maximum value or the average value of these data points in the cell are two possible fusion schemes.

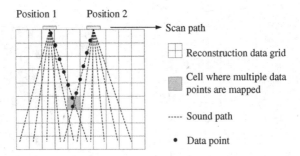

Fig. 2 Illustration of ultrasonic data reconstruction. The *shaded cell* exemplifies the case where multiple data points are mapped to one cell

2.2 Flaw Identification, Grouping, and Sizing

Flaw identification, grouping, and sizing procedure utilizes the reconstructed image. Automated or user-guided semi-automated operations are desired when a large amount of data need analysis.

Flaw identification can be made by iterating over the entire volumetric image and locating all image pixels whose intensity value is larger than a predefined threshold, e.g., $\alpha = 80\%$. Such pixels are identified as indication points. For each of those indication points, information about the physical location of the pixel, the indexes of the pixel in the image, and the normalized intensity is stored. It is also possible that only region of interest (ROI) is searched instead of the entire volumetric image. Using ROI can reduce the computation demands for flaw identification when the entire volumetric image is very large.

Multiple indication points may be connected (e.g., adjacent pixels) due to the scattering nature of ultrasonic signals and the dimensions of an actual flaw. Connected indication points can be clustered together since there is a large probability that those indication points are generated from an individual flaw. Region-growing methods (Adams and Bischof 1994) or similar algorithms can be used to cluster indication points according to the pixels' connectivity of the indication points. It is possible that an indication point group consists one isolated indication point. An indication point group will be considered as one flaw. Figure 3 shows an indication point group, where clustered indication points are marked as square dots. Flaw sizing for an indication point group is based on the method of Distance-Gain-Size (DGS) (Krautkrämer 1959). The basic idea of DGS is to evaluate the ultrasonic signal from an unknown reflector (e.g., a flaw) based upon the theoretical response of a flat-bottomed hole (FBH) reflector perpendicular to the beam axis. The signal intensity from a calibration FBH with a known size, e.g., a diameter of d_0, is calibrated to produce an echo amplitude of h_0 with the calibration signal intensity I_0. Given the base signal intensity I, the calibration gain is

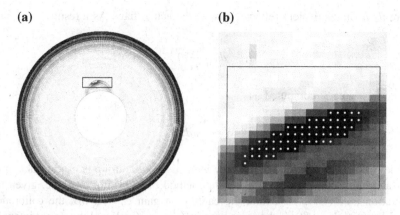

Fig. 3 A cluster of indication points marked as dots. **a** Global view, and **b** zoom to ROI

$$g_0 = 20 \log\left(\frac{I_0}{I}\right), \tag{1}$$

which leads to

$$I_0 = I10^{\frac{g_0}{20}}. \tag{2}$$

Assume the ultrasound inspection of an actual flaw gives an echo amplitude of h_1, the inspection gain is

$$g_1 = 20 \log\left(\frac{I_1}{I}\right), \tag{3}$$

leading to

$$I_1 = I10^{\frac{g_1}{20}}. \tag{4}$$

Denote the reflector area for the calibration hole as S_0 and the equivalent reflector area for the actual flaw as S_1. It is known that $S_0 I_0 \propto h_0$ and $S_1 I_1 \propto h_1$ and the following equations can be established:

$$\frac{S_1 I_1}{S_0 I_0} = \frac{h_1}{h_0} \tag{5}$$

and

$$\frac{\frac{1}{4}\pi d_1^2 I10^{\frac{g_1}{20}}}{\frac{1}{4}\pi d_0^2 I10^{\frac{g_0}{20}}} = \frac{h_1}{h_0}, \tag{6}$$

where d_1 is the equivalent reflector size of the actual flaw. As a result,

$$d_1 = d_0 \sqrt{\frac{h_1}{h_0}} 10^{\frac{g_0 - g_1}{40}}. \qquad (7)$$

The flaw area is computed as

$$S = \frac{1}{4} \pi d_1^2. \qquad (8)$$

The maximum intensity value of the indication point group is used to obtain an equivalent reflector size (ERS) using the method of DGS. For example, given the calibration hole size $d_0 = 2$ mm, the calibration gain $g_0 = 15$ dB, the calibration signal intensity $h_0 = 80\%$, the inspection gain $g_1 = 10$ dB, and the signal intensity of an indication $h_1 = 100\%$, the ERS of the indication is calculated as $d_1 = 2.98$ mm using Eq. (7). The reflector area is treated as the flaw area and is $\frac{1}{4}\pi d_1^2 = 6.98$ mm^2.

Fatigue reliability assessment relies on fracture mechanics. In particular, a fatigue crack propagation model is used to calculate the remaining useful life of a rotor. Fatigue crack propagation calculation takes the crack size of a chosen crack geometry. For embedded flaws identified by ultrasonic NDE, the flaw is assumed to be an embedded elliptical crack. Figure 4a presents the diagram of the embedded elliptical crack geometry, where **a** is the crack size. Since the actual flaw shape and orientation is unknown, an assumption must be made to transfer the ERS (i.e., the diameter of the reflector) to the size of an embedded elliptical crack. An accepted assumption is based on the idea that the reported reflector area is equal to the area of an embedded elliptical crack (Kern et al. 1998). The idea is shown in Fig. 4b. For example, an ultrasonic inspection reported a flaw with an ERS of 4 mm. With the assumption of $\mathbf{a}/c = 0.4$ for conservatism, the crack size is calculated from $\pi(4 \text{ mm})^2/4 = \pi a c$ and is $4 \text{ mm}/\sqrt{10}$. It should be noted that due to uncertainties from ultrasonic inspections, the dimensions of a flaw and its orientation, materials, etc., the actual flaw size or flaw area is different than ultrasonic inspection reported flaw size or flaw area. The uncertainties must be carefully included and

(a)   **(b)**

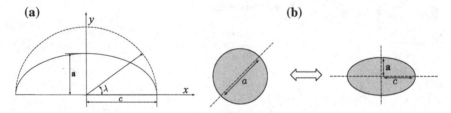

Fig. 4 **a** Embedded elliptical crack geometry. Semi-circle and semi-ellipse, and **b** illustration of converting the actual flaw size a (in terms of reflector diameter) to initial crack size **a** of an embedded elliptical crack

scientifically quantified to ensure a reliable assessment result of fatigue reliability. Uncertainty quantification is made using probabilistic modeling and is presented next.

3 Probabilistic Quantification of Flaws Using Probability of Detection Models

No inservice NDE method is perfect to produce precise results for detection, classification, and sizing of flaws (Georgiou 2006; Guan et al. 2014). The quality of ultrasonic NDE depends on many uncertain factors, including the capability of the ultrasonic probe, the service condition of the target structure being inspected, the variability of material properties, operation procedure and personnel, and so on. Scientific quantification of these uncertainties must be made in order to produce reliable and informative inspection results. Traditionally, deterministic treatment of the uncertainty uses safety factors (Freudenthal 1977; Kern et al. 1998). The determination of safety factors rely on expert judgment and long-term experiences. It is not a trivial task to find the optimal safety factors, and it may lead to risky or over conservative assessment results. Probabilistic modeling provides a rational approach for uncertainty management and quantification.

Two approaches are generally available for POD modeling (Achenbach 2000; Simola and Pulkkinen 1998). One approach uses hit/miss data, which only record whether a flaw was detected or not. This type of data is still in use for some nondestruction inspection methods, such as penetrant testing or magnetic particle testing. In other inspection systems additional information is available in testing data. For example, signal intensities and time indexes of ultrasonic NDE data, and voltage amplitudes and location information in electromagnetic responses. In those cases the flaw size or defect severity is closely correlated with signal responses, and thus the data are referred to as signal response data. Signal response data are usually continuous and denoted as $\hat{a}$. The variable of query is usually denoted as a. For example, a can be the actual flaw size and $\hat{a}$ is the ultrasonic inspection reported flaw size. It has been reported in many studies that $\ln \hat{a}$ and $\ln a$ is linearly correlated (Berens 1989; Schneider and Rudlin 2004) and can be expressed as

$$\ln \hat{a} = \alpha + \beta \ln a + \varepsilon, \tag{9}$$

where ε is a normal random variable with zero mean and standard deviation σ_ε. Both α and β are fitting parameters. A predefined threshold $\hat{a}_{th}$ is assumed according to the measurement noise and physical limits of measuring devices. It is also possible that $\hat{a}_{th}$ is specified by manufacturing criterion and standard. For example, a vendor may consider any flaw whose reported size is less than 1.0 mm is safe to be ignored. A flaw is regarded as identified if $\hat{a}$ exceeds the threshold value of $\hat{a}_{th}$, and the probability of detection of size a can be expressed as

$$\text{POD}(a) = \Pr(\ln \hat{a} > \ln \hat{a}_{\text{th}}), \tag{10}$$

where $\Pr(\cdot)$ represents the probability of an event $(\cdot)$. Using Eq. (9), the POD function is rewritten as

$$\text{POD}(a) = \Pr(\alpha + \beta \ln a + \varepsilon > \ln \hat{a}_{\text{th}}) = \Phi\left(\frac{\ln a - (\ln \hat{a}_{\text{th}} - \alpha)/\beta}{\sigma_\varepsilon/\beta}\right), \tag{11}$$

where $\Phi(\cdot)$ is the standard normal cumulative distribution function (CDF). Terms α and β are obtained using maximum likelihood estimator (MLE) from the calibration data. If the variable ε follows another probability distribution other than the standard normal distribution, Eq. (11) can still be established but the corresponding CDF of ε, instead of $\Phi(\cdot)$, should be used. Following the convention, random variables are denoted using capital letters (e.g., $\hat{A}$) and corresponding values are denoted using lower case letters (e.g., $\hat{a}$). Assume that a flaw is detected and the value of the reported flaw size is a', where a' is a positive real scalar. For convenience, the variable $\ln \hat{A}$ is used instead of $\hat{A}$. Denote the probability distributions for variables $\ln \hat{A}$, $\ln A$, and $\mathcal{E}$ as $p(\ln \hat{A})$, $p(\ln A)$, and $p(\mathcal{E})$, respectively. PDFs for $\ln \hat{A}$, $\ln A$, and $\mathcal{E}$ are represented by $f_{\ln \hat{A}}(\ln \hat{a})$, $f_{\ln A}(\ln a)$, and $f_{\mathcal{E}}(\varepsilon)$, respectively. The probability distribution of the actual flaw size, $p(\ln A)$, is of interest and its derivation is presented below.

Denote D as the event that a flaw is identified and $\bar{D}$ is the event that a flaw is not identified. The joint probability distribution $p(\ln A, \ln \hat{A}, \mathcal{E}|D)$ can be used to obtain $p(\ln A|D)$ as

$$p(\ln A|D) = \int \int p(\ln A, \ln \hat{A}, \mathcal{E}|D) \mathrm{d} \ln \hat{A} \mathrm{d} \mathcal{E}. \tag{12}$$

It can be further expressed, under the condition that $\ln \hat{A}$ and $\mathcal{E}$ are independent, as

$$p(\ln A|D) = \int \int p(\ln A| \ln \hat{A}, \mathcal{E}, D) p(\ln \hat{A}|D) p(\mathcal{E}|D) \mathrm{d} \ln \hat{A} \mathrm{d} \mathcal{E}. \tag{13}$$

Since $\ln \hat{a} = \alpha + \beta \ln a + \varepsilon$,

$$p(\ln A| \ln \hat{A}, \mathcal{E}, D) = \delta(\ln \hat{a} - \alpha - \beta \ln a - \varepsilon), \tag{14}$$

where $\delta(\cdot)$ is the Dirac delta function. Substitute Eq. (14) into Eq. (13) to obtain

$$p(\ln A|D) = \int_R f_{\ln \hat{A}}(\ln \hat{a})\mathrm{d}\ln \hat{a} \int_R f_{\mathcal{E}}(\varepsilon)\delta(\ln \hat{a} - \alpha - \beta \ln a - \varepsilon)\mathrm{d}\varepsilon$$

$$= \int_R f_{\ln \hat{A}}(\ln \hat{a})f_{\mathcal{E}}(\ln \hat{a} - \alpha - \beta \ln a)\mathrm{d}\ln \hat{a}. \tag{15}$$

3.1　Deterministic Conversion Model

Denote the raw signal feature, such as the signal intensity in the ultrasonic inspection data, as x, and the conversion is made through a mathematical model $m(x)$. It is clear that if the model is perfect $m(x) = \ln \hat{a}$ which leads to

$$f_{\ln \hat{A}}(\ln \hat{a}) = \delta(\ln \hat{a} - m(x)). \tag{16}$$

Substitution of Eq. (16) into Eq. (15) yields (recalling the actual value of $m(x)$ is now $\ln a'$)

$$p(\ln A|D) = \int_R \delta(\ln \hat{a} - m(x))f_{\mathcal{E}}(\ln \hat{a} - \alpha - \beta \ln a)\mathrm{d}\ln \hat{a} = f_{\mathcal{E}}(\ln a' - \alpha - \beta \ln a). \tag{17}$$

Recall $f_{\mathcal{E}}(\cdot)$ is a normal PDF with zero mean and standard deviation σ_ε. It is symmetric and $\alpha + \beta \ln a - \ln a'$ also follows a normal distribution with zero mean and standard deviation σ_ε. Recognize that $\ln A$ follows a normal PDF with mean $(\ln a' - \alpha)/\beta$ and standard deviation σ_ε/β, and thus A is a log-normal variable. The PDF of variable A is

$$p(A|D) = f_{A|D}(a) = \frac{1}{a(\sigma_\varepsilon/\beta)} \phi\left(\frac{\ln a - (\ln a' - \alpha)/\beta}{\sigma_\varepsilon/\beta}\right), \tag{18}$$

where $\phi(\cdot)$ is the standard normal PDF.

3.2　Probabilistic Conversion Model

If the conversion model is uncertain and the difference between the model output $m(x)$ and the estimated size $\ln \hat{a}$ is a random quantity e, the following equation can be established:

$$\ln \hat{a} = m(x) + e. \tag{19}$$

Denote the random variable as E, and the probability distribution function for E as $f_E(e)$. It is quite common to make the assumption that E is a normal variable with zero mean and standard deviation of σ_e.

$$p(\ln \hat{A}|D) = \int p(\ln \hat{A}, E|D)p(E|D)dE. \tag{20}$$

Since $\ln \hat{a} = m(x) + e$,

$$p(\ln \hat{A}, E|D) = \delta(\ln \hat{a} - m(x) - e). \tag{21}$$

and

$$p(\ln \hat{A}|D) = f_{\ln \hat{A}}(\ln \hat{a}) = \int_R \delta(\ln \hat{a} - m(x) - e)f_E(e)de = f_E(\ln \hat{a} - m(x)). \tag{22}$$

Substitute Eq. (22) into Eq. (15), with the actual value of $m(x) = \ln a'$, to obtain

$$p(\ln A|D) = \int_R f_E(\ln \hat{a} - \ln a')f_\varepsilon(\ln \hat{a} - \alpha - \beta \ln a)d\ln \hat{a}. \tag{23}$$

Recognizing Eq. (23) is a convolution of two normal probability distributions, it is well-known that the result is another normal distribution given by

$$p(\ln A|D) = \frac{1}{\sqrt{2\pi}\sqrt{\sigma_e^2 + \sigma_\varepsilon^2}}\exp\left\{-\frac{1}{2}\frac{(\alpha + \beta \ln a - \ln a')^2}{\sigma_e^2 + \sigma_\varepsilon^2}\right\}. \tag{24}$$

Again, $\ln A$ is a normal variable and A is a log-normal variable with the following PDF,

$$p(A|D) = f_{A|D}(a) = \frac{1}{a\left(\sqrt{\sigma_e^2 + \sigma_\varepsilon^2}/\beta\right)}\phi\left(\frac{\ln a - (\ln a' - \alpha)/\beta}{\sqrt{\sigma_e^2 + \sigma_\varepsilon^2}/\beta}\right). \tag{25}$$

It can be seen that if the degree of the uncertainty associated with the conversion model is approaching zero, i.e., $\sigma_e \to 0$, Eq. (25) reduces to Eq. (18). In addition, if the uncertainty in variable $\mathcal{E}$ is approaching zero, i.e., $\sigma_\varepsilon \to 0$, $\ln a = (\ln a' - \alpha)/\beta$.

3.3   No Indication Found in NDE Data

A clean NDE data does not indicate the target structure is completely free of flaws due to the inherent uncertainty and the inspection threshold $\hat{a}_{th}$. The distribution of A can readily be expressed using Bayes' theorem as

$$p(A|\bar{D}) = \frac{p(A,\bar{D})}{p(\bar{D})} = \frac{p(\bar{D}|A)p(A)}{p(\bar{D})}, \tag{26}$$

where $p(A)$ is the prior probability distribution of the flaw size, and $p(\bar{D}|A)$ is the probability of the event that no indication is found when a flaw actually exists. It should be noted that this usually requires some sort of NDE data base of prior inspections in order to have information about the flaw distribution of the fleet. Denote the prior PDF of A as $f_A(a)$. Using the concept of POD and Eq. (11), the probability of the event that the size of a flaw is not larger than a given value a conditional on clean inspection data is (Tang 1973; Zheng and Ellingwood 1998)

$$p(A \le a|\bar{D}) = \frac{\int_0^a [1 - POD(a)]f_A(a)da}{\int_0^\infty [1 - POD(a)]f_A(a)da}. \tag{27}$$

The probability distribution of a flaw with size a conditional on clean inspection data is

$$p(A|\bar{D}) = f_{A|\bar{D}}(a) = \frac{\partial[p(A \le a|\bar{D})]}{\partial a} = \frac{[1 - POD(a)]f_A(a)}{\int_0^\infty [1 - POD(a)]f_A(a)da}, \tag{28}$$

where $POD(a)$ is given in Eq. (11) and $f_A(a)$ is the prior PDF of the actual flaw size. If no prior information about the distribution of A is available, the assumption that the actual flaw size is uniformly distributed in the range of $(0, \hat{a}_{th})$ can be made. If information about $f_A(a)$ is available, the information can be directly incorporated. Based on the above discussions, the PDF of the actual flaw size with the NDE data can be summarized as

$$p(A|NDE) = \begin{cases} f_{A|D}(a) = \frac{1}{a}N\left(\frac{\ln a' - \alpha}{\beta}, \frac{\sigma_\varepsilon}{\beta}\right) & \text{Deterministic } m(x), \text{ size } a' \\ f_{A|D}(a) = \frac{1}{a}N\left(\frac{\ln a' - \alpha}{\beta}, \frac{\sqrt{\sigma_e^2 + \sigma_\varepsilon^2}}{\beta}\right) & \text{Probabilistic } m(x), \text{ size } a' \\ f_{A|\bar{D}}(a) = \frac{[1 - POD(a)]f_A(a)}{\int_0^\infty [1 - POD(a)]f_A(a)da} & \text{No indication} \end{cases}, \tag{29}$$

where $N(\cdot)$ represents a normal PDF. The first parameter of $N(\cdot)$ is the mean value and the second one is the standard deviation. All other symbols are defined as before.

4   Fatigue Reliability and Life Assessment

Fatigue reliability assessment involves the evaluation of an integral of the performance function over the failure domain. The integral can be expressed as

$$P_F \equiv P[g(x) < 0] = \int_{\forall x | g(x) < 0} p(x) dx, \tag{30}$$

where P_F is the probability of failure (POF), $x \in R^k$ is a real-valued k-dimensional uncertain variable, $g(x)$ is the performance function, such that $g(x) < 0$ represents the failure event, and $p(x)$ is the PDF of x. For fatigue reliability assessment, the failure event is usually defined as the flaw characterization (e.g., the crack size) being larger than a predefined threshold (e.g., the critical crack size). A fatigue crack may propagate subject to cyclic loads and eventually reaches the critical crack size. Paris' equation is one the most commonly used crack propagation models and is expressed as

$$\frac{d\mathbf{a}}{dN} = C(\Delta K)^m, \tag{31}$$

where $\mathbf{a}$ is the crack size, N is the number of load cycles, C and m are model parameters estimated from fatigue testing data, and ΔK is the stress intensity factor range during one load cycle. Calculation of the crack size given a number of load cycles N_t is made by cycle integration of Eq. (31). Based on the above crack propagation model, the performance function of a fatigue failure at cycle $N_t \in R^+$ is expressed as

$$g(x) := \mathbf{a}_c - \int_0^{N_t} \frac{d\mathbf{a}}{dN}, \quad \mathbf{a} = \mathbf{a}_0 \text{ at } N = 0. \tag{32}$$

where $\mathbf{a}_c$ is the critical crack size and $\mathbf{a}_0$ is the initial crack size. Considering model parameters and the initial crack size as uncertain variables, $x := (C, m, \mathbf{a}_0)$. It is also noticed that $g(x)$ is cycle-dependent or time-dependent since for an arbitrary given N_t, results of $g(x)$ and P_F depends on the actual value of N_t.

4.1   Monte Carlo-Based Estimation

Given the joint distribution of $(C, m, \mathbf{a}_0)$, simulation-based methods can be used to evaluate Eq. (30), such as Monte Carlo (MC) simulations and its variants (Billinton and Allan 1983). The basic procedure is described as follows:

1. Generate a sufficient number (n) of random instances from $p(x)$.
2. For each of the random instances, evaluate the crack size for a specified N_t and record the results.

The MC estimation of POF using n random samples is

$$\widehat{POF} = \frac{1}{n}\sum_{i=1}^{n} \mathbf{1}, \tag{33}$$

where $\mathbf{1}$ is a function taking value 1 if the calculation using ith random sample produces a failure event and taking value 0 otherwise. The variance of the estimator is estimated as

$$\text{Var}(\widehat{POF}) = \frac{1}{n-1}\left[\frac{1}{n}\sum_{i=1}^{n} \mathbf{1}^2 - \widehat{POF}^2\right]. \tag{34}$$

Extra attentions must be paid to error analysis in using MC simulations. This is due to the fact that in most realistic situations fatigue failure is considered as a rare event for high-reliability demanding systems, and thus POF is sometime a very small quantity. The minimal number of random instances required for the MC estimator depends on the risk level allowed for the system being evaluated. For example, a nuclear steam turbine may allow for a maximum 1 % relative error (RE) but 5 % RE may be enough for a wind turbine. It is well-known that the RE of an MC estimator can be obtained as (Rubinstein and Kroese 2007)

$$\text{RE} \overset{\text{def}}{=} \frac{\sqrt{\text{Var}(\widehat{POF})}}{POF} \approx \sqrt{\frac{1}{n \cdot POF}}, \tag{35}$$

and the minimal number of random instances required is

$$n \approx \frac{1}{\text{RE}^2 \cdot POF}. \tag{36}$$

For example, using an MC estimator to estimate a failure event with $POF = 10^{-6}$. Achieving 5 % RE requires at least $n = 4 \times 10^8$ random instances. In addition, from the Central Limit Theorem, a confidence interval $CI = [POF^-, POF^+]$ can be defined for the threshold $1 - 2\xi$. It is such that $\Pr(POF^- < \widehat{POF} < POF^+)$, and the CI can be expressed as

$$CI \approx \left[\widehat{POF} - z_\xi\sqrt{\text{Var}(\widehat{POF})}, \widehat{PoF} + z_\xi\sqrt{\text{Var}(\widehat{POF})}\right], \tag{37}$$

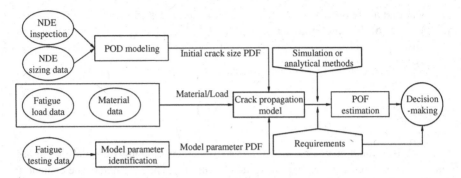

Fig. 5 Overall computational procedure of reliability assessment integrating ultrasonic NDE

where $z_\xi = \Phi^{-1}(1 - \xi)$ and $\Phi^{-1}(\cdot)$ is the inverse CDF of the standard normal variable. For instance, if the threshold is chosen as 95 % then $\xi = 2.5\%$ and $z_\xi \approx 1.96$.

Based on above discussions, the overall computational procedure of the reliability assessment using ultrasonic NDE data is shown in Fig. 5. Next, a realistic industrial example is presented to demonstrate the overall method and procedure.

4.2 An Application Example

A steam turbine rotor made of Cr–Mo–V material is of interest. For illustration purposes, historical ultrasonic NDE sizing data reported in Schwant and Timo (1985) are used to represent the actual detection features of the ultrasonic inspection system. Using the log-scale data shown in Fig. 6a and linear regression, the coefficient and intercept in Eq. (9) are estimated as $\beta = 0.658$, $\alpha = 0.381$, respectively. The standard deviation of ε is estimated as $\sigma_\varepsilon = 0.616$. The mean and 95 % bound prediction results are shown in Fig. 6a in solid and dash lines, respectively. For investigation purposes, three values of 0.5, 1.0, and 1.5 mm are considered as the threshold value $\hat{a}_{th}$. The POD model is established as the following equation:

$$\text{POD}(a) = \begin{cases} \Phi\left(\dfrac{\ln a + 1.6334}{0.9368}\right), & \hat{a}_{th} = 0.5\,\text{mm} \\[2mm] \Phi\left(\dfrac{\ln a + 0.5793}{0.9368}\right), & \hat{a}_{th} = 1.0\,\text{mm} \\[2mm] \Phi\left(\dfrac{\ln a - 0.0374}{0.9368}\right), & \hat{a}_{th} = 1.5\,\text{mm} \end{cases} \tag{38}$$

The three POD curves are shown in Fig. 6b. For an embedded elliptical crack as shown in Fig. 4a, the stress intensity factor of a point located at an angle λ with respect to the direction of the applied tensile stress σ is given by

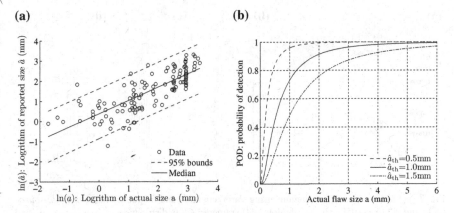

Fig. 6 **a** Sizing information of actual flaw size and ultrasonic NDE reported size in log-scale (*data source* Schwant and Timo 1985), mean and bound predictions using log-linear model, and **b** POD curves obtained using the sizing data with threshold values of 0.5, 1.0, and 1.5 mm

$$K = \sigma\sqrt{\pi \mathbf{a}M}/Q\left[\sin^2\lambda + \left(\frac{\mathbf{a}}{c}\right)^2\cos^2\lambda\right]^{1/4}, \qquad (39)$$

where M is the location factor, **a** the crack size defined before which is also the minor axis length of the semi-ellipse and c is the major axis length of the semi-ellipse. Term Q is the shape factor defined as $Q = \int_0^{\pi/2}\sqrt{1 - \left(\frac{c^2-\mathbf{a}^2}{c^2}\right)\sin^2\lambda}\,d\lambda$. For embedded crack geometry $M = 1.0$. It is worth mentioning that the most critical ratio value (**a**/c) depends on several factors and can not be easily predicted. For conservatism purposes in engineering applications, **a**/c takes 0.4 and K has a maximum value at $\lambda = \pi/2$. The joint distribution of $(\ln C, m)$ is estimated using the method of Bayesian parameter estimation and fatigue testing data of Cr–Mo–V material at 800 F reported in Shih and Clarke (1950). Details of Bayesian parameter estimation with Markov chain Monte Carlo (MCMC) simulations can be found in Guan et al. (2013). Considering the two parameters follow a multivariate normal distribution (MVN), the joint PDF is expressed as

$$f(\ln C, m) \sim \text{MVN}\left(\mu_{(\ln C,m)}, \Sigma_{(\ln C,m)}\right). \qquad (40)$$

The mean vector is $\mu_{(\ln C,m)} = [-22.23, 2.151]$ and the covariance matrix is $\Sigma_{(\ln C,m)} = \begin{bmatrix} 1.2143 & -0.17465 \\ -0.17465 & 0.02513 \end{bmatrix}$. The fitting performance of the estimation is visualized in Fig. 7a, where mean and 95 % bound predictions are shown.

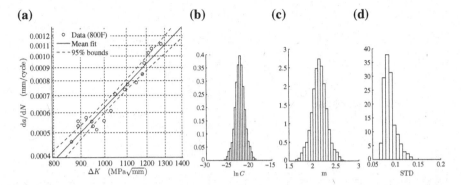

Fig. 7 **a** Mean and bound predictions using Paris' equation and fatigue testing data (*data source* Shih and Clarke 1950), **b** distribution of parameter C, **c** distribution of parameter m, and **d** distribution of standard deviation (STD) of the Gaussian likelihood

If no flaw was identified, the PDF of the actual flaw size is given by Eq. (28). Without loss of generality, $f_A(a)$ in Eq. (28) is assumed to be a uniform distribution between 0 and the detection threshold $\hat{a}_{th}$. Based on this assumption, PDFs of the actual flaw size and the corresponding POF evaluations are presented in Fig. 8a, b respectively.

The rotor was scheduled for an ultrasonic inspection after a 15-year service. The raw inspection data, reconstructed data, and the identified flaw is shown in Fig. 9. The raw inspection data were shown in Fig. 9a. The raw data were collected using a traditional 2 MHz monolithic probe. The probe was mounted to the surface of the outside diameter of the rotor. The scan schematic plot shown in Fig. 1b can be used to represent the actual acquisition setting. The step size is 0.188° along the moving path, and a total of 1921 positions are used to cover the entire section (e.g., $1921 \times 0.188 \approx 361°$). The reconstructed data were shown in Fig. 9b, where a flaw

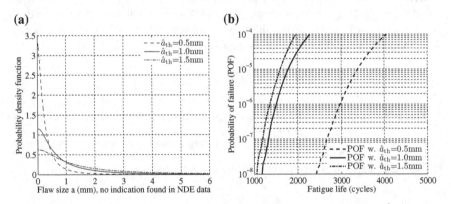

Fig. 8 Assessment results given that no flaw was found in NDE data. **a** PDFs of actual flaw size in terms of ERS, and **b** POF evaluations

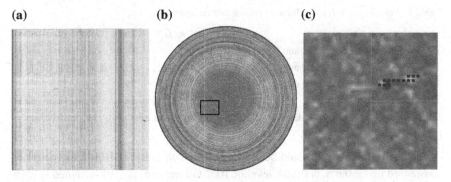

Fig. 9 **a** Raw ultrasonic inspection data, **b** reconstructed image, the flaw region is annotated by a rectangular box, and **c** the identified flaw as an indication point group denoted as *square dots*

indication was annotated by a rectangular box. Figure 9c presents the indication point group representing the flaw. Indication points are pixels whose intensity value is larger than a predefined threshold $\alpha = 50\%$. The concept of indication point can be found in Sect. 2.2. Using the method of DGS, the flaw size is reported as an ERS of 1.8 mm. The PDF of the actual flaw size is obtained from Eq. (18) and is presented in Fig. 10a.

Risk analysis depends on the interpretation of the safety parameters. For example, Nuclear Regulatory Commission (NRC) approved risk levels for nuclear power plants is 10^{-4} failures/year for rotor disk bursts under favorable configurations, and 10^{-5} failures/year for rotor disk bursts under unfavorable configurations (U.S. Nuclear Regulatory Commission 1987). A favorable configuration refers to the case when the pressure vessel and nuclear fuel rods are not to the side of the turbine. In addition, the risk level approved by NRC accounts for the entire rotor disk. When multiple forging parts are considered, uniformly assigning the overall

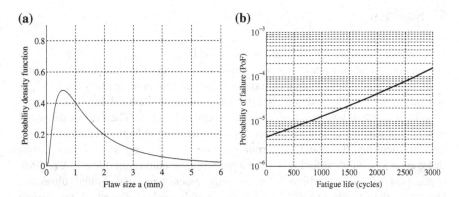

Fig. 10 Assessment results with a flaw reported as an ERS of 1.8 mm. **a** The PDF of the actual flaw size in terms of ERS, and **b** POF evaluations

Table 1 Fatigue life based on different failure rate requirements

Component risk (failures/year)	2.5×10^{-6}	5×10^{-6}	7.5×10^{-6}	1.0×10^{-5}
Life (years): no flaw, $\hat{a}_{th} = 0.5$ mm	53.3	55.5	56.8	57.8
Life (years): no flaw, $\hat{a}_{th} = 1.0$ mm	27.5	28.9	29.9	30.7
Life (years): no flaw, $\hat{a}_{th} = 1.5$ mm	24.9	26.1	26.9	27.5
Life (years): 1.8 mm flaw	0	1.8	8.4	13.2

Duration of each cycle is considered as 150 h. Results are based on 4×10^8 MC simulations

risk to each of the parts should be considered. For instance, if 10 forging parts are considered under the overall risk level of 10^{-4} failure/year, the risk assigned to each of the 10 parts is 10^{-5} failures/year. This component risk could be used to evaluate the remaining useful life for the part containing a flaw. The uniform distribution of risk to the number of rotor components, when only evaluating the most critical ones, can be considered as conservative as the total risk for the entire rotor is approximately the sum of the individual risks (low risk approximation). More accurate evaluation would be the calculation of the total rotor risk and its comparison to the acceptable risk limit of 10^{-4} per year. More details on this topic can be found in (U.S. Nuclear Regulatory Commission 1987). Based on different recommended failure rates for the rotor, a more detailed table can be built for decision-making, as shown in Table 1. For example, consider the cases that only one flaw is found in the entire rotor. Using the recommended failure rate of 10^{-5} assigned to the forging part having a flaw of 1.8 mm, the rotor can still be safely used for about 13.2 years and another NDE testing can be scheduled after another 10-year service. However using the recommended failure rate of 5×10^{-6} assigned to the forging part having a flaw of 1.8 mm, the rotor can only be used for about 1.8 years. It is possible that the rotor component might need a replacement after the evaluation.

5   Summary

This chapter presents a systematic method for fatigue reliability assessment integrating automatic ultrasonic nondestructive inspections. Uncertainties from ultrasonic inspections, flaw characterization, and fatigue model parameters are included and quantified. POD modeling is established using a classical log-linear model coupling the actual flaw size and the NDE reported flaw size. The PDF of the actual flaw size is derived considering three typical scenarios of NDE data: (1) No flaw, (2) with identified flaws and a deterministic model converting NDE signals to flaw sizes, and (3) with identified flaws and a probabilistic conversion model. A realistic application of steam turbine rotor integrity assessment with actual ultrasonic inspection data is used to demonstrate the overall method. Based on the current study, the following can be summarized: (1) An automatic ultrasonic inspection system is introduced. Ultrasonic data reconstruction method is developed. Flaw identification, grouping, and sizing methods are also proposed to characterize flaws, (2) a general methodology for fatigue reliability assessment is formulated by developing uncertainty quantification models for ultrasonic nondestructive inspections, sizing, and model parameters, (3) probability distributions of the actual flaw size under different NDE data scenarios are developed based on probabilistic modeling and Bayes' theorem. The derivation and results are general and can readily be adapted to different NDE applications, and (4) a realistic application example of steam turbine rotor with ultrasonic inspection data is presented to demonstrate the overall methodology. The interpretation of the results using risk parameters based on NRC recommendations is suggested.

References

Abbasi, W., & Metala, M. (2008). Recent advances in NDE technologies for turbines and generators. In *17th World Conference on Nondestructive Testing*.

Achenbach, J. (2000). Quantitative nondestructive evaluation. *International Journal of Solids and Structures, 37*(1), 13–27.

Adams, R., & Bischof, L. (1994). Seeded region growing. *IEEE Transactions on Pattern Analysis and Machine Intelligence, 16*(6), 641–647.

Berens, A. P. (1989). NDE reliability data analysis. In: *ASM Handbook* (vol. 17, 9th edn., pp. 689–701). ASM International.

Billinton, R., & Allan, R. (1983). *Reliability evaluation of engineering systems: concepts and techniques*. New York: Plenum Press.

Deng, W., Shark, L., Matuszewski, B., Smith, J., & Cavaccini, G. (2004). CAD model-based inspection and visualisation for 3D non-destructive testing of complex aerostructures. *Insight-Non-Destructive Testing and Condition Monitoring, 46*(3), 157–161.

Drinkwater, B., & Wilcox, P. (2006). Ultrasonic arrays for non-destructive evaluation: a review. *NDT and E International, 39*(7), 525–541.

Freudenthal, A. (1977). The scatter factor in the reliability assessment of aircraft structures. *Journal of aircraft, 14*(2), 202–208.

Geng, R. (2006). Modern acoustic emission technique and its application in aviation industry. *Ultrasonics, 44*, e1025–e1029.

Georgiou, G. (2006). *Probability of Detection (POD) curves: derivation, applications and limitations*. Technical Report 454, Jacobi Consulting Limited, London, UK.

Guan, X., Zhang, J., Kadau, K., & Zhou, S. K. (2013). Probabilistic fatigue life prediction using ultrasonic inspection data considering equivalent initial flaw size uncertainty. In D. O. Thompson, & D. E. Chimenti (Eds.), *AIP Conference Proceedings* (vol. 1511, pp. 620–627). AIP.

Guan, X., Zhang, J., Rasselkorde, E. M., Abbasi, W. A., & Kevin Zhou, S. (2014). Material damage diagnosis and characterization for turbine rotors using three-dimensional adaptive ultrasonic NDE data reconstruction techniques. *Ultrasonics, 54*(2), 516–525.

Kern, T., Ewald, J., & Maile, K. (1998). Evaluation of NDT-signals for use in the fracture mechanics safety analysis. *Materials at High Temperatures, 15*(2), 107–110.

Krautkrämer, J. (1959). Determination of the size of defects by the ultrasonic impulse echo method. *British Journal of Applied Physics, 10*(6), 240–245.

Rubinstein, R., & Kroese, D. (2007). *Simulation and the Monte Carlo method* (vol. 707). Wiley-interscience.

Schneider, C., & Rudlin, J. (2004). Review of statistical methods used in quantifying NDT reliability. *Insight-Non-Destructive Testing and Condition Monitoring, 46*(2), 77–79.

Schwant, R., & Timo, D. (1985). Life assessment of general electric large steam turbine rotors. In *Life assessment and improvement of turbo-generator rotors for fossil plants* (pp. 1–8). New York: Pergamon Press.

Shih, T., & Clarke, G. (1979). Effects of temperature and frequency on the fatigue crack growth rate properties of a 1950 vintage CrMoV rotor material. In *Fracture Mechanics: Proceedings of the Eleventh National Symposium on Fracture Mechanics* (vol. 700, p. 125). ASTM International.

Simola, K., & Pulkkinen, U. (1998). Models for non-destructive inspection data. *Reliability Engineering & System Safety, 60*(1), 1–12.

Sposito, G., Ward, C., Cawley, P., Nagy, P., & Scruby, C. (2010). A review of non-destructive techniques for the detection of creep damage in power plant steels. *NDT and E International, 43*(7), 555–567.

Tang, W. (1973). Probabilistic updating of flaw information(flaw prediction and control in welds). *Journal of Testing and Evaluation, 1*, 459–467.

U.S. Energy Information Administration. (2013). Annual energy outlook 2013. Retrieved from http://www.eia.gov.

U.S. Nuclear Regulatory Commission. (1987). Standard review plan for the review of safety analysis reports for nuclear power plants, LWR edn. Washington, D.C.: US Nuclear Regulatory Commission, Office of Nuclear Reactor Regulation.

Zheng, R., & Ellingwood, B. (1998). Role of non-destructive evaluation in time-dependent reliability analysis. *Structural Safety, 20*(4), 325–339.

Fusing Wavelet Features for Ocean Turbine Fault Detection

Janell Duhaney, Taghi M. Khoshgoftaar and Randall Wald

1 Introduction

As systems and machines continue to become increasingly complex, the demand for automated tools to monitor them has also increased as ensuring the reliability of these machines plays an important role in maximizing availability and minimizing maintenance costs. Machine condition monitoring (MCM) systems provide such a means for continuous and intelligent problem detection, while prognostics and health monitoring (PHM) systems work to predict the life expectancy of these machines. MCM/PHM systems have been widely used in many industries, including aeronautics, robotics, and energy production.

This paper concerns an MCM/PHM system for one specific machine that has been the subject of recent research efforts into renewable energy production—the ocean turbine. Ocean turbines provide a clean, renewable power alternative to fossil fuels by converting the kinetic energy in ocean currents to electricity. These machines will operate autonomously below the ocean's surface in varying and sometimes harsh environmental conditions which raises numerous reliability concerns, including susceptibility to bio-fouling (accumulation of flora or fauna on the surface of the turbine), corrosion of the cables connecting the turbine to its topside components due to salinity, as well as the possibility that animals or underwater debris may strike and/or damage the turbine (Beaujean et al. 2009).

J. Duhaney (✉) · T.M. Khoshgoftaar · R. Wald
Computer and Electrical Engineering and Computer Science, Florida Atlantic University,
Boca Raton, Florida
e-mail: fjduhane1@fau.edu

T.M. Khoshgoftaar
e-mail: khoshgof@fau.edu

R. Wald
e-mail: rwald1g@fau.edu

© Springer-Verlag London 2016
H. Pham (ed.), *Quality and Reliability Management and Its Applications*,
Springer Series in Reliability Engineering, DOI 10.1007/978-1-4471-6778-5_15

435

Frequent manual inspections are infeasible due to high expeditionary costs to access the machines, and problems which occur between inspection intervals can go undetected until the next maintenance visit. Also, ocean turbines are highly intolerant of false alarms due to the high costs associated with equipment retrieval. Thus, reliable and timely detection of problems is a must. A well-designed MCM/PHM system can provide these and other capabilities, such as predicting the likelihood of future faults and estimating time to failure under given operational conditions. Another added benefit of using an MCM/PHM system is that it allows for predictive maintenance of a machine (where maintenance tasks are performed as needed) versus routine (or time-based) preventative maintenance (where maintenance activities are scheduled at predetermined intervals). In a preventative maintenance scheme, one challenge is ascertaining how frequent such maintenance activities should be performed. Too frequent visits could end up wasting money and resources, while infrequent visits run the risk of faults developing between scheduled visits which may damage the machine and cause unnecessary downtime.

In the absence of a live ocean turbine, we rely on data from its test bed—a 22 kW dynamometer—to guide the design of the MCM/PHM system. The vibration signals gathered from this dynamometer test bed are expected to be similar to what we will observe in a deployed ocean turbine; thus, we utilize data we collect from this test bed to provide input to a data-driven MCM system being designed for the turbine.

This paper focuses on employing feature-level sensor fusion to enable machine learners to reliably detect changes in the operational state of the dynamometer. To the author's knowledge, this is the second study related to feature-level fusion of experimental data for enabling condition monitoring of an ocean turbine. In the prior study (Duhaney et al. 2011)—also conducted by our group—feature-level fusion was applied to vibration data gathered from two sensors attached to a box fan. Preliminary findings from that study encouraged future research in this direction. The study presented later in this paper was performed on data gathered from six sensors (as compared to two sensors in the earlier study) mounted on an ocean turbine's dynamometer, which, compared to the box fan, produces vibration signatures more closely related to that which would be expected from the turbine. One reason for this is the rotational velocity at which the dynamometer is operated which is similar to those of an ocean turbine; the blades of a box fan rotate at a much faster rate.

Additional backgrounds on data fusion and on the ocean turbine are provided in Sect. 2. The feature extraction and feature-level fusion techniques are described in Sect. 3. The case study discussed in Sect. 4 demonstrates how various data mining and machine learning techniques combined with the feature-level fusion approach allow for reliable classification of operational state. Results of the study are given in Sect. 5, while final comments and plans for future studies appear in Sect. 6.

Sensor fusion is the process of combining data originating from multiple sensor sources with the goal of achieving a more complete and/or accurate view. Through a case study involving experimental data, we demonstrate how feature-level sensor fusion—which is sensor fusion applied after descriptive features are extracted from the raw sensor data—can enable reliable fault detection in a condition monitoring system for an ocean tuine. Sensor fusion is needed in the condition monitoring context to combine readings from sensors measuring different physical phenomena and/or different components of a machine. This study assesses the abilities of six well-known machine learners to detect changes in the state of the turbine from the fused sensor data. Analysis of the performance of these classifiers showed more stable performances for the six classifiers in detecting the state of the machine from the fused data versus from the data from the individual sensor channels.

2 Related Work

Predictive maintenance involves using the condition of a machine to determine when and how often maintenance should be performed. One major component of predictive maintenance, called condition monitoring, focuses on monitoring the normal state of the machinery, allowing for failures to be detected based on significant changes in one or more of the parameters being monitored. Machine condition monitoring and prognostics health monitoring (MCM/PHM) systems allow for predictive maintenance of a machine by recording, processing, and interpreting data collected from sensors installed on the machine being monitored.

One machine that would greatly benefit from the automated monitoring capabilities of an MCM/PHM system is the ocean turbine. Research into developing a 20-kilowatt ocean turbine for harvesting ocean current energy from the Gulf Stream is underway by the Southeast National Marine Renewable Energy Center at the Florida Atlantic University (Beaujean et al. 2009). An MCM system will be employed to perform continuous, automated self-checking of this turbine to increase its reliability and to aid in maintenance planning.

An MCM system for an ocean turbine could include sensors to measure oil level and quality, temperature, turbidity, electrical output, rotational velocity, and vibration. Some sensors such as oil and temperature sensors produce a single reading at regular intervals. Others, like the vibration sensor, are capable of continuously emitting waveform measurements in a data stream. All the data gathered from these sensors must be combined or fused and then interpreted to provide accurate information about the turbine's environment, current state, and future health.

During the construction of the ocean turbine, vibration data collected from accelerometers mounted on a 20 kW dynamometer (designed to test components that will eventually be incorporated into the turbine) are used to guide the development of the MCM/PHM system (Duhaney et al. 2010). This dynamometer is shown in Fig. 1. In this diagram, we see four main components: a motor (MTRX), two gearboxes (GBXA and GBXB), and a generator (GENX). The motor (MTRX) is a 3-phase induction motor which simulates the effects of ocean current on the machine. This component is powered by the electrical signal from the grid which is conditioned by two variable frequency input drives (not shown in the figure). The motor MTRX is connected to a gearbox GBXA which reduces the rotational speed

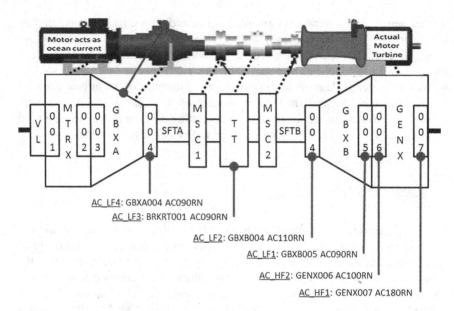

Fig. 1 Dynamometer diagram showing sensor locations

from MTRX by a 1:21.8 reduction ratio. The dynamometer side drive shaft SFTA couples to turbine drive shaft SFTB via coupling MSCX, both of which therefore rotate at this reduced rate. Planetary step-up gearbox GBXB has a 1:0.04 gear ratio and supplies an increased rotation speed to the motor GENX. By supplying an input rotation to GENX, an electrical current is produced.

Components labeled 001 through 004 in Fig. 1 to the left of driveshaft SFTA and 004 through 007 to the right of driveshaft SFTB in the same figure represent the bearing assemblies. These are numbered relative to the prime mover. Torque transducer TT records the torque of the driveshaft. Velocity transducer VL (i.e., encoder) measures the rotational velocity of the input force to the MTRX. Motor shaft coupler MSC1 joins the dynamometer to a portion of the driveshaft that functions as a test sleeve for the torque transducer. MSC2 joins the test sleeve to drive shaft SFTB for the machine under test.

There are four low-frequency accelerometers, AC_LF1, AC_LF2, AC_LF3, and AC_LF4, located closest to the prime mover, which are labeled as channels 4, 3, 2, and 1, respectively. AC_HF1 and AC_HF2 (channels 6 and 5, respectively) are the high-frequency vibration sensors which record the vibration from more rapid rotations by GENX. All sensors—except the one at channel 6 which was placed at 180°—were installed at or around 90° relative to axial view with prime mover in the foreground (i.e., MIMOSA Convention).

These accelerometers are used to record data, while the turbine operates in different states. Changes in vibration signatures from such rotating machinery typically indicate the presence of a problem such as a shift in its orientation or mechanical impact from an animal or object in its environment. Since the sensors are mounted on different components, each sensor provides unique information about the state of the machine. Although there may be some overlap (or redundancy) in the information provided by the sensors, it is necessary to analyze data from all the sensors to get a complete view of the system and to ensure that otherwise useful observations are not ignored during the state detection process. So some means of combining, or fusing, the data are needed.

Techniques for combining, or fusing, data are required at different stages of the monitoring process, including state detection, health assessment, and advisory generation; this makes data fusion a cross-cutting concern within an MCM system (Duhaney et al. 2010). Approaches to data fusion can be divided into categories based on the level at which they are performed. These levels are as follows:

(a) *Data-Level Fusion*: Data-level fusion techniques (Samadzadegan 2004) involves combining raw sensor signals prior to performing any data transformations, feature extraction, or data manipulation. In order to combine sensor signals at the data level, they must have originated from sources which produce the same type of signal.

(b) *Feature-Level Fusion:* Feature-level fusion (Samadzadegan 2004; Zhang et al. 2006; Sharma and Davis 2006; Gunatilaka and Baertlein 2001; Bedworth 1999; Kong et al. 2006; Frigui et al. 2010) is performed after features or attributes, which are individual measurable characteristics of the data, are extracted from the raw data. After extracting the feature set, the features from each signal are combined to produce a fused signal. In cases where the feature set may be large or features are correlated, feature selection methods (Khoshgoftaar et al. 2009) may be applied to select the most descriptive or informative features and reduce the set of features prior to data fusion. Unlike data-level fusion, feature-level fusion can be applied to data from both homogeneous and heterogeneous sensor types.

(c) *Decision-Level Fusion:* Decision-level fusion (Samadzadegan 2004; Gunatilaka and Baertlein 2001; Chen et al. 1997; Degara-Quintela et al. 2009; Veeramachaneni et al. 2008) occurs while or after a local decision has been made from each source. It involves making a local decision from each source or signal s_i and then combining the decisions to get the final output, usually with the aid of a mathematical model. This can be done by combining the decisions themselves or by combining the probabilities of class membership for each class and selecting the class with the highest probability (Veeramachaneni et al. 2008).

In this study, a feature-level approach is chosen to combine the sensor data. Future work will involve experimenting with data-level and decision-level fusion techniques and empirically comparing these techniques to determine which is best for combining data at different stages within the MCM/PHM system. A prerequisite to a feature-level fusion approach is some means of extracting usable features from the data. The feature extraction and feature-level fusion techniques are described in greater detail in Sect. 3.

3   Methodology

The following two sections discuss the feature extraction and feature-level fusion techniques in greater detail.

3.1   Feature Extraction

For our application, features are derived from the raw sensor signals by applying the short time wavelet transform with baselining (STWTB) methodology, which is fully defined in (Wald et al. 2010). This is a two-step process where the first step is the application of a traditional short time wavelet transformation (STWT) algorithm which converts the time series of amplitude readings into a time–frequency representation of the signal. The second step is the use of "baseline-differencing" to normalize the data relative to a given operating condition.

Wavelet transforms are favored over other vibration analysis techniques including the short time Fourier transform (after which it was modeled) for several reasons. One such advantage is the ability to show both which wave patterns are present in the raw signal as well as how much these waves change over time. The STWT algorithm applied in this study uses a Haar wavelet, which is simple and requires less computational effort. Each instance or observation in the file output by this algorithm is a vector of n features where each feature says how much of the original signal can be represented by oscillations at a given wavelength.

The second step in the STWTB process is to generate a baseline from the data gathered during a specific "normal" operational condition, and then to subtract the baseline from the current observations made by that source regardless of the current operational conditions of the machine. This baselining step was deemed necessary to remove those portions of the vibration signals that are characteristic of the machine's environment and/or operational conditions (in this case, its rotational velocity), so that the remaining signal only depicts the vibrations caused by actual abnormalities in the machine (and not its operating conditions).

3.2   Feature-Level Fusion

Our case study investigates feature-level fusion of wavelet transformed vibration data. To fuse the sensor data at the feature level, a set union of the features produced by the wavelet transform from all channels was performed, which, intuitively, should improve a classifier's ability to perform state detection since all the available data are being taken into account during the data mining process.

For a given experimental setup and a set $S = \{S_1, S_2, \ldots, S_m\}$ of sensor sources, n features are extracted from each source S_i. We will use S_{ij} to denote the jth feature extracted from the data from source S_i. The fused output from all m sources is given by

$$F(S) = \bigcup_{j=1}^{n} \bigcup_{i=1}^{m} S_{ij}$$

Simply put, the output of the fusion process will be a file containing all the features from all sources, or a total of $n \times m$ features. For simplicity, each source is considered equally reliable and thus its observations are considered no more or less important than any of the other sources.

In the case study discussed in Sect. 4, a combination of data mining and sensor fusion is used to identify abnormal states as they occur. Data mining and machine learning, which collectively refer to techniques for inferring knowledge from raw data by analyzing patterns, provide an avenue for automated interpretation of the sensor data and problem classification, while sensor fusion techniques are needed to combine data from multiple sources to get a complete, more accurate picture. These techniques will work collaboratively within an MCM/PHM system to allow for automatic interpretation of the raw data.

4 Experimental Setup

Vibration readings of rotating machinery contain distinct signatures which can be used to determine the state of a machine. By comparing acquired signals against a known baseline signal, we can determine whether the machine is operating in an abnormal state. In the case study described below, we show how we can achieve stable classification performance when distinguishing between 2° of resistive load (which constitute the states of interest). In other words, our system is attempting to detect when the resistive load being applied to the dynamometer has increased from 45 to 90 %, where 45 % load is considered to be normal. This is similar to having twice the normal load applied to the ocean turbine, as could occur if the on-board electrolysis device or other such devices meant to consume or utilize the power being generated by the turbine is unexpectedly drawing more energy.

Data collection and time synchronous averaging (Lebold et al. 2000) of the vibration signals is performed by a wavebook data acquisition unit. The baseline signal for this experiment was acquired with 45 % resistive load and SFTA (the dynamometer drive shaft) turning at 25 RPM. Data were also acquired with a 90 % resistive load applied at the same RPM and then with 45 and 90 % load at 50 RPM. Data gathered at 25 RPM will be used to generate the classification models used in this study; these models will then be tested against the data collected at 50 RPM. Again, we stress that the change in load is what we aim to detect. We test against a different RPM here to ensure the robustness of these techniques in varying operational conditions (such as speed) since typically the speed of the turbine will vary during operation based on the flow velocity of the ocean currents.

For the first of two experiments (denoted *Experiment A*), data from the accelerometers were sampled at 5,000 Hz for 32 s for each of the four setups (i.e., configurations of load and RPM), producing a total of 160,000 readings per setup and sensor. These accelerometers are denoted as channels 1 through 6 (CH1, CH2,..., CH6) throughout this case study. The resulting dataset consisted of 24 files = 6 channels × 2 loads × 2 speeds.

Experiment B, the second experiment, was conducted using the same hardware but the data were collected separately. 4 s of data were sampled at 5,000 Hz from all six sensors, producing 20,000 instances or readings per sensor per setup. The same naming convention is used for the sensors, meaning that CHx in Experiment A is the same sensor as CHx in Experiment B. For Experiment B, there are also a total of 24 files in the dataset. This second batch of data was acquired using the same procedure as was used for Experiment A to replicate the data and test the consistency and stability of the feature-level fusion approach employed here.

The STWTB technique described in 3.1 was applied to each of the 48 files (the 24 files from Experiment A and the 24 files from Experiment B) separately. As previously described, this transform converts the time series of raw amplitude readings to a time–frequency representation of the signal. For each RPM, the baseline was generated from the 45 % load data for each source and RPM, implying that there are 12 baselines, and then subtracting the baseline for a given source and RPM from both the 45 and 90 % data observed by that source at the specified RPM. This is done twice, once for each RPM. This baselining step was deemed necessary to remove those portions of the vibration signals that are characteristic of the speed of the system, so that the remaining signal only depicts the vibrations caused by actual abnormalities in the machine (and not its operating conditions). The output of this algorithm is 13 numeric features, with lower values for a specific frequency and instance indicating a close similarity to the baseline signal.

The overall goal of this study is to produce a state detection system which is capable of distinguishing between two states (45 % load and 90 % load). This is a simplified version of the real problem in which there will be multiple possible system states. Multi-class classification, or the problem involving detecting more than two system states, will be considered in future studies. To prepare the datasets from each experiment for binary classification (determining which of two states a particular set of observations belong to), the data gathered at 45 % load are appended to the corresponding file containing the data gathered at 90 % load for the same channel and RPM. This means that there are now 12 files per experiment.

In this study, feature-level fusion is done by performing a set union of the features across all sensor channels for that RPM as described in Sect. 3.2. So, after fusing the data, there are two files per experiment: one for building the classification model (the 25 RPM data) and the other to be used as the test set (the 50 RPM data). Each of the two files contained 78 wavelet features = 13 wavelet features from each of the six channels. The test set is a group of observations or instances that have the same characteristics which are the training set (the one used to build the model). These test data are not involved in generating the models; instead, they are used to evaluate the model's ability to label new data. Since the data in the test set are also gathered under controlled circumstances, the state that each instance in the test set actually represents is already known. The performance of each model can therefore be evaluated by comparing its state prediction for an instance in the test set against the known state for that instance.

4.1 Classifiers

Six (6) machine learning techniques were trained to detect the underlying patterns in the vibration signatures and to predict the state of the machine. Seven models were built for each machine learner (or classifier) per experiment by training the classifier on the 25 RPM data from each accelerometer independently (producing one model per sensor channel) and then from the fused data. The classifiers used in this case study are all available in WEKA[1] data mining software package (Witten and Frank 2005).

- Naive Bayes (NB)—A simplified form of a Bayesian network, the Naive Bayes learner (Frank et al. 2000) applies Bayes' rule of conditional probability and an assumption of independence among the features to predict the probability that an instance belongs to a specific class.
- Multi-Layer Perceptron (MLP)—The MLP neural network (Charalampidis and Muldrey 2009) is a form of feed-forward neural network which uses a learning technique known as back propagation to continually updating the weights it assign to individual connections within the neural network based on the amount of error in the output compared to the expected outcome. Two parameters were changed: the *hiddenLayers* parameter (the number of intermediate, or hidden, layers in the network) was set to '3' and the *validationSetSize* (percentage of the training dataset reserved for validating the MLP model during back propagation) was set to '10'.
- k-Nearest Neighbor (5-NN)—The k-Nearest Neighbor algorithm (Fraiman et al. 2010) classifies a new instance by taking a majority vote of the classes of the *k* instances in the training dataset that are closest to the new instance within the feature space. Default values were selected for all parameters of the IBk algorithm—the WEKA implementation of the k-nearest neighbor algorithm—with the exception of the value of *k* which was set to 5 and the *distanceWeighting* (which tells the learner to use an inverse distance weighting to determine how to classify an instance) which was set to 'Weight by 1/distance.'
- Support Vector Machine (SVM)—The simplest form of the SVM is a hyperplane which divides a set of instances into two classes with maximum margin. Its support vectors are a subset of instances which are used to find this hyperplane. Two parameters for John Platt's SMO algorithm (Platt 1998), which is the WEKA implementation of the SVM, were changed: *c* (representing the

---

[1]Available for download from http://www.cs.waikato.ac.nz/ml/weka.

complexity constant of the SVM) was set to 5.0 and *buildLogisticModels* (which allows the SVM to obtain proper probability estimates) was set to 'true.'

- Decision Tree (C4.5)—The decision tree is a tree-like machine learning model. Decision rules comprised comparisons of attributes to numeric thresholds are coded as branches, and the predicted values (which in a classification problem would be the class labels) are coded as leaves. The C4.5 algorithm (Quinlan 1993) used here generates decision trees recursively, computing each of its comparison thresholds based on the information gain (i.e., the difference in entropy) of the independent attributes at each level. We built this classifier using default values for J48, the WEKA implementation of the C4.5 decision tree algorithm.
- Logistic Regression (LR)—Also known as the maximum entropy classifier, the logistic regression learner (Witten and Frank 2005) labels an instance in a binary classification problem based on the measured probability of the class of interest (which in our case is the faulty scenario), similar to Naive Bayes. The probabilities are based on a parametric model having parameters estimated from the training data. The logistic regression learner is implemented in WEKA using a multinomial regression model for minimizing error.

The results for all six learners are presented in the next section. Default parameter values were used unless otherwise noted. Non-default parameter values were used only where experimentation indicated an overall improvement in classification performance for all channels. Details and results of individual experiments were excluded due to space limitations.

Because an equal amount of data was collected during the normal state as with the abnormal state, the data distribution are said to be balanced. In a typical real-world application, the number of faulty instances in a dataset will be significantly less than the number of fault-free instances, leading to a data imbalance. In imbalanced dataset problems where the abnormal class has significantly fewer instances than the normal class, a classifier can simply label all examples as belonging to the normal class to seemingly achieve a high ratio of correctly versus incorrectly classified instances (Seliya et al. 2009). Since we have a balanced class distribution in this case study, it will suffice to use the classifier's accuracy—the percentage of correctly classified instances—as a measure of its performance. The accuracy is calculated by dividing the number of correctly labeled/classified instances by the total number of instances.

5 Results

The results of Experiment A are shown in the table in Fig. 2, while those for Experiment B are shown in Fig. 3. Each table shows the performances of the six classifiers on distinguishing between the two loads from data from independent channels (CH1, CH2,..., CH6) and the fused channel (denoted Fused). The results

Learners	CH1	CH2	CH3	CH4	CH5	CH6	Average	Fused
5-NN	0.99163	0.93996	0.99324	0.98835	0.51468	0.50035	0.82137	**0.99884**
SVM	0.99069	0.93126	0.98410	0.98433	0.99998	0.83752	0.95465	**0.98892**
MLP	0.98595	0.92455	0.97222	0.97787	0.96930	0.82624	0.94269	**0.98549**
LR	0.94208	0.93245	0.90995	0.93204	0.79277	0.78931	0.88310	**0.92807**
C4.5	0.83221	0.93536	0.91488	0.85016	0.74312	0.53322	0.80149	**0.83641**
NB	0.67726	0.59974	0.79989	0.48615	0.73241	0.64474	0.65670	**0.73642**

Fig. 2 Results of experiment A

Learners	CH1	CH2	CH3	CH4	CH5	CH6	Average	Fused
5-NN	0.92719	1.00000	0.98720	0.63494	0.98331	0.50000	0.83877	**1.00000**
SVM	0.85004	1.00000	0.99686	0.99969	0.94932	0.50000	0.88265	**1.00000**
MLP	0.85073	0.99903	0.98711	0.99717	0.95514	0.50000	0.88153	**0.99997**
LR	0.82813	1.00000	1.00000	1.00000	1.00000	1.00000	0.97135	**0.99943**
NB	0.83206	0.61158	0.65276	0.63255	0.73035	0.99921	0.74308	**0.88475**
C4.5	0.89606	0.89434	0.84598	0.82096	0.91518	0.50000	0.81209	**0.82096**

Fig. 3 Results of experiment B

from the fused data are shown in the last column. The average accuracy across all six channels is also presented (second column from the right).

The higher of the two accuracies (i.e., the average accuracy of the individual models versus the accuracy of the fused models) is bolded for emphasis and the learners are ranked in order of the fused accuracies with the best learner shown first. The fused results are not compared to the results for the individual channels separately because the goal is to consider the input from all channels together. Basing decisions on multiple sensors versus on a single sensor is favored for several reasons. Aside from the fact that the sensors all measure different components, there are also two different types of sensors (four low-frequency sensors and two high-frequency sensors) which can capture information that the other sensor type would have missed.

The results on the individual channels are presented here to show that the machine learner performances differ across individual channels. Note, first, that there is no individual or group of channels that consistently yield optimal results for the learners across both experiments. Also, none of the learners produced models having an average accuracy across the six channels greater than 98 % for either of the two experiments.

In the first experiment, Experiment A, the 5-nearest neighbor algorithm (5-NN) generated near perfect models on the fused data. Its accuracy on the fused data (99.884 %) is dramatically greater than the average from the individual channels (82.137 %), meaning that the models built on the fused data were roughly 20 % better than the counterparts built on data from individual channels. The support vector machine (SVM) and multi-layer perceptron (MLP) learners are close runners-up in performance, which accuracy values exceeding 98 %. Both of these learners realize greater than a 4 % increase in accuracy meaning that there were at least 6,400 (4 % of 160,000) more observations that were correctly identified on the fused data than from the individual channels on average.

Experiment B, which was a repetition of Experiment A, was performed to ensure that these techniques yielded consistent results across multiple runs under the same conditions. The amount of data collected for Experiment B was 25 % of that collected for Experiment A. This was done to confirm that the specified feature extraction, feature-level fusion, and machine learning techniques were effective even with shorter bursts of data. 5-NN and SVM models were now perfect for distinguishing between the two loads regardless of the dynamometer's operational speed.

From both tables, it can be easily seen that the performance on the fused data exceeded the average performance of that learner across the individual channels for both experiments. In other words, by applying feature-level fusion, classification performance on distinguishing between two states was better than the average performance on each sensor channel. Overall, the 5-NN learner gives the best and most consistent results than the other 5 learners. Using the 5-NN with our feature extraction and feature-level fusion techniques, reliable detection between normal, and abnormal operational states of the ocean turbine is possible.

6 Conclusion

Machine condition monitoring (MCM) systems allow for automated detection of changes in the state of a machine being monitored. A data-driven MCM system could employ a combination of vibration analysis, machine learning, and data fusion techniques to interpret and process readings from sensors installed on a machine to provide insight into the health of the machine. Feature-level fusion can be applied within an MCM system to obtain more reliable problem detection capabilities, as we demonstrate in this paper. Through a case study, we showed how feature-level fusion provided more stable classifier performances than were obtained on the individual channels. Future work includes analyzing and comparing different data-level, feature-level, and decision-level algorithms to determine the best approach to combining data within each layer of the MCM system.

Acknowledgements The work discussed here grew from collaborations within the Prognostics and Health Monitoring (PHM) working group of the Southeast National Marine Renewable Energy Center (SNMREC) at Florida Atlantic University and was funded through SNMREC by the State of Florida.

References

Beaujean, P.-P., Khoshgoftaar, T. M., Sloan, J. C., Xiros, N., & Vendittis, D. (2009). Monitoring ocean turbines: a reliability assessment. In *Proceedings of the 15th ISSAT International Reliability and Quality in Design Conference*, 2009 (pp. 367–371).

Duhaney, J., Khoshgoftaar, T. M., & Sloan, J. C. (2011). Feature level sensor fusion for improved fault detection in MCM systems for ocean turbines. In *Proceedings of the 24th Florida Artificial Intelligence Research Society Conference (FLAIRS'24)*.

Duhaney, J., Khoshgoftaar, T. M., Cardei, I., Alhalabi, B., & Sloan, J. C. (2010). Applications of data fusion in monitoring inaccessible ocean machinery. In *Proceedings of the 16th ISSAT International Reliability and Quality in Design Conference* (Washington D.C., USA) (pp. 308—313).

Samadzadegan, F. (2004). Data integration related to sensors, data and models. In *Proceedings of the ISPRS, no. 4 in 35, Natural Resources Canada*, 2004 (pp. 569–574).

Zhang, S.-W., Pan, Q., Zhang, H.-C., Shao, Z.-C., & Shi, J.-Y. (2006). Prediction of protein homo-oligomer types by pseudo amino acid composition: Approached with an improved feature extraction and naive bayes feature fusion. *Amino Acids, 30*, 461–468. doi:10.1007/s00726-006-0263-8.

Sharma, V., & Davis, J. W. (2006). Feature-level fusion for object segmentation using mutual information. In *Proceedings of the IEEE International Workshop on Object Tracking and Classification Beyond the Visible Spectrum*, 2006.

Gunatilaka, A. H., & Baertlein, B. A. (2001). Feature-level and decision-level fusion of noncoincidently sampled sensors for land mine detection. *IEEE Transactions on Pattern Analysis and Machine Intelligence, 23*, 577–589.

Bedworth, M. D. (1999). Source diversity and feature-level fusion. In R. Evans, L. White, D. McMichael, & L. Sciacca (Eds.), *Proceedings of Information Decision and Control 99*, (Adelaide, Australia) (pp. 597–602). Institute of Electrical and Electronic Engineers, Inc.

Kong, A., Zhang, D., & Kamel, M. (2006). Palmprint identification using feature-level fusion. *Pattern Recognition, 39*(3), 478–487.

Frigui, H., Zhang, L., & Gader, P. D. (2010). Context-dependent multisensor fusion and its application to land mine detection. *IEEE Transactions on Geoscience and Remote Sensing, 99*, 1–16.

Khoshgoftaar, T., Bullard, L., & Geo, K. (2009) Attribute selection using rough sets in software quality classification. *International Journal of Reliability, Quality, and Safety Engineering, 16* (1), 73–89.

Chen, K., Wang, L., & Chi, H. (1997). Methods of combining multiple classifiers with different features and their applications to text-independent speaker identification. *International Journal of Pattern Recognition and Artificial Intelligence, 11*, 417–445.

Degara-Quintela, N., Pena, A., & Torres-Guijarro, S. (2009). A comparison of score-level fusion rules for onset detection in music signals. In *Proceedings of the 10th International Society for Music Information Retrieval Conference (ISMIR 2009)*.

Veeramachaneni, K., Osadciw, L., Ross, A., & Srinivas, N. (2008). Decision-level fusion strategies for correlated biometric classifiers. In *Proceedings of IEEE Computer Society Workshop on Biometrics at the Computer Vision and Pattern Recogniton (CVPR) Conference*, (Anchorage, USA) 2008 (pp. 1–6).

Wald, R., Khoshgoftaar, T. M., Beaujean, P. -P. J., & Sloan, J. C. (2010). A review of prognostics and health monitoring techniques for autonomous ocean systems. In *Proceedings of the 16th ISSAT International Reliability and Quality in Design Conference*, 2010 (pp. 308–313).

Lebold, M., McClintic, K., Campbell, R., Byington, C., & Maynard, K. (2000). Review of vibration analysis methods for gearbox diagnostics and prognostics. In *Proceedings of the 54th Meeting of the Society for Machinery Failure Prevention Technology*, May 2000 (pp. 623–634).

Witten, I. H., & Frank, E. (2005). *Data mining: Practical machine learning tools and techniques* (2nd ed.). San Francisco: Morgan Kaufmann.

Frank, E., Trigg, L., Holmes, G., & Witten, I. H. (2000). Technical note: Naive Bayes for regression. *Machine Learning, 41*(1), 5–25.

Charalampidis, D., & Muldrey, B. (2009). Clustering using multilayer perceptrons. *Nonlinear Analysis: Theory, Methods & Applications, 71*(12), e2807–e2813.

Fraiman, R., Justel, A., & Svarc, M. (2010). Pattern recognition via projection-based kNN rules. *Computational Statistics & Data Analysis, 54*(5), 1390–1403.

Platt, J. C. (1998). Sequential minimal optimization: A fast algorithm for training support vector machines. *Advances in Kernel MethodsSupport Vector Learning, 208*(MSR-TR-98-14), 1–21.

Quinlan, J. R. (1993). *C4.5: Programs for machine learning*. San Francisco, CA: Morgan Kaufmann Publishers Inc.

Seliya, N., Khoshgoftaar, T. M., & Hulse, J. V. (2009). A study on the relationships of classifier performance metrics. In *Proceedings of the 21st IEEE International Conference on Tools with Artificial Intelligence, ICTAI '09*, (Washington, DC, USA) (pp. 59–66). IEEE Computer Society.

Index

© Springer-Verlag London 2016
H. Pham (ed.), *Quality and Reliability Management and Its Applications*,
Springer Series in Reliability Engineering, DOI 10.1007/978-1-4471-6778-5

Printed in the United States
By Bookmasters